Lectures on Abstract Algebra

MICHAEL K. BUTLER

Version 25.1

m.k.butler@bolton.ac.uk

Published by:
Pendlebury Press Limited
20 May Road
Swinton, Manchester
M20 5FR
United Kingdom

ISBN: 978-1-9999846-5-6

CONTENTS

Introduction

This book is based on lectures given by the author at the University of Bolton in the UK between 1994 and 2019. These were attended by students from an exceptionally wide range of backgrounds. Over the years these notes were continuously refined in the light of the experience of teaching and in response to feedback from students, over many deliveries of the material.

There follows a brief outline of the contents of each chapter. If you are new to the subject, you may prefer to skip this on first reading. Each chapter concludes with a set of exercises designed to test and consolidate your understanding of the material. Solutions to many of the exercises are given at the back of the book. However, it is hoped that you will spend some time attempting and thinking about each exercise before reaching for the solution.

The first chapter introduces the algebraic structure known as a *group*. A group is a set, on which a binary operation $*$ has been defined, satisfying the axioms of *associativity*, *identity* and *inverses* :

$(a * b) * c = a * (b * c)$ for all $a, b, c \in G$

There is $e \in G$ such that $e * a = a = a * e$ for all $a \in G$

For each $a \in G$ there is $a^{-1} \in G$ such that $a * a^{-1} = e = a^{-1} * a$

The study of groups underlies all of the material in this book, in the sense that other algebraic structures such as vector spaces and rings may be thought of as groups with additional algebraic structure.

We look at examples of infinite group, such as groups of invertible matrices. We also introduce finite groups, including groups of symmetries and groups of permutations.

In chapter 2 we meet the idea of a *subgroup*, as a group that is contained inside another group. Subgroups are the prototypes for various other sub-objects that will be encountered, including *subspaces*, *subrings* and *subalgebras*. We also look at mappings or functions from one group to another that preserve structure. Such mappings are called *group homomorphisms*. Once again these are the prototypes for various other structure preserving mappings that will be encountered, including *linear transformations* and *ring homomorphisms*.

For the next five chapters we turn our attention to linear algebra, the study of *vector spaces*. You may have already met the idea of a *vector* as a quantity with both magnitude and direction. Examples of vector quantities include velocity and force. Vector spaces are a substantial generalisation of the idea of a vector, and these are introduced in chapter 3. Vector spaces are abelian groups under vector addition. They also have the extra structure of an operation called *multiplication by a scalar*: we may take a vector v and a number λ (known as a *scalar*) and multiply these together to give a new vector λv.

The most well-known example is $\boldsymbol{R}^2$ in which the operations are

$$\begin{pmatrix} x_1 \\ y_1 \end{pmatrix} + \begin{pmatrix} x_2 \\ y_2 \end{pmatrix} = \begin{pmatrix} x_1 + x_2 \\ y_1 + y_2 \end{pmatrix} \qquad \lambda \begin{pmatrix} x \\ y \end{pmatrix} = \begin{pmatrix} \lambda x \\ \lambda y \end{pmatrix}.$$

Chapter 4 introduces *subspaces* and *linear transformations* by analogy with the concepts of subgroup and group homomorphism encountered in the theory of groups. In particular, a linear transformation $\boldsymbol{R}^n \to \boldsymbol{R}^m$ may be carried out by multiplying a column vector in $\boldsymbol{R}^n$ by an $m \times n$ matrix to give a column vector in $\boldsymbol{R}^m$.

In chapter 5 we introduce the concept of a *basis* for a vector space. A basis is a set of vectors that may be thought of as the basic building blocks for the vector space. For example, for $\boldsymbol{R}^2$ we may take the two vectors

$$\begin{pmatrix} 1 \\ 0 \end{pmatrix}, \begin{pmatrix} 0 \\ 1 \end{pmatrix}$$

as a basis. Any vector in $\boldsymbol{R}^2$ may be written as a *linear combination* of the basis vectors as

$$\begin{pmatrix} x \\ y \end{pmatrix} = x \begin{pmatrix} 1 \\ 0 \end{pmatrix} + y \begin{pmatrix} 0 \\ 1 \end{pmatrix}$$

We investigate the effects of changing the basis, both for the domain and for the codomain of a linear transformation, and look at how to calculate the matrix representing the linear transformation with respect to our choice of bases.

In chapter 6 we study *endomorphisms,* which are linear transformations f with the same vector space as both the domain and the codomain. It can happen that for a vector v we have $f(v) = \lambda v$ for some scalar λ, so that $f(v)$ is precisely a multiple of the original vector v. Such vectors are called *eigenectors* and the corresponding scalars λ are called *eigenvalues.* Often we are able to obtain eigenvectors which form a basis for the vector space. In this case, relative to the basis of eigenvectors, f is represented by a diagonal matrix. The matrix features the eigenvalues on the leading diagonal.

In the vector space $\boldsymbol{R}^2$ the *scalar product,* sometimes known as the "dot product", is given by

$$\begin{pmatrix} x_1 \\ y_1 \end{pmatrix} \cdot \begin{pmatrix} x_2 \\ y_2 \end{pmatrix} = x_1 x_2 + y_1 y_2$$

Notice that the product of a pair of vectors is a scalar. In chapter 7 we introduce a generalisation of the scalar product, known as an *inner product.* Such a product assigns to a pair of vectors u and v a scalar $\langle u, v \rangle$. We introduce the concept of *orthogonality*: a pair of non-zero vectors are orthogonal if $\langle u, v \rangle = 0$. This leads to the idea of an *orthogonal basis* in which any pair of distinct vectors are orthogonal. We study *symmetric* linear transformations which are represented with respect to the standard basis by symmetric matrices. We show that where it is possible to obtain a basis of eigenvectors for a symmetric linear transformation, it may be arranged that this basis be orthogonal.

In chapter 8 we return to the study of groups. Let H be a subgroup of a group G and $a \in G$. The set $aH = \{ah : h \in H\}$ is known as a *coset.* We explore the properties of cosets, and show how the cosets of H form a group under the operation $aH \cdot bH = abH$, provided that H is a special sort of subgroup known as a *normal* subgroup. These groups of cosets are known as *quotient groups.*

In chapter 9 we explore the idea of a group *acting* on a set. This leads us to the proof of the First Sylow Theorem, which is useful in predicting the existence of subgroups of particular order.

In chapter 10 we introduce the idea of a *simple* group. In group theory the simple groups play a role analogous to prime numbers in number theory. The classification of the finite simple groups, which was rigorously completed around 2004, is a story in itself. One family of simple groups are the integers modulo p, where p is a prime number. A second family is provided by certain groups of permutations. Yet a third family arise as groups of matrices.

Pressing the analogy of simple groups with prime numbers further, we can decompose a group into *composition factors* by means of what is known as a *composition series*. In chapter 11 this leads to the idea of a *soluble group*, in which all composition factors are abelian, which is the key to exploring the solvability of polynomial equations.

In chapter 12 we turn our attention to fields. A field is, roughly speaking, a commutative ring in which we are able to carry out division (with the usual exception of dividing by zero). For example, the rational numbers $\boldsymbol{Q}$ and the real numbers $\boldsymbol{R}$ are fields. If a field F lies inside a field E, then we say that E is an *extension* of F. For example, the field $\boldsymbol{C}$ of complex numbers is an extension of the field $\boldsymbol{R}$ of real numbers. We may think of an extension field E as a vector space over the base field F. This insight allows us the make use of the machinery of linear algebra in studying extensions of fields.

Chapter 13 builds on this work on fields to study the solvability of polynomial equations. For a quadratic equation

$$x^2 + bx + c = 0 \text{ with } b, c \in \boldsymbol{Q}$$

it is well known that the roots are given by

$$x = \frac{-b \pm \sqrt{b^2 - 4c}}{2}.$$

Hence in order to solve such an equation, all that is required beyond the arithmetic operations of adding, subtracting, multiplying and dividing is the extraction of square roots. It is also possible to express the roots of *cubic* (degree 3) and *quartic* (degree 4) polynomials in terms of the coefficients using just arithmetic and nth roots. However, this cannot be done for many examples of *quintic* (degree 5) polynomial. This chapter culminates in showing why this is so. When we wish to solve a polynomial equation over a given field, we often need to extend the field to obtain roots. For example, $x^2 + 1 = 0$ is a polynomial equation over $\boldsymbol{R}$, but to obtain roots we need to extend to the field $\boldsymbol{C}$. To a given choice of base field F and extension field E we may associate a group, known as the Galois group $Gal(E/F)$. The connection of solvability of polynomial equations with group theory is at last revealed: a polynomial equation is solvable if and only if the corresponding Galois group is a soluble group.

In chapter 14 we turn our attention to *rings*. Whilst a group has just a single binary operation, a ring has two binary operations. Usually, these are *addition* and *multiplication*. Familiar number systems such as the integers $\boldsymbol{Z}$, the rational numbers $\boldsymbol{Q}$, and the real numbers $\boldsymbol{R}$ are rings under the usual arithmetic operations. We can also construct rings of other algebraic objects, such as polynomials and matrices.

Chapter 15 introduces *subrings* and *ring homomorphisms* by analogy with the concepts of subgroup and group homomorphism encountered in chapter 2. We also introduce a special type of subring, known as an *ideal*. In ring theory, ideals play a role analogous to the role of normal subgroups in group theory. This leads to the construction of *quotient rings* in chapter 16.

In chapter 17 we study *integral domains* and return to the study of fields. An integral domain is a commutative ring with unity in which there are no divisors of zero, so that $xy = 0$ only if $x = 0$ or $y = 0$. This is true, for example, in the ring of integers. However, in the integers modulo 10 this is not true because, for example, $2 \times 5 = 0$. We show how integral domains and fields may be constructed as quotient rings. This is applied in chapter 18 to the construction of finite fields. It turns out that there is one finite field for each positive integer power of each prime p.

The fundamental theorem of arithmetic tells us that each positive integer (except 1) is either a prime, or may be factorised uniquely as a product of primes. An integral domain in which a similar result holds is called a *unique factorisation domain* (UFD), and these are studied in chapter 19. Not every integral domain is a UFD. For example, the ring $\boldsymbol{Z}[\sqrt{-3}\,]$ comprises all complex numbers of the form $a + b\sqrt{-3}$ where a and b are integers. In this ring, 4 may be factorised in the usual way as 2×2. However, 4 also has a different factorisation as $(1 + \sqrt{-3}\,)(1 - \sqrt{-3}\,)$.

In chapter 20 we introduce the concept of a *real algebra* as a real vector space that also has a *product* or *multiplication* of vectors. A well-known example is the *vector product* on $\boldsymbol{R}^3$, sometimes known as the "cross product". Under this multiplication of vectors, $\boldsymbol{R}^3$ becomes an *algebra.*

In any algebra, we require

$$(u+v)\cdot w = u\cdot w + v\cdot w$$

$$u\cdot(v+w) = u\cdot v + u\cdot w$$

$$\lambda u\cdot v = u\cdot \lambda v = \lambda(u\cdot v)$$

for all vectors u, v and w and all $\lambda \in \boldsymbol{R}$. These three conditions may be summed up by saying that the product is *bilinear*, so that $u\cdot v$ is linear both in u and in v.

The other conditions to be satisfied vary from case to case. Many algebras are *associative*, so that $(u\cdot v)\cdot w = u\cdot(v\cdot w)$. We shall also meet important examples of algebras that are not associative.

In chapter 21 we study the *exterior algebra* of a vector space. In such algebras, the product of vectors u and v is written $u\wedge v$. Suppose that the vector space $\boldsymbol{R}^3$ has basis $\{e_1, e_2, e_3\}$. The exterior algebra of $\boldsymbol{R}^3$ has basis

$$\{1, e_1, e_2, e_3, e_1\wedge e_2, e_1\wedge e_3, e_2\wedge e_3, e_1\wedge e_2\wedge e_3\}$$

and so is of dimension $2^3 = 8$. More generally, the exterior algebra of $\boldsymbol{R}^N$ is of dimension 2^N. Exterior algebras are interesting because they give us a substantial generalisation of the vector product of $\boldsymbol{R}^3$. They also give us very convenient methods for defining and manipulating determinants and traces of matrices, and elegant proofs of many standard results from linear algebra.

In chapter 22 we introduce *Lie algebras*, in which the product of vectors u and v is written $[u, v]$. Unlike our previous examples, Lie algebras are *not* generally associative. Instead, they satisfy the *Jacobi identity*:

$$[[u,v],w] + [[v,w],u] + [[w,u],v] = \mathbf{0}$$

We have already met one example of a Lie algebra: $\boldsymbol{R}^3$ with the vector product. Many other important examples arise as algebras of matrices, where the Lie product is defined by

$$[A,B]=AB-BA.$$

Chapter 23 introduces the study of *matrix groups*. We have already met the *general linear groups* comprising all invertible matrices of a given size and the *special linear groups* comprising all matrices with determinant 1. Other examples include the group of *orthogonal matrices*. For a matrix group G, by applying calculus we can determine the set of all matrices that are tangential to differentiable curves through the identity matrix in G. This set of matrices is a real vector space, known as the *tangent space* of G. It also has the structure of a Lie algebra. Since vector spaces are generally easier to study than groups, the tangent spaces of matrix groups are a powerful tool for studying and classifying matrix groups.

In chapter 24 we build upon the idea of a *real inner product space* from chapter 7. A *norm* on a real vector space V is a mapping $V \to \boldsymbol{R}$. For an inner product $\langle u, v\rangle$ we define a norm by $\|v\| = \sqrt{\langle v, v\rangle}$. We say that A is a *normed algebra* if its norm satisfies $\|xy\| = \|x\| \cdot \|y\|$. We study the properties of normed real algebras and show that there are only three associative normed real algebras with unity: the real numbers $\boldsymbol{R}$, the complex numbers $\boldsymbol{C}$ and the quaternions $\boldsymbol{H}$.

Finally in chapter 25 we turn our attention to Clifford algebras. These are important generalisations of the complex numbers. We begin with a real vector space V with a basis $\{e_1, e_2, ..., e_n\}$. We define a product in such a way that $e_i^2 = e_i \cdot e_i = -1$ for each i. For example, if V is a one-dimensional real vector space then its Clifford algebra $Cl(1)$ is the algebra of complex numbers, $\boldsymbol{C}$. If V is a two-dimensional real vector space then its Clifford algebra $Cl(2)$ is the algebra of quaternions, $\boldsymbol{H}$. The Clifford algebras exhibit an interesting periodicity phenomenon: if $Cl(n)$ is an algebra A, then $Cl(n+8) = Mat(16, A)$, the algebra of 16×16 matrices with entries in A.

Chapter 0: Preliminaries

The concept of a *set* is our starting point, and is an undefined notion; the intuition is that of a collection of distinct objects known as the *elements*. We write $x \in S$ where x is an element of a set S.

We may display a set by listing the elements within parentheses:

$$\{x, y, z, \ldots\}$$

Alternatively, we may display the set of all elements x that satisfy some condition P as

$$\{x : Px\}$$

The *empty set*, denoted $\varnothing$, has no elements.

For a pair of sets S and T, if each element of S is also an element of T, then we call S a *subset* of T, and write $S \subseteq T$. Where S is a subset of T, but not equal to T, we call S a *proper subset* of T, and write $S \subset T$.

Let S and T be sets. Their *intersection* is

$$S \cap T = \{x : x \in S \text{ and } x \in T\}.$$

The sets are *disjoint* if $S \cap T = \varnothing$. Their *union* is

$$S \cup T = \{x : x \in S \text{ or } x \in T\},$$

where 'or' is used in the inclusive sense. The Cartesian product is the set of ordered pairs

$$S \times T = \{(x, y) : x \in S, y \in T\}.$$

A *function* or *mapping* $f : S \to T$ assigns to each $x \in S$ a unique $f(x) \in T$. S is called the *domain* of the mapping, and T is called the *codomain*. The mapping f is called a *surjection* if for each $y \in T$ there is one or more $x \in S$ such that $y = f(x)$. It is called an *injection* if when $x_1 \neq x_2$ in S we also have $f(x_1) \neq f(x_2)$ in T. Equivalently, if $f(x_1) = f(x_2)$ then $x_1 = x_2$. A *bijection* is a mapping that is both a surjection and an injection. Note that a mapping f has an inverse mapping $f^{-1} : T \to S$ only in the case where f is a bijection.

For mappings $f : S \to T$ and $g : T \to U$ the *composite mapping* $g \circ f : S \to U$ is defined by $g \circ f(x) = g(f(x))$.

A *binary operation* $*$ on a set S takes a pair of elements $x, y \in S$ and returns a result $x * y \in S$.

A binary operation $*$ is said to be *commutative* if

$$x * y = y * x \text{ for all } x, y \in S$$

and *associative* if

$$(x * y) * z = x * (y * z) \text{ for all } x, y, z \in S.$$

A *relation* on a set S is a subset $R \subseteq S \times S$; where $(x, y) \in R$ we write xRy. We are interested in three properties of relations:

Reflexive:	xRx for all $x \in S$
Symmetric:	If xRy then yRx for all $x, y \in S$
Transitive:	If xRy and yRz then xRz for all $x, y, z \in S$.

A relation which is reflexive, symmetric and transitive is called an *equivalence relation.*

Four standard number systems are denoted by bold capitals throughout the book:

$\boldsymbol{Z}$	the integers
$\boldsymbol{Q}$	the rational numbers
$\boldsymbol{R}$	the real numbers
$\boldsymbol{C}$	the complex number

For a number system R, we write R^* for the set of non-zero elements, so that for example $\boldsymbol{Q}^*$ denotes the set of non-zero rationals. We shall write $n\boldsymbol{Z}$ for the set of integer multiples of a natural number n.

For a number system R, we write $R[x]$ for the set of polynomials with one variable x and coefficients in R. We write $Mat(n, R)$ for the set of $n \times n$ matrices with entries in R. For such a matrix A, we write $\det A$ for the determinant of A, and $\operatorname{tr} A$ for the trace of A.

For positive $a, b \in \boldsymbol{Z}$ we write $\gcd(a, b)$ for the greatest common divisor of a and b, and $\operatorname{lcm}(a, b)$ for the lowest common multiple.

Chapter 1: The Group Axioms and Examples

The set of integers $\boldsymbol{Z}$ under the binary operation of addition satisfies:

(i) $a+b=b+a$ for all $a,b\in\boldsymbol{Z}$

(ii) $(a+b)+c=a+(b+c)$ for all $a,b,c\in\boldsymbol{Z}$

(iii) The integer 0 is the *identity*:

$0+a=a=a+0$ for all $a\in\boldsymbol{Z}$

(iv) For each $a\in\boldsymbol{Z}$ the integer $-a$ is the *inverse:*

$a+(-a)=0=(-a)+a.$

Various binary operations on various sets satisfy similar properties to these. This leads us to define a general algebraic structure called a *group*, which consists of a set together with a binary operation on that set satisfying properties similar to those above. The advantage of taking this step of abstraction is that whatever constructions we make for groups in general we may make for any particular example of a group, and whatever we prove to be true of groups in general will also be true of any particular example of a group.

1.1 Definition A *group* is a set G with a binary operation $*$ on G satisfying:

G1: *Associativity:* $(a*b)*c=a*(b*c)$ for all $a,b,c\in G$

G2: *Identity:* There is an element $e\in G$ such that $e*a=a=a*e$ for all $a\in G$

G3: *Inverses:* For each $a\in G$ there is an element $a^{-1}\in G$ such that $a*a^{-1}=e=a^{-1}*a$

These three statements are known as the *group axioms*. If we also have

Commutativity: $a*b=b*a$ for all $a,b\in G$

then we say that the group is *abelian*.

For a group $(G, *)$, where it is clearly understood what the binary operation $*$ is, we often simply refer to "the group G".

Examples (a) $(\boldsymbol{Z},+)$, $(\boldsymbol{Q},+)$ and $(\boldsymbol{R},+)$ are all abelian groups, each with identity element 0. In each case the inverse of x is $-x$

(b) $(\boldsymbol{Q}^*,\cdot)$ and $(\boldsymbol{R}^*,\cdot)$ are abelian groups, where $\boldsymbol{Q}^*$ and $\boldsymbol{R}^*$ denote the non-zero rationals and the non-zero real numbers respectively and $\cdot$ denotes multiplication. In each case, the identity element is 1, and the inverse of x is $\frac{1}{x}$.

(c) The set of invertible $n \times n$ matrices with entries in $\boldsymbol{Q}$ is a group under matrix multiplication known as a *general linear group*, denoted $GL(n,\boldsymbol{Q})$. Similarly, the set of invertible $n \times n$ matrices with entries in $\boldsymbol{R}$ is a group denoted $GL(n,\boldsymbol{R})$. These groups are *not* abelian, except for the trivial case $n = 1$.

In each case the identity element is the identity matrix, for example for $n = 2$

$$e = \begin{pmatrix} 1 & 0 \\ 0 & 1 \end{pmatrix}.$$

(d) The set of vectors in 3-dimensional space form an abelian group under the binary operation of vector addition defined by

$$\begin{pmatrix} x_1 \\ y_1 \\ z_1 \end{pmatrix} + \begin{pmatrix} x_2 \\ y_2 \\ z_2 \end{pmatrix} = \begin{pmatrix} x_1 + x_2 \\ y_1 + y_2 \\ z_1 + z_2 \end{pmatrix}.$$

The identity is the zero vector $\begin{pmatrix} 0 \\ 0 \\ 0 \end{pmatrix}$.

The inverse of a vector $\begin{pmatrix} x \\ y \\ z \end{pmatrix}$ is $\begin{pmatrix} -x \\ -y \\ -z \end{pmatrix}$.

In a non-abelian group it is usual to omit the operation symbol and write $a * b$ as ab. In an additive abelian group we often prefer to write $a + b$.

1.2 Proposition Any group G has only one identity element. Each element $a \in G$ has only one inverse.

Proof: Suppose that G has a pair of identity elements e_1 and e_2. Then

$$\begin{aligned} e_1 * e_2 &= e_1 \text{ because } e_2 \text{ is an identity} \\ \text{and } e_1 * e_2 &= e_2 \text{ because } e_1 \text{ is an identity.} \end{aligned}$$

Hence $e_1 = e_2$ and so there is only one identity element.

Now suppose that $a \in G$ has a pair of inverses a^{-1} and a^*.

$$\begin{aligned} \text{Then } a^{-1} &= a^{-1} * e && \text{by G2} \\ &= a^{-1} * (a * a^*) && \text{by G3} \\ &= (a^{-1} * a) * a^* && \text{by G1} \\ &= e * a^* && \text{by G3} \\ &= a^* && \text{by G2} \end{aligned}$$

Hence $a^{-1} = a^*$ and so $a \in G$ has only one inverse.

1.3 Proposition Suppose that G is a group and $a, b, c \in G$.

Then (i) if $a * b = a * c$ then $b = c$

(ii) if $b * a = c * a$ then $b = c$

Proof: (i) Suppose $a * b = a * c$

By G3, the element a has an inverse a^{-1} which we may combine on the left of both sides of the equation

$$a^{-1} * (a * b) = a^{-1} * (a * c)$$

Then by G1 $(a^{-1} * a) * b = (a^{-1} * a) * c$, so that $e * b = e * c$ by G3, and finally $b = c$ by G2.

The proof of (ii) is left as an exercise.

These are called *cancellation laws.*

1.4 Proposition Suppose that G is a group and $a, b \in G$.

Then (i) $(a^{-1})^{-1} = a$

(ii) $(a * b)^{-1} = b^{-1} * a^{-1}$

Part (i) tells us that the inverse of the inverse of an element is the element itself. Part (ii) tells us that the inverse of a product of elements is the product of the inverses of the individual elements *with the order reversed.* It is important to remember this when working with matrices, for example.

Proof: (i) a and $(a^{-1})^{-1}$ are each the inverse of a^{-1}. By 1.2 they must be equal.

(ii)

$$\begin{aligned} (b^{-1} * a^{-1}) * (a * b) &= b^{-1} * e * b \\ &= b^{-1} * b \\ &= e \end{aligned}$$

$$\text{and } \begin{aligned} (a * b) * (b^{-1} * a^{-1}) &= a * e * a^{-1} \\ &= a * a^{-1} \\ &= e \end{aligned}$$

so that $b^{-1} * a^{-1}$ is the inverse of $a * b$

A *finite group* is a group which has a finite number of elements. If G has n elements then we write $|G| = n$ and say that the group is of *order* n. For the remainder of this chapter we introduce some important families of finite group.

1.5 Definition Suppose that Π is a geometric figure in 2 or 3 dimensional space. A *symmetry* of Π is a bijection from Π to itself that preserves distances in Π.

Examples (a) The set of rotational symmetries of a regular n-gon form a group under the binary operation of composition of rotations. For example, for a square the set of rotations $\{0°, 90°, 180°, 270°\}$ about the centre forms a group. The identity element is the rotation of $0°$. The group is of order 4.

(b) Consider an equilateral triangle:

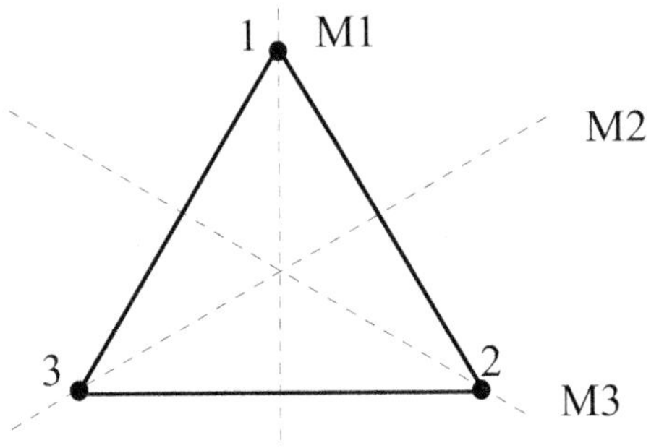

The symmetries are rotations through 120° and 240° clockwise, reflections in the lines M1, M2 and M3, and the identity mapping.

1.6 Definition The group of symmetries of the regular n-gon is called the *dihedral group* and denoted D_n.

The regular n-gon has $2n$ symmetries: n reflections and n rotations. Hence each group D_n is of order $2n$.

A finite group may be exhibited by constructing its *Cayley table*, where the entry at the intersection of row labelled X and the column labelled Y gives the result $X * Y$.

Example For the equilateral triangle above, let I be the identity symmetry and let R be the symmetry of rotation through 120° clockwise. Then R^2 is the symmetry of rotation through 240° and $R^3 = I$. Let M_1, M_2 and M_3 be the symmetries of reflection in the three lines as shown.

The Cayley table for D_3 is as follows:

•	I	R	R^2	M_1	M_2	M_3
I	I	R	R^2	M_1	M_2	M_3
R	R	R^2	I	M_3	M_1	M_2
R^2	R^2	I	R	M_2	M_3	M_1
M_1	M_1	M_2	M_3	I	R	R^2
M_2	M_2	M_3	M_1	R^2	I	R
M_3	M_3	M_1	M_2	R	R^2	I

1.7 The Klein Four-group

The Klein four-group, or *viergruppe V*, has four elements. Each element combined with itself gives the identity. The Cayley table is as follows:

	e	a	b	c
e	e	a	b	c
a	a	e	c	b
b	b	c	e	a
c	c	b	a	e

1.8 The Quaternion group, Q_8:

The elements of this group are the identity, 1, an element -1 such that $-1 \times -1 = 1$ and three elements i, j, and k which satisfy

$$i^2 = j^2 = k^2 = -1$$

and furthermore that

$$ij = k \qquad jk = i \qquad ki = j$$

Notice that $ji = j(jk) = j^2k = -1 \times k = -k$. Similarly $kj = -i$, and $ik = -j$ giving a further three elements.

We obtain the following Cayley table:

$\cdot$	1	i	j	k	-1	$-i$	$-j$	$-k$
1	1	i	j	k	-1	$-i$	$-j$	$-k$
i	i	-1	k	$-j$	$-i$	1	$-k$	j
j	j	$-k$	-1	i	$-j$	k	1	$-i$
k	k	j	$-i$	-1	$-k$	$-j$	i	1
-1	-1	$-i$	$-j$	$-k$	1	i	j	k
$-i$	$-i$	1	$-k$	j	i	-1	k	$-j$
$-j$	$-j$	k	1	$-i$	j	$-k$	-1	i
$-k$	$-k$	$-j$	i	1	k	j	$-i$	-1

There are also *generalised quaternion groups* Q_{12}, Q_{16}, Q_{20} and so on.

We also mention the *trivial group*, which has only one element - the identity.

The following fact is useful when completing Cayley tables:

1.9 Proposition - The Latin Square Property

In the Cayley table of a finite group, each element appears exactly once in each row and exactly once in each column.

Proof: The proof uses the cancellation properties 1.3 and is left as an exercise.

1.10 Definition For a set S, a bijection $S \to S$ is called a *permutation* of S.

Let G be the set of all permutations of S and let $\circ$ be the binary operation of composition, where $\sigma \circ \pi$ means carry out permutation σ and then permutation π. Then $(G, \circ)$ is a group, called the *symmetric group* on S.

WARNING: Some textbooks write $\sigma \circ \pi$ to mean carry out permutation π and then σ.

In particular, for $S = \{1, 2, 3, ..., n\}$ the group of permutations of S is called the *symmetric group of degree n*, denoted S_n.

A permutation $\sigma : S \to S$ may be represented by a $2 \times n$ matrix:

$$\begin{pmatrix} 1 & 2 & 3 & . & . & . & n \\ \sigma(1) & \sigma(2) & \sigma(3) & . & . & . & \sigma(n) \end{pmatrix}.$$

For example, the permutation of $\{1, 2, 3\}$ which maps $1 \mapsto 1$, $2 \mapsto 3$ and $3 \mapsto 2$ may be represented by

$$\begin{pmatrix} 1 & 2 & 3 \\ 1 & 3 & 2 \end{pmatrix}.$$

The other five elements of S_3 are

$$\begin{pmatrix} 1 & 2 & 3 \\ 1 & 2 & 3 \end{pmatrix}, \begin{pmatrix} 1 & 2 & 3 \\ 3 & 2 & 1 \end{pmatrix}, \begin{pmatrix} 1 & 2 & 3 \\ 2 & 1 & 3 \end{pmatrix}, \begin{pmatrix} 1 & 2 & 3 \\ 3 & 1 & 2 \end{pmatrix} \text{and} \begin{pmatrix} 1 & 2 & 3 \\ 2 & 3 & 1 \end{pmatrix}.$$

The identity element of each group S_n is

$$\begin{pmatrix} 1 & 2 & 3 & . & . & . & n \\ 1 & 2 & 3 & . & . & . & n \end{pmatrix}.$$

An inverse for each permutation is obtained by swapping the two rows of the matrix and then reordering the columns. For example in S_6

$$\text{if } \sigma = \begin{pmatrix} 1 & 2 & 3 & 4 & 5 & 6 \\ 3 & 1 & 5 & 2 & 6 & 4 \end{pmatrix} \text{then } \sigma^{-1} = \begin{pmatrix} 1 & 2 & 3 & 4 & 5 & 6 \\ 2 & 4 & 1 & 6 & 3 & 5 \end{pmatrix}.$$

Notice that each group S_n has $n!$ elements.

The groups S_n are non-abelian for $n \geq 3$.

Example If $\sigma = \begin{pmatrix} 1 & 2 & 3 & 4 \\ 2 & 4 & 3 & 1 \end{pmatrix}$ and $\pi = \begin{pmatrix} 1 & 2 & 3 & 4 \\ 4 & 1 & 2 & 3 \end{pmatrix}$ in S_4

then $\sigma \circ \pi = \begin{pmatrix} 1 & 2 & 3 & 4 \\ 1 & 3 & 2 & 4 \end{pmatrix}$ whereas $\pi \circ \sigma = \begin{pmatrix} 1 & 2 & 3 & 4 \\ 1 & 2 & 4 & 3 \end{pmatrix}$

so $\sigma \circ \pi \neq \pi \circ \sigma$, showing that S_4 is not an abelian group.

1.11 Definition Suppose that $a_1, a_2, ..., a_r$ are distinct elements of $\{1, 2, 3, ..., n\}$.

The permutation which maps a_1 to a_2, a_2 to a_3, a_3 to $a_4, ...$ a_{r-1} to a_r and a_r to a_1 leaving all other elements unchanged is called a *cycle of length* r and is denoted by $(a_1 a_2 a_3 ... a_r)$.

Example In S_5 the cycle of length 3 written (1 3 5) is the permutation

$$\begin{pmatrix} 1 & 2 & 3 & 4 & 5 \\ 3 & 2 & 5 & 4 & 1 \end{pmatrix}.$$

Notice that this cycle could also be written as (3 5 1) or (5 1 3).

1.12 Proposition Any non-identity permutation of S_n can be uniquely factorised as a product of disjoint cycles in S_n.

Example Consider the following permutation in S_{12}:

$$\begin{pmatrix} 1 & 2 & 3 & 4 & 5 & 6 & 7 & 8 & 9 & 10 & 11 & 12 \\ 6 & 3 & 5 & 11 & 4 & 12 & 1 & 2 & 9 & 10 & 8 & 7 \end{pmatrix}.$$

This may be factorised as (1 6 12 7)(2 3 5 4 11 8). The elements 9 and 10 are left unaltered, and by convention cycles of length 1 may be omitted from the notation.

Returning briefly to the dihedral groups, by numbering the corners of a regular *n*-gon each symmetry may be identified with that permutation in S_n which permutes the corners accordingly. In this way, for each $n \geq 3$ we may think of each group D_n sitting inside the corresponding S_n. This is an example of a *subgroup*, and we study these in chapter 2. In particular, for $n = 3$ we have $D_3 = S_3$. This is an example of an *isomorphism*, also introduced in chapter 2.

1.13 Modular Arithmetic Groups

Suppose that we count in such a way that after the number 3, instead of proceeding to 4, we return to 0 and begin again. Thus we count 0, 1, 2, 3, 0, 1, 2, 3, 0, ...

Then, for example, 1 + 1 = 2 as usual, but 2 + 2 = 0. We draw a Cayley table for this new binary operation of addition on the set of integers $\{0,1,2,3\}$:

+	0	1	2	3
0	0	1	2	3
1	1	2	3	0
2	2	3	0	1
3	3	0	1	2

This is a group, called a *modular arithmetic group*, working *modulo* 4. It is denoted $\boldsymbol{Z}_4$. We may similarly construct a modular arithmetic group working modulo any natural number. We show how these groups may be constructed in a more mathematically rigorous way.

1.14 Congruence of Integers

Suppose that n is a natural number. For a pair of integers a and b we say that a is *congruent to* b *modulo* n if $a-b$ is divisible by n, so that

$$a-b=\lambda n \text{ for some } \lambda \in \boldsymbol{Z}.$$

When this is the case, we write $a \equiv b \pmod{n}$.

Examples (a) 4 is congruent to 10 modulo 2, since 4 – 10 is divisible by 2. We write $4 \equiv 10 \pmod{2}$.

(b) 16 is congruent to 6 modulo 5, since 16 – 6 is divisible by 5. We write $16 \equiv 6 \pmod{5}$.

(c) 4 is *not* congruent to 10 modulo 5, since 4 – 10 is *not* divisible by 5. We write $4 \not\equiv 10 \pmod{5}$.

1.15 Proposition Congruence modulo *n* is an equivalence relation on the set of integers.

Proof: We must verify that congruence modulo *n* is reflexive, symmetric and transitive.

Reflexive: $a-a=0\cdot n$ and so $a\equiv a(\operatorname{mod} n)$ for each $a\in \boldsymbol{Z}$.

Symmetric: Suppose that $a\equiv b(\operatorname{mod} n)$ so that $a-b=\lambda n$ for some $\lambda\in \boldsymbol{Z}$.

Then $b-a=(-\lambda)n$ for $-\lambda\in \boldsymbol{Z}$ so that $b\equiv a(\operatorname{mod} n)$

Transitive: Suppose that $a\equiv b(\operatorname{mod} n)$ so that $a-b=\lambda_1 n$ for some $\lambda_1\in \boldsymbol{Z}$ and $b\equiv c(\operatorname{mod} n)$ so that $b-c=\lambda_2 n$ for some $\lambda_2\in \boldsymbol{Z}$.

Then $a-c=(a-b)+(b-c)=\lambda_1 n+\lambda_2 n=(\lambda_1+\lambda_2)n$ for $\lambda_1+\lambda_2\in \boldsymbol{Z}$ and hence $a\equiv c(\operatorname{mod} n)$.

The equivalence classes of integers under the relation of congruence are called *congruence classes.*

Examples (a) For $n = 2$ there are two congruence classes, one comprising all even integers and the other comprising all odd integers:

$$\{\ldots-4,-2,0,2,4,6,8,\ldots\}$$
$$\{\ldots-3,-1,1,3,5,7,9,\ldots\}.$$

(b) For $n = 4$ there are four congruence classes:

$$\{\ldots-8,-4,0,4,8,12,\ldots\}$$
$$\{\ldots-7,-3,1,5,9,13,\ldots\}$$
$$\{\ldots-6,-2,2,6,10,14,\ldots\}$$
$$\{\ldots-5,-1,3,7,11,15,\ldots\}.$$

Under congruence modulo *n* there are *n* congruence classes. Notice that each integer belongs to one and only one congruence class. We let $[a]$ denote the congruence class to which the integer a belongs, and say that a is a *representative* of the class $[a]$. We now develop arithmetic of congruence classes.

1.16 Definition For a pair of congruence classes $[a]$ and $[b]$, specified by representatives a and b respectively, the sum of the congruence classes is given by

$$[a]+[b]=[a+b]$$

Example For the congruence classes modulo 4 we may obtain

$$[6]+[7]=[6+7]=[13]$$

This procedure appears to depend upon the *choice* of representatives for the congruence classes. Clearly we could have made different choices, for example $[6]=[10]$ and $[7]=[11]$. Then we would have obtained

$$[10]+[11]=[10+11]=[21]$$

Whilst the answer *appears* to be different, it is actually the same as before since $[13]=[21]$. This is no lucky chance. It turns out that the result will be the same regardless of the choice of representative for each congruence class. We need to show that this is true in general, or that the binary operation $+$ is *well-defined.*

1.17 Proposition Suppose that a and a' are a pair of representatives for a congruence class, so that $[a]=[a']$. Similarly, b and b' are a pair of representatives for another congruence class, so that $[b]=[b']$. Then we have

$$[a+b]=[a'+b']$$

Proof:

$[a]=[a']$ so that $a\equiv a'(\bmod n)$ and $a-a'=\lambda_1 n$ for some $\lambda_1\in\boldsymbol{Z}$.

Similarly $[b]=[b']$ so that $b\equiv b'(\bmod n)$ and $b-b'=\lambda_2 n$ for some $\lambda_2\in\boldsymbol{Z}$.

Then

$$(a+b)-(a'+b')=(a-a')+(b-b')=\lambda_1 n+\lambda_2 n=(\lambda_1+\lambda_2)n$$

with $\lambda_1+\lambda_2\in\boldsymbol{Z}$.

Hence $a+b\equiv a'+b'(\bmod n)$ and so $[a+b]=[a'+b']$.

We denote the set of congruence classes modulo n by $\boldsymbol{Z}_n$.

1.18 Proposition $(\boldsymbol{Z}_n, +)$ is an abelian group.

Proof: We must verify that the group axioms hold for $(\boldsymbol{Z}_n, +)$:

$$\begin{aligned} [a]+[b] &= [a+b] \\ &= [b+a] \\ &= [b]+[a]. \end{aligned}$$

$$\begin{aligned} ([a]+[b])+[c] &= [a+b]+[c] = [(a+b)+c] \\ &= [a+(b+c)] = [a]+[b+c] \\ &= [a]+([b]+[c]). \end{aligned}$$

The identity element is [0]:

$$\begin{aligned} [a]+[0] &= [a+0] = [a] \\ [0]+[a] &= [0+a] = [a]. \end{aligned}$$

For a congruence class $[a]$ the inverse is $[-a]$:

$$\begin{aligned} [a]+[-a] &= [a+(-a)] = [0] \\ [-a]+[a] &= [(-a)+a] = [0]. \end{aligned}$$

Example The group $(\boldsymbol{Z}_4, +)$ has the following Cayley table:

+	[0]	[1]	[2]	[3]
[0]	[0]	[1]	[2]	[3]
[1]	[1]	[2]	[3]	[0]
[2]	[2]	[3]	[0]	[1]
[3]	[3]	[0]	[1]	[2]

Notice that apart from the presence of the brackets, this is the same as the Cayley table that we obtained via our simplistic approach. From now on we omit the brackets around numbers in Cayley tables of modular arithmetic groups.

In a similar way, we may define multiplication of congruence classes:

1.19 Definition The product of a pair of congruence classes $[a]$ and $[b]$ is

$$[a] \cdot [b] = [a \cdot b]$$

Once again we must show that this binary operation is well-defined:

1.20 Proposition If $[a] = [a']$ and $[b] = [b']$ then $[a \cdot b] = [a' \cdot b']$.

Proof:

$[a] = [a']$ so that $a \equiv a' \pmod{n}$ and $a - a' = \lambda_1 n$ for some $\lambda_1 \in \boldsymbol{Z}$.

Similarly, $[b] = [b']$ so that $b \equiv b' \pmod{n}$ and $b - b' = \lambda_2 n$ for some $\lambda_2 \in \boldsymbol{Z}$.

Then

$$\begin{aligned} a \cdot b - a' \cdot b' &= a \cdot b - a \cdot b' + a \cdot b' - a' \cdot b' \\ &= a \cdot (b - b') + (a - a') \cdot b' \\ &= a\lambda_1 n + \lambda_2 n b' \\ &= (\lambda_1 a + \lambda_2 b')n \end{aligned}$$

Hence $a \cdot b \equiv a' \cdot b' \pmod{n}$ so that $[a \cdot b] = [a' \cdot b']$.

We use the notation $\boldsymbol{Z}_p^*$ for the set of non-zero elements of $\boldsymbol{Z}_p$.

1.21 Proposition $(\boldsymbol{Z}_p^*, \cdot)$ is an abelian group iff p is prime.

Proof: Verification of commutativity and associativity is routine. The identity is $[1]$. All that remains is to show the existence of inverses.

Suppose that p is prime and $[a]$ is a non-zero congruence class of $\boldsymbol{Z}_p$.

Then $\gcd(a, p) = 1$. By the Euclidean algorithm there are integers r and s such that $ra + sp = 1$.

Then $[r] \cdot [a] = [ra] = [1 - sp] = [1]$, and hence $[a]$ has an inverse.

If p is *not* prime then there are natural numbers $a, b < p$ such that $p = ab$ so that

$$[a] \cdot [b] = [ab] = [p] = [0]$$

so $(\boldsymbol{Z}_p^*, \cdot)$ is not get a group in this case.

Examples (a) $(\boldsymbol{Z}_5^*, \cdot)$ is a group. Here is its Cayley table:

$\cdot$	1	2	3	4
1	1	2	3	4
2	2	4	1	3
3	3	1	4	2
4	4	3	2	1

(b) $(\boldsymbol{Z}_4^*, \cdot)$ is *not* a group:

$\cdot$	1	2	3
1	1	2	3
2	2	0	2
3	3	2	1

The element 2 has no inverse.

In fact, $\cdot$ is not a binary operation on $\boldsymbol{Z}_4^*$ because $2 \cdot 2 = 0$.

In subsequent chapters, we write $\boldsymbol{Z}_n$ for the additive group $(\boldsymbol{Z}_n, +)$ and we write $\boldsymbol{Z}_p^*$ for the multiplicative group $(\boldsymbol{Z}_p^*, \cdot)$ where p is a prime.

Exercises 1

1. Prove that in a group G if $b * a = c * a$ then $b = c$ for all $a, b, c \in G$.

2. Prove that in the Cayley table of a finite group, each element appears exactly once in each row and exactly once in each column.

3. Write each element of S_3 in cycle notation.

4. In S_5 let σ and π be the permutations

$$\sigma = \begin{pmatrix} 1 & 2 & 3 & 4 & 5 \\ 3 & 2 & 5 & 1 & 4 \end{pmatrix} \qquad \pi = \begin{pmatrix} 1 & 2 & 3 & 4 & 5 \\ 4 & 3 & 1 & 2 & 5 \end{pmatrix}.$$

Write down the permutations $\sigma \circ \pi$, $\pi \circ \sigma, \sigma^{-1}$ and π^2

5. Express the following permutation in S_{12} as a product of cycles:

$$\begin{pmatrix} 1 & 2 & 3 & 4 & 5 & 6 & 7 & 8 & 9 & 10 & 11 & 12 \\ 3 & 9 & 5 & 8 & 7 & 11 & 1 & 12 & 2 & 6 & 10 & 4 \end{pmatrix}.$$

6. Express the following product of cycles in S_{10} as a permutation:

$$(\ 1\ 4\ 6\ )(\ 2\ 3\ 9\ 8\ )(\ 5\ 7\ ).$$

7. In S_6, find the permutation σ in each of the following:

(i) $$\begin{pmatrix} 1 & 2 & 3 & 4 & 5 & 6 \\ 2 & 5 & 1 & 4 & 3 & 6 \end{pmatrix} \circ \sigma = \begin{pmatrix} 1 & 2 & 3 & 4 & 5 & 6 \\ 6 & 1 & 2 & 5 & 4 & 3 \end{pmatrix}$$

(ii) $$\sigma \circ \begin{pmatrix} 1 & 2 & 3 & 4 & 5 & 6 \\ 4 & 2 & 1 & 3 & 6 & 5 \end{pmatrix} = \begin{pmatrix} 1 & 2 & 3 & 4 & 5 & 6 \\ 6 & 5 & 4 & 3 & 2 & 1 \end{pmatrix}.$$

8. Consider the symmetries of a rectangle: I is the identity, and R is a rotation of 180 degrees. Reflections in the horizontal and vertical centre lines are denoted M_1 and M_2.

Draw the Cayley table for the group of symmetries of a rectangle.

Consider the symmetries of a square: I is the identity, and R is a rotation of 90 degrees clockwise. Reflections in the four lines of symmetry (horizontal, vertical and the two diagonals) are denoted M_1, M_2, M_3 and M_4, labelled in clockwise sequence.

Draw the Cayley table for the group D_4 of symmetries of the square.

9. Write down the congruence classes of the integers modulo 5.

10. Draw the Cayley table for each of the groups $(\boldsymbol{Z}_5,+),(\boldsymbol{Z}_6,+)$ and $(\boldsymbol{Z}_7^*,\cdot)$

Chapter 2: Subgroups and Group Homomorphisms

2.1 Definition Suppose that $(G,*)$ is a group and $H \subseteq G$. We say that H is a *subgroup* of G if $(H,*)$ is itself a group under the *same* binary operation as G.

Notice that for any group G, the trivial group $\{e\}$ consisting of just the identity element and the group G itself are both subgroups of G. A subgroup which is not G itself is called a *proper subgroup* of G.

Examples (a) $(2\boldsymbol{Z},+)$ is a subgroup of $(\boldsymbol{Z},+)$.

In fact, for any natural number n the group $(n\boldsymbol{Z},+)$ is a subgroup of $(\boldsymbol{Z},+)$.

(b) $(\boldsymbol{Q}^*,\cdot)$ is a subgroup of $(\boldsymbol{R}^*,\cdot)$, and $(\boldsymbol{Q},+)$ is a subgroup of $(\boldsymbol{R},+)$.

(c) The set of $n \times n$ matrices with entries in $\boldsymbol{Q}$ and determinant 1 is a group under matrix multiplication known as the *special linear group*, denoted $SL(n,\boldsymbol{Q})$. For each natural number n, $SL(n,\boldsymbol{Q})$ is a subgroup of $GL(n,\boldsymbol{Q})$.

2.2 Test for a Subgroup Suppose that $(G,*)$ is a group and H is a non-empty subset of G.

Then H is a subgroup of G if and only if

(i) if $a,b \in H$ then $a*b \in H$

(ii) if $a \in H$ then $a^{-1} \in H$

Proof: "$\Rightarrow$" We have (i) because $*$ must also be a binary operation on H.

Since $(H,*)$ is a group, any element $a \in H$ must have an inverse a^{-1} in H so we have (ii).

"$\Leftarrow$" We need to verify G1, G2 and G3. G1 is immediate - since it holds for the whole of G it must hold for the subset H.

By G3 for G, any element $a \in H$ has an inverse a^{-1}. By (ii) this inverse is an element of H. For $a \in H$ we have $a^{-1} \in H$ such that $a * a^{-1} = e \in G$, using G2 for G. By (i) we have $e \in H$.

Examples (a) If $a, b \in n\boldsymbol{Z}$ then $a = \lambda n$ and $b = \mu n$ for some $\lambda, \mu \in \boldsymbol{Z}$. Then

$$a + b = \lambda n + \mu n = (\lambda + \mu)n \text{ with } \lambda + \mu \in \boldsymbol{Z}$$

Hence $a + b \in n\boldsymbol{Z}$. Also

$$-a = -(\lambda n) = (-\lambda)n \text{ with } -\lambda \in \boldsymbol{Z}$$

Hence $-a \in n\boldsymbol{Z}$. By 2.2 $(n\boldsymbol{Z}, +)$ is a subgroup of $(\boldsymbol{Z}, +)$.

(b) Let $H = \left\{ \begin{pmatrix} a & b \\ 0 & a \end{pmatrix} : a, b \in \boldsymbol{Q}, a \neq 0 \right\}$

$$\begin{pmatrix} a_1 & b_1 \\ 0 & a_1 \end{pmatrix} \begin{pmatrix} a_2 & b_2 \\ 0 & a_2 \end{pmatrix} = \begin{pmatrix} a_1 a_2 & a_1 b_2 + a_2 b_1 \\ 0 & a_1 a_2 \end{pmatrix} \in H$$

$$\begin{pmatrix} a & b \\ 0 & a \end{pmatrix}^{-1} = \frac{1}{a^2} \begin{pmatrix} a & -b \\ 0 & a \end{pmatrix} = \begin{pmatrix} \frac{1}{a} & -\frac{b}{a^2} \\ 0 & \frac{1}{a} \end{pmatrix} \in H.$$

Hence by 2.2, H is a subgroup of $GL(2, \boldsymbol{Q})$.

2.3 Proposition Suppose that H_1 and H_2 are subgroups of a group G.

Then their intersection $H_1 \cap H_2$ is also a subgroup of G.

Proof: Suppose that $a, b \in H_1 \cap H_2$. Then $a, b \in H_1$ and $a, b \in H_2$. We have $ab \in H_1$ because H_1 is a subgroup and $ab \in H_2$ because H_2 is a subgroup. Hence $ab \in H_1 \cap H_2$.

Also, $a^{-1} \in H_1$ and $a^{-1} \in H_2$ because H_1 and H_2 are subgroups and hence $a^{-1} \in H_1 \cap H_2$.

By 2.2 it follows that $H_1 \cap H_2$ is a subgroup of G.

Example $(2\boldsymbol{Z},+)$ and $(3\boldsymbol{Z},+)$ are subgroups of $(\boldsymbol{Z},+)$.

By 2.3, $2\boldsymbol{Z}\cap 3\boldsymbol{Z}=6\boldsymbol{Z}$ is also a subgroup of $(\boldsymbol{Z},+)$.

2.4 Definition Suppose that G is a group and $a\in G$. The *centraliser* of a in G is the set of all those elements of G which commute with a:

$$C(a)=\{x\in G : xa=ax\}$$

Examples

We shall find the centralisers of three matrices in $GL(2,\boldsymbol{Q})$.

(a) $A=\begin{pmatrix}1 & -1\\ 0 & 1\end{pmatrix}$

If $\begin{pmatrix}a & b\\ c & d\end{pmatrix}$ belongs to the centraliser then

$$\begin{pmatrix}1 & -1\\ 0 & 1\end{pmatrix}\begin{pmatrix}a & b\\ c & d\end{pmatrix}=\begin{pmatrix}a & b\\ c & d\end{pmatrix}\begin{pmatrix}1 & -1\\ 0 & 1\end{pmatrix}$$

Hence $\begin{pmatrix}a-c & b-d\\ c & d\end{pmatrix}=\begin{pmatrix}a & -a+b\\ c & -c+d\end{pmatrix}$

By comparing the top left entries we have $a-c=a$ and so $c=0$.

Similarly, from the top right entries $b-d=-a+b$ and so $a=d$.

Hence $C(A)=\left\{\begin{pmatrix}a & b\\ 0 & a\end{pmatrix} : a,b\in\boldsymbol{Q}, a\neq 0\right\}$.

(b) $B=\begin{pmatrix}2 & 1\\ -1 & 3\end{pmatrix}$ $\qquad \begin{pmatrix}2 & 1\\ -1 & 3\end{pmatrix}\begin{pmatrix}a & b\\ c & d\end{pmatrix}=\begin{pmatrix}a & b\\ c & d\end{pmatrix}\begin{pmatrix}2 & 1\\ -1 & 3\end{pmatrix}$

Hence $\begin{pmatrix}2a+c & 2b+d\\ -a+3c & -b+3d\end{pmatrix}=\begin{pmatrix}2a-b & a+3b\\ 2c-d & c+3d\end{pmatrix}$

Top left: $2a+c=2a-b$ and so $c=-b$.

Top right: $2b+d=a+3b$ and so $d=a+b$.

Hence $C(B)=\left\{\begin{pmatrix} a & b \\ -b & a+b \end{pmatrix} : a,b\in \boldsymbol{Q}, a^2+ab+b^2\neq 0\right\}.$

(c) $I=\begin{pmatrix} 1 & 0 \\ 0 & 1 \end{pmatrix}$

$C(I)=GL(2,\boldsymbol{Q})$, since any 2×2 matrix will commute with I.

2.5 Proposition For any group G and any $a\in G$ the centraliser $C(a)$ is a subgroup of G.

Proof: Suppose that $x,y\in C(a)$ so that $xa=ax$ and $ya=ay$.

Then $(xy)a=x(ya)=x(ay)=(xa)y=(ax)y=a(xy)$

so that we have $xy\in C(a)$. Also

$$x^{-1}a=(x^{-1}a)e=(x^{-1}a)(xx^{-1})=x^{-1}(ax)x^{-1}$$
$$=x^{-1}(xa)x^{-1}=(x^{-1}x)(ax^{-1})=e(ax^{-1})=ax^{-1}$$

so that we have $x^{-1}\in C(a)$. Hence by 2.2 $C(a)$ is a subgroup of G.

For $a\in G$ and n a natural number we write

a^n for $\underbrace{a*a*\ldots*a}_{n}$ and a^{-n} for $\underbrace{a^{-1}*a^{-1}*\ldots*a^{-1}}_{n}$

By convention, we take $a^0=e$, the identity.

2.6 Definition Suppose that G is a group and $a\in G$. The group

$$\{a^n : n\in \boldsymbol{Z}\}$$

is called the *subgroup of G generated by* a, denoted $\langle a\rangle$.

Examples (a) Consider the group of rotational symmetries of a square

$$\{0°, 90°, 180°, 270°\}.$$

We have $\langle 0° \rangle = \{0°\}, \langle 180° \rangle = \{0°, 180°\}$ and $\langle 90° \rangle = \{0°, 90°, 180°, 270°\}$.

(b) In $(\boldsymbol{Z}, +)$ we have $\langle 2 \rangle = \{\ldots, -2, 0, 2, 4, 6, 8, \ldots\}$.

In $(\boldsymbol{Q}^*, \cdot)$ we have $\langle 2 \rangle = \{\ldots, \frac{1}{4}, \frac{1}{2}, 1, 2, 4, 8, \ldots\}$

2.7 Definition Let a be an element of a group G. If $a^n \neq e$ for all positive n then we say that a has an *infinite order*. Otherwise, the smallest positive n for which $a^n = e$ is called the *order* of a.

Examples (a) Consider the group of rotational symmetries of a regular hexagon, $\{0°, 60°, 120°, 180°, 240°, 300°\}$.

0° is of order 1

60° is of order 6

120° is of order 3

180° is of order 2

240° is of order 3

300° is of order 6.

(b) In the group $(\boldsymbol{Z}, +)$ *every* element, apart from 0, is of infinite order.

Notice that in any group the identity element is the *only* element of order 1.

2.8 Definition A group G is said to be *cyclic* if it is generated by a single element, so that $G = \langle a \rangle$ for some $a \in G$.

Examples (a) Let $G = \{0°, 60°, 120°, 180°, 240°, 300°\}$ be the group of rotational symmetries of a regular hexagon. Then G is cyclic since $G = \langle 60° \rangle$.

(b) $(\boldsymbol{Z}, +)$ is cyclic, since it is generated by the integer 1.

(c) $(\boldsymbol{Q},+)$ is *not* cyclic. Can you prove this?

2.9 Proposition Any cyclic group is abelian.

Proof: The proof is left as an exercise.

2.10 Definition Suppose that $(G,*)$ and $(H,\circ)$ are groups. A mapping $f\colon G \to H$ is called a *homomorphism* of groups if

$$f(a*b) = f(a)\circ f(b)$$

for every pair of elements $a,b \in G$.

If f is surjective then it is called an *epimorphism* of groups; if f is injective then it is called a *monomorphism* of groups; if f is surjective and injective then it is called an *isomorphism* of groups. If there is an isomorphism $f\colon G \to H$ then we say that the groups G and H are *isomorphic* and we write $G \cong H$.

Examples (a) The mapping $f\colon GL(2,\boldsymbol{R}) \to (\boldsymbol{R}^*,\cdot)$ given by $f(A) = \det A$ is a homomorphism, since for $A,B \in GL(2,\boldsymbol{R})$ we have $\det AB = \det A \det B$. (This well-known fact is proved in 21.8) It is surjective since for any non-zero $x \in \boldsymbol{R}$ we may take

$$A = \begin{pmatrix} x & 0 \\ 0 & 1 \end{pmatrix}$$

so that

$$f(A) = \det\begin{pmatrix} x & 0 \\ 0 & 1 \end{pmatrix} = x \times 1 - 0 \times 0 = x.$$

Hence f is an epimorphism. However f is not injective since, for example, for

$$A = \begin{pmatrix} 2 & 0 \\ 0 & 2 \end{pmatrix} \text{ and } B = \begin{pmatrix} 5 & 1 \\ 1 & 1 \end{pmatrix}$$

we have both $f(A) = 4$ and $f(B) = 4$. Hence f is not a monomorphism and not an isomorphism.

(b) The mapping $f: (\boldsymbol{Z}, +) \to (\boldsymbol{Z}, +)$ given by $f(n) = 2n$ is a homomorphism, since for $m, n \in \boldsymbol{Z}$ we have

$$f(m+n) = 2(m+n) = 2m + 2n = f(m) + f(n).$$

It is injective since for $m, n \in \boldsymbol{Z}$, if $f(m) = f(n)$ then $2m = 2n$ so that $m = n$.

Hence f is a monomorphism. However f is not surjective since, for example, there is no integer n such that $f(n) = 3$. Hence f is not an epimorphism and not an isomorphism.

(c) Let $\boldsymbol{R}^+$ denote the set of positive real numbers. The mapping $f: (\boldsymbol{R}, +) \to (\boldsymbol{R}^+, \cdot)$ given by $f(x) = 10^x$ is a homomorphism, since for $x, y \in \boldsymbol{R}$ we have

$$f(x+y) = 10^{x+y} = 10^x \cdot 10^y = f(x) \cdot f(y).$$

It is surjective, since for any $y \in \boldsymbol{R}^+$ we may take $x = \log_{10} y \in \boldsymbol{R}$ so that $f(x) = y$. Hence f is an epimorphism. It is also injective since

$$\text{if } f(x) = f(y) \text{ so that } 10^x = 10^y \text{ then } x = y.$$

Hence f is a monomorphism and also an isomorphism, and so we may write $(\boldsymbol{R}, +) \cong (\boldsymbol{R}^+, \cdot)$.

(d) The mapping $f: (\boldsymbol{Z}, +) \to (\boldsymbol{Z}, +)$ given by $f(n) = n^2$ is *not* a homomorphism.

For example $f(1+1) = f(2) = 2^2 = 4$ whereas

$$f(1) + f(1) = 1^2 + 1^2 = 2.$$

2.11 Proposition For each prime p the group $(\boldsymbol{Z}_p^*, \cdot)$ is isomorphic to the group $(\boldsymbol{Z}_{p-1}, +)$

Examples (a) $(\boldsymbol{Z}_5^*, \cdot) \cong (\boldsymbol{Z}_4, +)$ via the following isomorphism:

$$1 \leftrightarrow 0$$
$$2 \leftrightarrow 1$$
$$3 \leftrightarrow 3$$
$$4 \leftrightarrow 2$$

(b) By observing that $(\mathbf{Z}_{10}, +)$ is cyclic and generated by 1, and that $(\mathbf{Z}^*_{11}, \cdot)$ is cyclic and generated by 2, we may obtain the following isomorphism:

$\mathbf{Z}_{10}$	0	1	2	3	4	5	6	7	8	9
$\mathbf{Z}^*_{11}$	1	2	4	8	5	10	9	7	3	6

There is no algorithm for finding a generator for the multiplicative group - it is just a matter of trial and error.

2.12 Definition The *(external) direct product* of a pair of groups G and H is the set of ordered pairs $G \times H$ under the binary operation given by

$$(g_1, h_1) \cdot (g_2, h_2) = (g_1 g_2, h_1 h_2)$$

for $g_1, g_2 \in G$ and $h_1, h_2 \in H$. It is routine to check that the group axioms are satisfied.

Examples (a) Consider the group of integers $\mathbf{Z}$ under addition. The direct product of $\mathbf{Z}$ with itself is $\mathbf{Z} \times \mathbf{Z}$ under the binary operation

$$(m_1, n_1) + (m_2, n_2) = (m_1 + m_2, n_1 + n_2)$$

for integers m_1, m_2, n_1 and n_2.

(b) $\mathbf{Z}_2 \times \mathbf{Z}_2$ is isomorphic to the Klein four-group, V.

(c) $\mathbf{Z}_2 \times \mathbf{Z}_3$ is isomorphic to $\mathbf{Z}_6$. To see this, observe that $\mathbf{Z}_2 \times \mathbf{Z}_3$ is cyclic, with generator $(1,1)$. The following table gives an isomorphism:

$\mathbf{Z}_6$	0	1	2	3	4	5
$\mathbf{Z}_2 \times \mathbf{Z}_3$	$(0,0)$	$(1,1)$	$(0,2)$	$(1,0)$	$(0,1)$	$(1,2)$

More generally, if m and n are coprime then $\mathbf{Z}_m \times \mathbf{Z}_n \cong \mathbf{Z}_{mn}$.

(d) $Z_2 \times Z_4$ is not isomorphic to Z_8. To see this, observe that the element 1 in Z_8 is of order 8, whereas in $Z_2 \times Z_4$ all non-identity elements are of order 2 or order 4.

2.13 Proposition Suppose that $f: G \to H$ is a homomorphism of groups and that e_G and e_H are the identity elements of G and H respectively. Then we have:

(i) $f(e_G) = e_H$

(ii) $f(g^{-1}) = f(g)^{-1}$ for all $g \in G$.

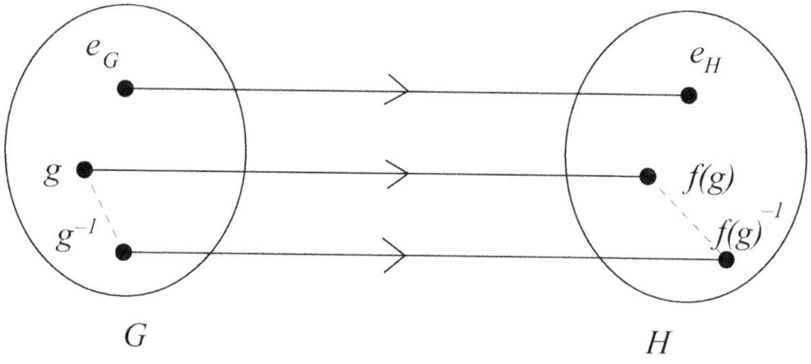

Proof: (i) For any $g \in G$ we have $e_G \cdot g = g$ and so $f(e_G \cdot g) = f(g)$. But f is a homomorphism and so $f(e_G) \cdot f(g) = f(g) = e_H \cdot f(g)$

Cancelling using 1.3 we have $f(e_G) = e_H$.

(ii) For any $g \in G$ we have $gg^{-1} = e_G$, and so $f(gg^{-1}) = f(e_G) = e_H$ by (i) above.

But f is a homomorphism and so $f(g)f(g^{-1}) = e_H$.

Similarly, $f(g^{-1})f(g) = e_H$. We have shown that $f(g)$ and $f(g^{-1})$ are inverses in H.

Hence by uniqueness of inverses from 1.2 we have $f(g^{-1}) = f(g)^{-1}$.

2.14 Definition Suppose $f: G \to H$ is a homomorphism of groups. The *image* of f is $\mathrm{im} f = \{f(g) : g \in G\}$. The *kernel* of f is $\ker f = \{g \in G : f(g) = e_H\}$.

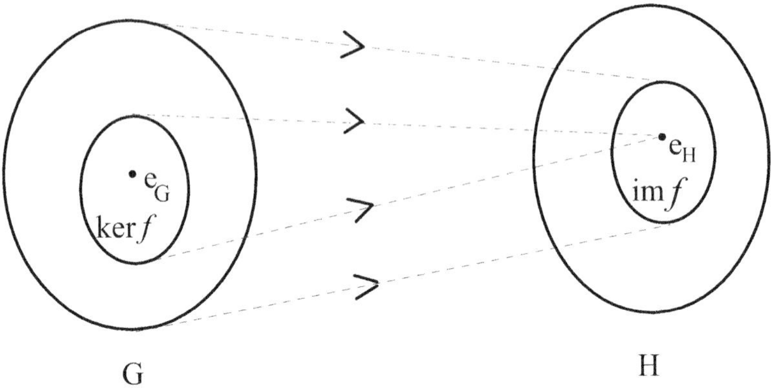

Examples (a) For $f: GL(2, \boldsymbol{R}) \to (\boldsymbol{R}^*, \cdot)$ given by $f(A) = \det A$ we have

$$\mathrm{im} f = \boldsymbol{R}^*$$

$$\ker f = \{A \in GL(2, \boldsymbol{R}) : \det A = 1\} = SL(2, \boldsymbol{R}).$$

The group $SL(2, \boldsymbol{R})$ is an example of a *special linear group.*

(b) For $f: (\boldsymbol{Z}, +) \to (\boldsymbol{Z}, +)$ given by $f(n) = 2n$ we have

$$\mathrm{im} f = 2\boldsymbol{Z}$$

$$\ker f = \{n \in \boldsymbol{Z} : 2n = 0\} = \{0\}$$

2.15 Proposition If $f : (G, *) \to (H, \circ)$ is a homomorphism of groups then we have the following:

(i) im f is a subgroup of H

(ii) f is an epimorphism iff im $f = H$

(iii) ker f is a subgroup of G

(iv) f is a monomorphism iff ker $f = \{e_G\}$.

Proof: (i) If $h_1, h_2 \in \text{im} f$ then there are $g_1, g_2 \in G$ such that $f(g_1) = h_1$ and $f(g_2) = h_2$.

Then $f(g_1 * g_2) = f(g_1) \circ f(g_2) = h_1 \circ h_2$ so that $h_1 \circ h_2 \in \text{im} f$.

Also, by 2.13(ii), we have $f(g_1^{-1}) = f(g_1)^{-1} = h_1^{-1}$ so that $h_1^{-1} \in \text{im} f$.

Hence by 2.2 we conclude that im f is a subgroup of H.

(ii) f is an epimorphism iff im $f = H$ is an immediate consequence of the definitions.

(iii) If $g_1, g_2 \in \ker f$ then $f(g_1) = e_H$ and $f(g_2) = e_H$.

Then $f(g_1 * g_2) = f(g_1) \circ f(g_2) = e_H \circ e_H = e_H$ so that $g_1 * g_2 \in \ker f$.

Also, using 2.13(ii), we have $f(g_1^{-1}) = f(g_1)^{-1} = e_H^{-1} = e_H$ so that $g_1^{-1} \in \ker f$.

Hence by 2.2 we conclude that ker f is a subgroup of G.

(iv) Finally we show that f is a monomorphism iff ker $f = \{e_G\}$:

" $\Rightarrow$ " Suppose f is a monomorphism and suppose $a \in \ker f$ so that $f(a) = e_H$.

But also $f(e_G) = e_H$ by 2.13(i)

If $f(a) = f(e_G)$ then $a = e_G$. Hence ker $f = \{e_G\}$.

"$\Leftarrow$" Suppose $\ker f = \{e_G\}$ and suppose $f(a) = f(b)$.

Then $f(a)f(b)^{-1} = f(b)f(b)^{-1} = e_H$ and so $f(ab^{-1}) = e_H$.

It follows that $ab^{-1} \in \ker f$, so that $ab^{-1} = e_G$ and so $a = b$.

Hence f is a monomorphism.

Examples For the two examples in 2.14 above:

(a) $\operatorname{im} f = \boldsymbol{R}^*$, so f is an epimorphism

$\ker f = SL(2, \boldsymbol{R})$, so f is *not* a monomorphism.

(b) $\operatorname{im} f = 2\boldsymbol{Z}$, so f is *not* an epimorphism.

$\ker f = \{0\}$, so f is a monomorphism.

We return to the study of groups in chapter 8.

Exercises 2

1. Find the order of each of the elements in the group of rotational symmetries of a regular octagon.

2. List all of the proper non-trivial subgroups of the group of rotational symmetries of

   (i) the square

   (ii) the regular hexagon

   (iii) the regular octagon

   (iv) the regular pentagon.

3. Explain why

   (i) $(\boldsymbol{Z},+)$ is a subgroup of $(\boldsymbol{R},+)$.

   (ii) $(\boldsymbol{R}^+,\cdot)$ is a subgroup of $(\boldsymbol{R}^*,\cdot)$, where $\boldsymbol{R}^+$ is the set of positive real numbers.

   (iii) $SL(2,\boldsymbol{R})$ is a subgroup of $GL(2,\boldsymbol{R})$.

4. List the proper subgroups of

   (i) the symmetry group of a rectangle

   (ii) the groups D_4 and D_5.

5. Let $H=\left\{\begin{pmatrix} a & b \\ -b & a+b \end{pmatrix} : a, b \in \boldsymbol{Q}, a^2+ab+b^2 \neq 0\right\}$.

   Using 2.2, verify that H is a subgroup of $GL(2,\boldsymbol{Q})$.

6. Find the order of each element in

   (i) D_4 (ii) Q_8 (iii) $\boldsymbol{Z}_8$

7. Prove that every cyclic group is abelian.

Is the converse, that any abelian group is cyclic, true?

8. Find the centralisers of the following matrices in $GL(2, \boldsymbol{Q})$:

(i) $\begin{pmatrix} 1 & 1 \\ 0 & 1 \end{pmatrix}$ (ii) $\begin{pmatrix} 1 & 2 \\ 3 & 4 \end{pmatrix}$.

9. State, with reasons, whether or not each of the following mappings is an epimorphism, a monomorphism or an isomorphism:

(a) $f: (\boldsymbol{Q}^*, \cdot) \to (\boldsymbol{Q}^*, \cdot)$ given by $f(x) = 2x$.

(b) $f: (\boldsymbol{Z}, +) \to (2\boldsymbol{Z}, +)$ given by $f(n) = 4n$.

(c) $f: GL(2, \boldsymbol{Q}) \to (\boldsymbol{Q}^*, \cdot)$ given by $f(M) = \det M$.

(d) $f: (\boldsymbol{R}, +) \to (\boldsymbol{R}, +)$ given by $f(x) = 3x$.

10. Find the order of each element in $(\boldsymbol{Z}_6, +)$ and in $(\boldsymbol{Z}_5, +)$. Find all of the proper non-trivial subgroups of $(\boldsymbol{Z}_6, +)$ and of $(\boldsymbol{Z}_5, +)$.

11. Consider the set of complex numbers $G = \{1, i, -1, -i\}$.

Draw the Cayley table for multiplication on G and show that

$$G \cong \boldsymbol{Z}_4.$$

12. Show that $(\boldsymbol{Z}_6, +) \cong (\boldsymbol{Z}_7^*, \cdot)$.

13. Verify the group axioms for the external direct product $G \times H$ of groups G and H.

14. Show that the Klein four-group is isomorphic to $\boldsymbol{Z}_2 \times \boldsymbol{Z}_2$.

15. Show that $\boldsymbol{Z}_2 \times \boldsymbol{Z}_5 \cong \boldsymbol{Z}_{10}$ and $\boldsymbol{Z}_3 \times \boldsymbol{Z}_5 \cong \boldsymbol{Z}_{15}$

16. Is $\boldsymbol{Z}_3 \times \boldsymbol{Z}_6 \cong \boldsymbol{Z}_{18}$? Why, or why not?

Chapter 3: Vector Spaces

3.1 Vectors in N-dimensional space

We begin by discussing the distinction between *vector quantities* and *scalar quantities* which arises early in the study of physics and applied mathematics. A *vector quantity* is one which has both magnitude and direction. Examples include velocity, acceleration and force. For example, a force pushes with a certain strength in a particular direction. By contrast, a *scalar quantity* is one which has only magnitude. Examples include mass, area, volume and energy.

Shortly we shall introduce a substantial generalisation of the concept of vector when we introduce structures called *vector spaces*. However, we begin by considering vectors representing quantities having direction in 2 or 3-dimensional space (the precise definition of *dimension* is given in chapter 5).

Consider the plane equipped with Cartesian coordinates. Each point P in the plane may be specified by giving its coordinates (x, y). The origin has coordinates $(0, 0)$. A vector quantity in 2 dimensions may represented by a directed line segment in the plane, beginning at the origin and ending at the point P. The length of the line segment represents the magnitude of the quantity, and the direction of the line segment represents the direction of the quantity. Such a vector quantity v may be written as the matrix

$$v = \begin{pmatrix} x \\ y \end{pmatrix}$$

We call such matrices *column vectors* or, where no ambiguity arises, simply *vectors*. The set of such vectors is denoted $\boldsymbol{R}^2$. The *magnitude* of a vector, written $|v|$, is given by the length of its line segment. This may be calculated using the well-known theorem of Pythagoras:

$$|v| = \sqrt{x^2 + y^2}\,.$$

Examples If $v = \begin{pmatrix} 3 \\ 4 \end{pmatrix}$ then $|v| = \sqrt{3^2 + 4^2} = \sqrt{25} = 5$.

If $v = \begin{pmatrix} 12 \\ -5 \end{pmatrix}$ then $|v| = \sqrt{12^2 + (-5)^2} = \sqrt{169} = 13$.

There are two important operations which may be carried out on vectors. The first is *vector addition*:

For a pair of vectors $u = \begin{pmatrix} x_1 \\ y_1 \end{pmatrix}$ and $v = \begin{pmatrix} x_2 \\ y_2 \end{pmatrix}$ their sum is the vector

$$u + v = \begin{pmatrix} x_1 + x_2 \\ y_1 + y_2 \end{pmatrix}.$$

Example If $u = \begin{pmatrix} 3 \\ 4 \end{pmatrix}$ and $v = \begin{pmatrix} 5 \\ 2 \end{pmatrix}$ then their sum is $u + v = \begin{pmatrix} 8 \\ 6 \end{pmatrix}$.

The second operation is *multiplication by a scalar.* We may multiply a vector by a real number to give a new vector.

For a vector $v = \begin{pmatrix} x \\ y \end{pmatrix}$ and a real number λ the multiplication of v by the scalar λ is $\lambda v = \begin{pmatrix} \lambda x \\ \lambda y \end{pmatrix}$.

Example If $v = \begin{pmatrix} 3 \\ 4 \end{pmatrix}$ and $\lambda = 10$ then $\lambda v = \begin{pmatrix} 30 \\ 40 \end{pmatrix}$.

In particular, notice that for $\lambda = -1$ we have $\lambda v = \begin{pmatrix} -x \\ -y \end{pmatrix}$, which is usually written as $-v$.

The set of all 2 dimensional vectors is an *abelian group* under vector addition, because we have:

Commutativity: $u + v = v + u$ for all vectors u and v.

Associativity: $(u + v) + w = u + (v + w)$ for all vectors u, v and w.

Identity: The zero vector $\mathbf{0} = \begin{pmatrix} 0 \\ 0 \end{pmatrix}$ satisfies $v + \mathbf{0} = v = \mathbf{0} + v$ for all vector v

Inverses: For each vector v the vector $-v$ satisfies $v + (-v) = \mathbf{0} = (-v) + v$

When the generalisation of the concept of vector is introduced shortly, these will also be abelian groups.

Another operation on vectors is the *scalar product,* which combines a pair of vectors to give a real number:

3.2 Definition For vectors u and v in $\boldsymbol{R}^2$ the *scalar product* is

$$u.v = |u||v|\cos\theta$$

where θ is the angle between the two vectors.

The scalar product may be conveniently calculated in terms of the components of column vectors, as the following result shows:

3.3 Proposition If $u = \begin{pmatrix} x_1 \\ y_1 \end{pmatrix}$ and $v = \begin{pmatrix} x_2 \\ y_2 \end{pmatrix}$ then the scalar product is given by

$$u.v = x_1x_2 + y_1y_2.$$

Proof: Suppose that $u = \begin{pmatrix} x_1 \\ y_1 \end{pmatrix}$ and $v = \begin{pmatrix} x_2 \\ y_2 \end{pmatrix}$ and that the angle between u and v is θ.

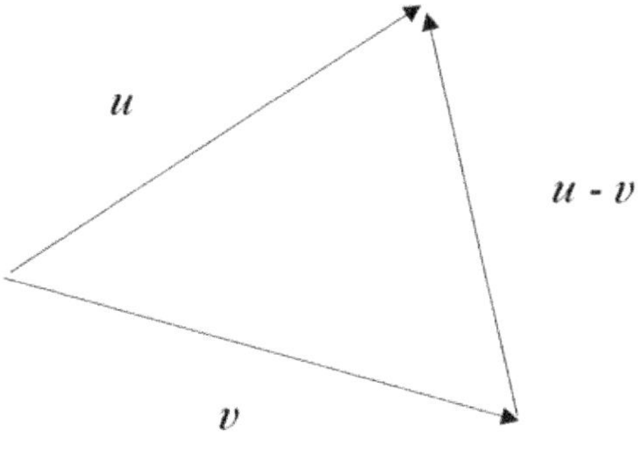

Applying the cosine rule to the triangle of vectors above, we have

$$|u - v|^2 = |u|^2 + |v|^2 - 2|u||v|\cos\theta$$

Meanwhile, using the definition of magnitude we obtain

$$\begin{aligned} |u - v|^2 &= (x_1 - x_2)^2 + (y_1 - y_2)^2 \\ &= x_1^2 + x_2^2 + y_1^2 + y_2^2 - 2(x_1x_2 + y_1y_2) \\ &= |u|^2 + |v|^2 - 2(x_1x_2 + y_1y_2) \end{aligned}$$

By comparing these two formulae we see that

$$|u||v|\cos\theta = x_1x_2 + y_1y_2$$

Example If $u = \binom{3}{4}$ and $v = \binom{5}{2}$ then the scalar product is given by

$$u.v = 3\times 5 + 4\times 2 = 23.$$

We may rearrange 3.2 to obtain $\cos\theta = \frac{u.v}{|u||v|}$ to find the angle between vectors written as column vectors.

Example For u and v as in the example above, we have

$$u.v = 23,\ |u| = 5 \text{ and } |v| = \sqrt{29}\,.$$

Hence $\cos\theta = \frac{23}{5\sqrt{29}} = 0.854$ and $\theta = 31^\circ$, approximated to the nearest degree.

All of our definitions and results for vectors in 2 dimensions may be matched with similar definitions and results for vectors in 3 dimensions. A vector quantity in 3 dimensions may be represented by a column vector

$$\begin{pmatrix} x \\ y \\ z \end{pmatrix}$$

Vector addition is defined as follows:

$$\text{if } u = \begin{pmatrix} x_1 \\ y_1 \\ z_1 \end{pmatrix} \text{ and } v = \begin{pmatrix} x_2 \\ y_2 \\ z_2 \end{pmatrix} \text{ then } u+v = \begin{pmatrix} x_1+x_2 \\ y_1+y_2 \\ z_1+z_2 \end{pmatrix}.$$

The set of three dimensional vectors, denoted $\boldsymbol{R}^3$, is an abelian group under vector addition. Multiplication by a scalar is as follows:

$$\text{If } v = \begin{pmatrix} x \\ y \\ z \end{pmatrix} \text{ and } \lambda \in \boldsymbol{R} \text{ then } \lambda v = \begin{pmatrix} \lambda x \\ \lambda y \\ \lambda z \end{pmatrix}.$$

The magnitude of v is given by $|v| = \sqrt{x^2 + y^2 + z^2}$.

As before, the scalar product is $u.v = |u||v| \cos\theta$, where θ is the angle between the two vectors, and for column vectors

$$u = \begin{pmatrix} x_1 \\ y_1 \\ z_1 \end{pmatrix} \text{ and } v = \begin{pmatrix} x_2 \\ y_2 \\ z_2 \end{pmatrix}$$

the scalar product may be calculated as $u.v = x_1x_2 + y_1y_2 + z_1z_2$.

Indeed, we may define vectors in N dimensions, for each natural number N, and treat these as column vectors with N entries. We define vector addition, multiplication by a scalar, magnitude and scalar product in ways similar to above to give a vector space $\boldsymbol{R}^N$.

3.4 Abstract Vector Spaces

We shall now introduce the concept of a vector space. We motivate this by returning to vectors in $\boldsymbol{R}^2$. Recall that we add together a pair of vectors in $\boldsymbol{R}^2$ as follows:

$$\begin{pmatrix} x_1 \\ y_1 \end{pmatrix} + \begin{pmatrix} x_2 \\ y_2 \end{pmatrix} = \begin{pmatrix} x_1 + x_2 \\ y_1 + y_2 \end{pmatrix}$$

This binary operation on $\boldsymbol{R}^2$ is commutative and associative. The zero vector provides an identity, and inverses may be obtained by changing the signs of both entries. Hence $\boldsymbol{R}^2$ is an abelian group under vector addition.

Furthermore, multiplication by a scalar given by

$$\lambda \begin{pmatrix} x \\ y \end{pmatrix} = \begin{pmatrix} \lambda x \\ \lambda y \end{pmatrix}$$

satisfies the following:

$$\textbf{V1:}\quad (\lambda+\mu)\begin{pmatrix}x\\y\end{pmatrix}=\lambda\begin{pmatrix}x\\y\end{pmatrix}+\mu\begin{pmatrix}x\\y\end{pmatrix}$$

$$\textbf{V2:}\quad \lambda\left\{\begin{pmatrix}x_1\\y_1\end{pmatrix}+\begin{pmatrix}x_2\\y_2\end{pmatrix}\right\}=\lambda\begin{pmatrix}x_1\\y_1\end{pmatrix}+\lambda\begin{pmatrix}x_2\\y_2\end{pmatrix}$$

$$\textbf{V3:}\quad \lambda\left(\mu\begin{pmatrix}x\\y\end{pmatrix}\right)=(\lambda\mu)\begin{pmatrix}x\\y\end{pmatrix}$$

$$\textbf{V4:}\quad 1\begin{pmatrix}x\\y\end{pmatrix}=\begin{pmatrix}x\\y\end{pmatrix}$$

where λ and μ are real numbers. It is left as an exercise to verify these. Since $\boldsymbol{R}^2$ is an abelian group under vector addition, and satisfies V1 through V4 under multiplication by scalars, we say that $\boldsymbol{R}^2$ is a *vector space.*

Using $\boldsymbol{R}^2$ as our motivating example, we now give the definition of a vector space. We begin by defining an algebraic structure known as a *field,* as a set F together with a pair of binary operations + and $\cdot$ on F such that

(i) $(F,+)$ is an abelian group

(ii) $\cdot$ is commutative and associative, has a *unity* element, written 1, satisfying $1\cdot a=a=a\cdot 1$ and each non-zero element $a\in F$ has a multiplicative inverse, written a^{-1}, satisfying

$$a\cdot a^{-1}=1=a^{-1}\cdot a$$

(iii) $\cdot$ *distributes* over +, so that

$$a\cdot(b+c)=a\cdot b+a\cdot c \text{ for all } a,b,c\in F.$$

Familiar examples of fields are the rational numbers $\boldsymbol{Q}$, the real numbers $\boldsymbol{R}$, the complex numbers $\boldsymbol{C}$, and the modular integers $\boldsymbol{Z}_p$ for each prime number p. Fields are studied in depth in chapters 12, 17 and 18.

3.5 Definition Suppose that F is a field and V is a non-empty set. We shall call the elements of F *scalars* and the elements of V *vectors.*

Suppose also that we have a binary operation + on V (called *vector addition*) and an operation which combines a scalar $\lambda\in F$ with a vector $v\in V$ to give a vector $\lambda v\in V$ (called *multiplication by a scalar*).

Then V is a *vector space over* F if V is an abelian group under + and we have:

V1: $(\lambda+\mu)v = \lambda v + \mu v$

V2: $\lambda(u+v) = \lambda u + \lambda v$

V3: $\lambda(\mu v) = (\lambda\mu)v$

V4: $1v = v$

for all vectors $u, v \in V$ and all scalars $\lambda, \mu \in F$.

Examples (a) We have already seen that $\boldsymbol{R}^2$ is a vector space over $\boldsymbol{R}$. It is not difficult to see that $\boldsymbol{R}^3, \boldsymbol{R}^4$ and so on are also vector spaces over $\boldsymbol{R}$ with respect to the usual operations of vector addition and multiplication by a scalar. Also, in a rather trivial way, $\boldsymbol{R}$ may be thought of as a vector space over itself.

(b) In real analysis, a *sequence* is an infinite set of real numbers arranged in some particular order. A sequence is usually written like this:

$$x_1, x_2, x_3, \ldots$$

We call x_1 the first *term* of the sequence, x_2 the second term, and so on. A short notation for a sequence is (x_n). Usually a sequence will be defined by a formula, for example

$$(3n) = 3, 6, 9, 12, \ldots$$

$$(\tfrac{1}{n}) = 1, \tfrac{1}{2}, \tfrac{1}{3}, \tfrac{1}{4}, \ldots$$

We denote the set of all sequences of real numbers by $\boldsymbol{R}^\infty$. This is a vector space over $\boldsymbol{R}$ under suitable operations. We first define vector addition. Suppose that we have a pair of sequences

$$(x_n) = x_1, x_2, x_3, \ldots$$

$$(y_n) = y_1, y_2, y_3, \ldots$$

Their sum is defined by

$$(x_n) + (y_n) = x_1 + y_1, x_2 + y_2, x_3 + y_3, \ldots$$

This operation is commutative and associative. The identity is the sequence

$$(0) = 0, 0, 0, \ldots$$

Each sequence $x_1, x_2, x_3, \ldots$ has inverse $-x_1, -x_2, -x_3, \ldots$

Hence $\boldsymbol{R}^\infty$ is an abelian group under addition of sequences.

Multiplication of a sequence by a real number is defined by

$$\lambda(x_n) = \lambda x_1, \lambda x_2, \lambda x_3, \ldots$$

It may be verified that the properties V1 through V4 hold for these operations on sequences.

(c) Consider the set of all functions $\boldsymbol{R} \longrightarrow \boldsymbol{R}$, which we shall denote [**R**, **R**]. Suppose that f and g are functions in [**R**, **R**]. Their sum $f+g$ is defined by

$$(f+g)(x) = f(x) + g(x)$$

and scalar multiplication λf is defined by

$$(\lambda f)(x) = \lambda f(x)$$

For example, if $f(x) = x^2$, $g(x) = \sin x$ and $\lambda = 3$ then

$$(f+g)(x) = x^2 + \sin x \qquad\qquad (\lambda f)(x) = 3x^2.$$

Once again the set is an abelian group under vector addition, and V1 through V4 hold, so that [**R**, **R**] is a vector space over **R**.

(d) The set of column vectors with complex entries

$$\mathbf{C}^2 = \left\{ \begin{pmatrix} w + xi \\ y + zi \end{pmatrix} : w, x, y, z \in \boldsymbol{R} \right\}$$

is a vector space over $\boldsymbol{C}$.

(e) The field $\boldsymbol{Q}(\sqrt{2}\,) = \{a + b\sqrt{2} \ : a, b \in \boldsymbol{Q}\}$ is a vector space over $\boldsymbol{Q}$ under the operations

$$(a_1 + b_1\sqrt{2}\,) + (a_2 + b_2\sqrt{2}\,) = (a_1 + a_2) + (b_1 + b_2)\sqrt{2}$$

$$\lambda(a + b\sqrt{2}\,) = \lambda a + \lambda b\sqrt{2} \ \text{ for } \lambda \in \boldsymbol{Q}.$$

Similarly

$$\boldsymbol{Q}(\sqrt{2}, \sqrt{3}) = \{w + x\sqrt{2} + y\sqrt{3} + z\sqrt{6} \ : w, x, y, z \in \boldsymbol{Q}\}$$

is a vector space over the field of rational numbers $\boldsymbol{Q}$. Examples of this type are important in chapter 12 when we study fields and their extensions.

A vector space over $\boldsymbol{R}$ is called a *real vector space.* A vector space over $\boldsymbol{C}$ is called a *complex vector space.*

3.6 Proposition Let V be a vector space over a field F. The zero vector $\mathbf{0}$ is unique. Furthermore, each vector v has a unique inverse $-v$.

Proof: The proof is left as an exercise, and is very similar to the result for groups proved in 1.2.

Note that we distinguish the scalar 0 from the vector $\mathbf{0}$ in our notation by rendering the latter in bold type.

3.7 Proposition Let V be a vector space over a field F. Then

(i) $0v = \mathbf{0}$ for each $v \in V$

(ii) $\lambda\mathbf{0} = \mathbf{0}$ for each $\lambda \in F$.

Proof: (i) $(0+0)v = 0v$ and so $0v + 0v = 0v$.

Hence $0v + 0v = 0v + \mathbf{0}$, and so by cancellation $0v = \mathbf{0}$.

(ii) $\lambda(\mathbf{0}+\mathbf{0}) = \lambda\mathbf{0}$ and so $\lambda\mathbf{0} + \lambda\mathbf{0} = \lambda\mathbf{0}$.

Hence $\lambda\mathbf{0} + \lambda\mathbf{0} = \lambda\mathbf{0} + \mathbf{0}$, and so by cancellation $\lambda\mathbf{0} = \mathbf{0}$.

Exercises 3

1. Evaluate the following:

(i) $2\begin{pmatrix} 3 \\ -4 \end{pmatrix}$ (ii) $5\begin{pmatrix} 1 \\ -2 \\ 3 \end{pmatrix}$

(iii) $\begin{pmatrix} 7 \\ 2 \end{pmatrix}+\begin{pmatrix} 4 \\ 6 \end{pmatrix}$ (iv) $\begin{pmatrix} 9 \\ 5 \end{pmatrix}-\begin{pmatrix} 3 \\ -2 \end{pmatrix}$

(v) $\begin{pmatrix} 6 \\ -5 \\ 2 \end{pmatrix}+\begin{pmatrix} -1 \\ 8 \\ -5 \end{pmatrix}$ (vi) $\begin{pmatrix} -3 \\ 6 \\ 5 \end{pmatrix}-\begin{pmatrix} -2 \\ 3 \\ 10 \end{pmatrix}$

2. Find the magnitude of each of the following vectors:

(i) $\begin{pmatrix} 5 \\ 8 \end{pmatrix}$ (ii) $\begin{pmatrix} 6 \\ -2 \end{pmatrix}$

(iii) $\begin{pmatrix} 1 \\ -8 \\ 5 \end{pmatrix}$ (iv) $\begin{pmatrix} 7 \\ -3 \\ 6 \\ 10 \end{pmatrix}$

3. Calculate the following scalar products of vectors:

(i) $\begin{pmatrix} 2 \\ 1 \end{pmatrix}\cdot\begin{pmatrix} 1 \\ 4 \end{pmatrix}$ (ii) $\begin{pmatrix} 3 \\ 5 \\ -2 \end{pmatrix}\cdot\begin{pmatrix} 1 \\ -2 \\ 0 \end{pmatrix}$

(iii) $\begin{pmatrix} 1 \\ 4 \\ 3 \end{pmatrix}\cdot\begin{pmatrix} 2 \\ -1 \\ 7 \end{pmatrix}$ (iv) $\begin{pmatrix} 1 \\ 2 \\ 3 \\ 4 \end{pmatrix}\cdot\begin{pmatrix} 4 \\ 3 \\ 2 \\ 1 \end{pmatrix}$

In each case, calculate the angle between the two vectors, approximated to the nearest degree.

4. Verify that the set of vectors $\boldsymbol{R}^2$ is an abelian group under addition. Verify also that for multiplication by a scalar the properties V1, V2, V3 and V4 hold for $\boldsymbol{R}^2$.

5. Verify that the set of functions $\boldsymbol{R} \rightarrow \boldsymbol{R}$, with vector addition and multiplication by a scalar as defined in the chapter satisfy the properties V1, V2, V3 and V4.

Chapter 4: Subspaces and Linear Transformations

We have already met the concepts of subset and subgroup. Similarly, a *subspace* is a vector space contained within another vector space.

4.1 Definition Suppose that V is a vector space over a field F, and that U is a subset of V. We say that U is a *subspace* of V if U is itself a vector space over F under the same operations of vector addition and multiplication by a scalar of V.

Example

$$U = \left\{ \begin{pmatrix} x \\ y \\ 0 \end{pmatrix} : x, y \in \boldsymbol{R} \right\} \text{ is a subspace of } \boldsymbol{R}^3$$

Note that for any vector space V, the trivial vector space $\{\mathbf{0}\}$ and V itself are subspaces of V.

To check whether or not a given subset forms a subspace we apply the following test:

4.2 Test for a Subspace Suppose that V is a vector space over F and U is a non-empty subset of V.

U is a subspace if and only if

(i) if $u, v \in U$ then $u + v \in U$

(ii) if $v \in U$ and $\lambda \in F$ then $\lambda v \in U$

Examples (a) For the example above we see that

(i) $$\begin{pmatrix} x_1 \\ y_2 \\ 0 \end{pmatrix} + \begin{pmatrix} x_2 \\ y_2 \\ 0 \end{pmatrix} = \begin{pmatrix} x_1 + x_2 \\ y_1 + y_2 \\ 0 \end{pmatrix} \in U$$

(ii) $$\lambda\begin{pmatrix} x \\ y \\ 0 \end{pmatrix} = \begin{pmatrix} \lambda x \\ \lambda y \\ 0 \end{pmatrix} \in U$$

confirming that U is indeed a subspace of $\boldsymbol{R}^3$.

(b) $U = \left\{ \begin{pmatrix} x \\ y \end{pmatrix} : 3x + 5y = 0 \right\}$ is a subspace of $\boldsymbol{R}^2$.

To see this, suppose that $\begin{pmatrix} x_1 \\ y_1 \end{pmatrix}, \begin{pmatrix} x_2 \\ y_2 \end{pmatrix} \in U$ so that

$$3x_1 + 5y_1 = 0,\ 3x_2 + 5y_2 = 0.$$

Then we have

$$3(x_1 + x_2) + 5(y_1 + y_2) = (3x_1 + 5y_1) + (3x_2 + 5y_2) = 0 + 0 = 0$$

so that

$$\begin{pmatrix} x_1 \\ y_1 \end{pmatrix} + \begin{pmatrix} x_2 \\ y_2 \end{pmatrix} = \begin{pmatrix} x_1 + x_2 \\ y_1 + y_2 \end{pmatrix} \in U.$$

Also $3(\lambda x_1) + 5(\lambda y_1) = \lambda(3x_1 + 5y_1) = \lambda \cdot 0 = 0$ so that

$$\lambda\begin{pmatrix} x_1 \\ y_1 \end{pmatrix} = \begin{pmatrix} \lambda x_2 \\ \lambda y_2 \end{pmatrix} \in U.$$

(c) A sequence of real numbers (x_n) is said to *converge* to the *limit* $a \in \boldsymbol{R}$ if the terms of the sequence approach arbitrarily close to a as n becomes large (the precise definition of convergence belongs to the study of *real analysis*). We write $x_n \to a$ as $n \to \infty$. For example, $\frac{1}{n} \to 0$ as $n \to \infty$. A sequence that converges to 0 is called a *null sequence*. It may be shown that

$$\text{if } x_n \to 0 \text{ and } y_n \to 0 \text{ then } x_n + y_n \to 0 \text{ as } n \to \infty$$

and

$$\text{if } x_n \to 0 \text{ then } \lambda x_n \to 0 \text{ as } n \to \infty$$

Hence the set of null sequences is a subspace of $\boldsymbol{R}^\infty$.

(d) A function $f: \boldsymbol{R} \to \boldsymbol{R}$ is said to be *bounded* if there are real numbers b and B such that

$$b \leq f(x) \leq B \text{ for all values of } x.$$

For example, the function $\cos x$ is bounded: a lower bound is $b = -1$ and an upper bound is $B = 1$. By contrast, the function x^3 is not bounded.

It may be shown that if f and g are bounded functions then so are $f + g$ and λf. Hence the set of bounded functions is a subspace of $[\boldsymbol{R}, \boldsymbol{R}]$.

(e) $\boldsymbol{Q}(\sqrt{2}\,)$ is a subspace of $\boldsymbol{Q}(\sqrt{2}, \sqrt{3}\,)$.

(f) Consider $U = \left\{ \begin{pmatrix} x \\ 1 \end{pmatrix} : x \in \boldsymbol{R} \right\} \subseteq \boldsymbol{R}^2$.

U is not a subspace of $\boldsymbol{R}^2$.

For example $\begin{pmatrix} 0 \\ 1 \end{pmatrix} \in U$ but $2\begin{pmatrix} 0 \\ 1 \end{pmatrix} = \begin{pmatrix} 0 \\ 2 \end{pmatrix} \notin U$.

4.3 Proposition Suppose that U_1 and U_2 are subspaces of a vector space V.

Then their intersection $U_1 \cap U_2$ is also a subspace of V.

Proof: This is left as an exercise. It is similar to the proof that the intersection of a pair of subgroups is a subgroup, 2.3.

Example

$$U_1 = \left\{ \begin{pmatrix} x \\ y \\ 0 \end{pmatrix} : x, y \in \boldsymbol{R} \right\} \text{ and } U_2 = \left\{ \begin{pmatrix} 0 \\ y \\ z \end{pmatrix} : y, z \in \boldsymbol{R} \right\}$$

are subspaces of $\boldsymbol{R}^3$.

$$U_1 \cap U_2 = \left\{ \begin{pmatrix} 0 \\ y \\ 0 \end{pmatrix} : y \in \boldsymbol{R} \right\} \text{ is also a subspace of } \boldsymbol{R}^3$$

4.4 Proposition Suppose that U_1 and U_2 are subspaces of a vector space V.

Then $U_1 + U_2 = \{u_1 + u_2 : u_1 \in U_1 \text{ and } u_2 \in U_2\}$ is also a subspace of V.

Proof: This is left as an exercise.

Example

$$U_1 = \left\{ \begin{pmatrix} x \\ 0 \\ 0 \end{pmatrix} : x \in \boldsymbol{R} \right\} \text{ and } U_2 = \left\{ \begin{pmatrix} 0 \\ y \\ 0 \end{pmatrix} : y \in \boldsymbol{R} \right\}$$

are subspaces of $\boldsymbol{R}^3$.

$$U_1 + U_2 = \left\{ \begin{pmatrix} x \\ y \\ 0 \end{pmatrix} : x, y \in \boldsymbol{R} \right\} \text{ is also a subspace of } \boldsymbol{R}^3.$$

4.5 Definition Suppose that U_1 and U_2 are subspaces of a vector space V.

We say that V is the *direct sum* of U_1 and U_2 if:

(i) $V = U_1 + U_2$

(ii) $U_1 \cap U_2 = \{\mathbf{0}\}$.

We write $V = U_1 \oplus U_2$.

Example

$$U_1 = \left\{ \begin{pmatrix} x \\ 0 \end{pmatrix} : x \in \boldsymbol{R} \right\} \text{ and } U_2 = \left\{ \begin{pmatrix} 0 \\ y \end{pmatrix} : y \in \boldsymbol{R} \right\}$$

are subspaces of $\boldsymbol{R}^2$. For each vector in $\boldsymbol{R}^2$ we have

$$\begin{pmatrix} x \\ y \end{pmatrix} = \begin{pmatrix} x \\ 0 \end{pmatrix} + \begin{pmatrix} 0 \\ y \end{pmatrix}, \text{ and clearly } U_1 \cap U_2 = \left\{ \begin{pmatrix} 0 \\ 0 \end{pmatrix} \right\}.$$

Hence $\boldsymbol{R}^2 = U_1 \oplus U_2$.

4.6 Proposition $V = U_1 \oplus U_2$ if and only if each $v \in V$ can be written in a unique way as $v = u_1 + u_2$ for some $u_1 \in U_1$ and $u_2 \in U_2$.

Proof: This is left as an exercise.

We are interested in mappings between vector spaces that preserve structure. We explored a similar idea when we studied *homomorphisms* of groups in chapter 2. For vector spaces, as well as preserving vector addition we also need to preserve multiplication by a scalar.

4.7 Definition Suppose that V and W are vector spaces over the same field F. A mapping $f: V \to W$ is called a *linear transformation* if

(i) $f(u+v) = f(u) + f(v)$ for all $u, v \in V$

(ii) $f(\lambda v) = \lambda f(v)$ for all $\lambda \in F$ and all $v \in V$.

If f is injective then it is called a *monomorphism*;

If f is surjective then it is called an *epimorphism*;

If f is bijective (both injective and surjective) then it is called an *isomorphism*.

If there is an isomorphism between a pair of vector spaces V and W then we say that V is *isomorphic* to W, and we write $V \cong W$.

Examples (a) Consider $f: \boldsymbol{R}^2 \to \boldsymbol{R}^3$ given by

$$f\begin{pmatrix} x \\ y \end{pmatrix} = \begin{pmatrix} x \\ y \\ x+y \end{pmatrix}$$

We show that f is a linear transformation:

$$\text{(i)}\ f\left\{\begin{pmatrix} x_1 \\ y_1 \end{pmatrix} + \begin{pmatrix} x_2 \\ y_2 \end{pmatrix}\right\} = f\begin{pmatrix} x_1 + x_2 \\ y_1 + y_2 \end{pmatrix} = \begin{pmatrix} x_1 + x_2 \\ y_1 + y_2 \\ (x_1 + x_2) + (y_1 + y_2) \end{pmatrix}$$

$$= \begin{pmatrix} x_1 \\ y_1 \\ x_1 + y_1 \end{pmatrix} + \begin{pmatrix} x_2 \\ y_2 \\ x_2 + y_2 \end{pmatrix}$$

$$= f\begin{pmatrix} x_1 \\ y_1 \end{pmatrix} + f\begin{pmatrix} x_2 \\ y_2 \end{pmatrix}$$

$$\text{(ii)}\ f\left\{\lambda\begin{pmatrix} x \\ y \end{pmatrix}\right\} = f\begin{pmatrix} \lambda x \\ \lambda y \end{pmatrix} = \begin{pmatrix} \lambda x \\ \lambda y \\ \lambda x + \lambda y \end{pmatrix} = \lambda\begin{pmatrix} x \\ y \\ x + y \end{pmatrix} = \lambda f\begin{pmatrix} x \\ y \end{pmatrix}$$

It is a monomorphism but *not* an epimorphism, for example

$$\begin{pmatrix} 0 \\ 0 \\ 1 \end{pmatrix} \notin \operatorname{im} f.$$

(b) Consider $f: \boldsymbol{R}^3 \to \boldsymbol{R}^2$ given by

$$f\begin{pmatrix} x \\ y \\ z \end{pmatrix} = \begin{pmatrix} x + y \\ y + z \end{pmatrix}$$

It is left as an exercise to show that this is a linear transformation. It is an epimorphism since for any $\begin{pmatrix} x \\ z \end{pmatrix} \in \boldsymbol{R}^2$ we have $f\begin{pmatrix} x \\ 0 \\ z \end{pmatrix} = \begin{pmatrix} x \\ z \end{pmatrix}$.

It is *not* a monomorphism, since for example $f\begin{pmatrix} 1 \\ 1 \\ 1 \end{pmatrix} = f\begin{pmatrix} 2 \\ 0 \\ 2 \end{pmatrix}$.

(c) Let V be the real vector space of linear functions:

$$V = \{ax + b : a, b \in \boldsymbol{R}\}.$$

Consider the mapping $f : V \to \boldsymbol{R}^2$ given by

$$f(ax + b) = \begin{pmatrix} a \\ b \end{pmatrix}.$$

It is left as an exercise to show that this is a linear transformation. The mapping is an isomorphism, and hence $V \cong \boldsymbol{R}^2$.

(d) Consider $f : \boldsymbol{R}^\infty \to \boldsymbol{R}^3$ given by

$$f(x_1, x_2, x_3, \dots) = \begin{pmatrix} x_1 \\ 2x_3 \\ -x_5 \end{pmatrix}$$

This is a linear transformation. It is an epimorphism but not a monomorphism.

(e) Let V be the vector space of all differentiable functions $\boldsymbol{R} \to \boldsymbol{R}$ and let W be the vector space of all functions $\boldsymbol{R} \to \boldsymbol{R}$. Define a mapping $D : V \to W$ by $D(f) = f'$, the derivative.

For example, $D(x^3) = 3x^2$, $D(\sin x) = \cos x$.

Then D is a linear transformation by virtue of the following well known facts about differentiation:

$$(f + g)' = f' + g'$$
$$(\lambda f)' = \lambda f'.$$

We now prove some basic results about linear transformations.

4.8 Proposition Suppose that $f : V \to W$ is a linear transformation and that 0_V and 0_W are the zero vectors of V and W respectively. Then $f(0_V) = 0_W$

Proof: First observe that $0_V + 0_V = 0_V$ so that $f(0_V + 0_V) = f(0_V)$.

Since f is a linear transformation we have $f(0_V) + f(0_V) = f(0_V)$, and consequently we have

$f(0_V) + f(0_V) = f(0_V) + 0_W$. The result follows by cancellation.

4.9 Definition Suppose that $f: V \to W$ is a linear transformation. The set $\{f(v) : v \in V\}$ is called the *image* of f, written $\operatorname{im} f$

The set $\{v \in V : f(v) = 0_W\}$ is called the *kernel* of f, written $\ker f$.

These may be compared with the similar definitions made for group homomorphisms in 2.14.

Example Consider the linear transformation $f: \boldsymbol{R}^3 \to \boldsymbol{R}^3$ given by

$$f\begin{pmatrix} x \\ y \\ z \end{pmatrix} = \begin{pmatrix} 0 \\ y \\ z \end{pmatrix}$$

$$\operatorname{im} f = \left\{ \begin{pmatrix} x \\ y \\ z \end{pmatrix} : x = 0 \right\} \text{ and } \quad \ker f = \left\{ \begin{pmatrix} x \\ y \\ z \end{pmatrix} : y = 0 \text{ and } z = 0 \right\}$$

The kernel and image of a linear transformation have properties very similar to the kernel and image of a group homomorphism:

4.10 Proposition Suppose that $f: V \to W$ is a linear transformation. Then we have

(i) $\operatorname{im} f$ is a subspace of W

(ii) $\ker f$ is a subspace of V

(iii) f is a monomorphism if and only if $\ker f = \{0_V\}$.

(iv) f is an epimorphism if and only if $\operatorname{im} f = W$.

Proof: (i) Suppose $w_1, w_2 \in \operatorname{im} f$ so that $w_1 = f(v_1)$ and $w_2 = f(v_2)$ for some $v_1, v_2 \in V$.

Then $w_1 + w_2 = f(v_1) + f(v_2) = f(v_1 + v_2)$ so that $w_1 + w_2 \in \operatorname{im} f$.

Also $\lambda w_1 = \lambda f(v_1) = f(\lambda v_1)$ so that $\lambda w_1 \in \text{im} f$.

Hence $\text{im} f$ is a subspace of W.

(ii) Suppose $v_1, v_2 \in \ker f$ so that $f(v_1) = \mathbf{0}$ and $f(v_2) = \mathbf{0}$.

Then $f(v_1 + v_2) = f(v_1) + f(v_2) = \mathbf{0} + \mathbf{0} = \mathbf{0}$ so that we have $v_1 + v_2 \in \ker f$.

Also, $f(\lambda v_1) = \lambda f(v_1) = \lambda \mathbf{0} = \mathbf{0}$ so that we have $\lambda v_1 \in \ker f$.

Hence $\ker f$ is a subspace of V.

(iii) Suppose f is a monomorphism and $v \in \ker f$.

Then $f(v) = f(\mathbf{0})$ and so $v = \mathbf{0}$. Hence $\ker f = \{\mathbf{0}\}$.

Conversely, suppose $\ker f = \{\mathbf{0}\}$ and $f(v_1) = f(v_2)$.

Then $f(v_1 - v_2) = f(v_1) - f(v_2) = \mathbf{0}$ so that $v_1 - v_2 \in \ker f$ and consequently $v_1 - v_2 = \mathbf{0}$.

Hence $v_1 = v_2$ and so f is a monomorphism.

(iv) follows immediately from the definitions of epimorphism and image.

A linear transformation $f: \boldsymbol{R}^m \to \boldsymbol{R}^n$ can be represented by an $n \times m$ matrix.

Example Consider the linear transformation $f: \boldsymbol{R}^3 \to \boldsymbol{R}^2$ given by

$$f\begin{pmatrix} x \\ y \\ z \end{pmatrix} = \begin{pmatrix} x + 2y + 3z \\ 4x + 5y + 6z \end{pmatrix} = \begin{pmatrix} 1 & 2 & 3 \\ 4 & 5 & 6 \end{pmatrix}\begin{pmatrix} x \\ y \\ z \end{pmatrix}$$

f can be represented by the matrix $\begin{pmatrix} 1 & 2 & 3 \\ 4 & 5 & 6 \end{pmatrix}$.

Exercises 4

1. Show that the set

$$U = \left\{ \begin{pmatrix} x \\ y \end{pmatrix} : 2x + 3y = 0 \right\}$$

is a subspace of the vector space $\boldsymbol{R}^2$.

2. Show that the following sets are subspaces of $\boldsymbol{R}^3$:

$$U_1 = \left\{ \begin{pmatrix} x \\ 2x \\ -x \end{pmatrix} : x \in \boldsymbol{R} \right\} \qquad U_2 = \left\{ \begin{pmatrix} x \\ y \\ z \end{pmatrix} : x - 2y + 3z = 0 \right\}.$$

3. Consider the vector space of functions $[\boldsymbol{R}, \boldsymbol{R}]$. Show that the set

$$U = \{f \in [\boldsymbol{R}, \boldsymbol{R}] : f(0) = 0\}$$

is a subspace of $[\boldsymbol{R}, \boldsymbol{R}]$.

4. Prove that if U_1 and U_2 are subspaces of a vector space V, then $U_1 \cap U_2$ is also a subspace of V.

   Is $U_1 \cup U_2$ a subspace of V in general?

5. Prove that if U_1 and U_2 are subspaces of a vector space V, then $U_1 + U_2$ is also a subspace of V.

6. Verify that the mapping $f : \boldsymbol{R}^3 \rightarrow \boldsymbol{R}^2$ given by

$$f\begin{pmatrix} x \\ y \\ z \end{pmatrix} = \begin{pmatrix} x + y \\ y + z \end{pmatrix}$$

is a linear transformation.

7. Let $V = \{ax + b : a, b \in \boldsymbol{R}\}$. Show that the mapping $f : V \rightarrow \boldsymbol{R}^2$ given by

$$f(ax + b) = \begin{pmatrix} a \\ b \end{pmatrix}$$

is an isomorphism.

8. Let V be the *real* vector space of vectors

$$\begin{pmatrix} w+xi \\ y+zi \end{pmatrix}$$

with vector addition defined as for $\boldsymbol{C}^2$, and with multiplication by a scalar $\lambda \in \boldsymbol{R}$ defined by

$$\lambda \begin{pmatrix} w+xi \\ y+zi \end{pmatrix} = \begin{pmatrix} \lambda w + \lambda xi \\ \lambda y + \lambda zi \end{pmatrix}.$$

Show that $V \cong \boldsymbol{R}^4$. Explain why it would *not* be correct to say that $\boldsymbol{C}^2 \cong \boldsymbol{R}^4$.

9. Show that if $f: U \to V$ and $g: V \to W$ are linear transformations of vector spaces then $g \circ f: U \to W$ is also a linear transformation.

10. Write down the matrix representing the following linear transformation $f: \boldsymbol{R}^4 \to \boldsymbol{R}^5$:

$$f\begin{pmatrix} w \\ x \\ y \\ z \end{pmatrix} = \begin{pmatrix} w+2x+3y+4z \\ w-x+y-z \\ w+y \\ x-z \\ 3w-5x+y-6z \end{pmatrix}.$$

11. Prove that $V = U_1 \oplus U_2$ if and only if each $v \in V$ can be written in a unique way as

$$v = u_1 + u_2 \text{ for some } u_1 \in U_1 \text{ and } u_2 \in U_2.$$

Chapter 5: The Basis for a Vector Space

Suppose that V is a vector space over a field F, and that $\lambda_1, \lambda_2, ..., \lambda_n \in F$ and $v_1, v_2, ..., v_n \in V$.

A sum of the form $\lambda_1 v_1 + \lambda_2 v_2 + ... + \lambda_n v_n$ is called a *linear combination* of the vectors.

5.1 Definition A set of vectors $\{v_1, v_2, ..., v_n\}$ is *linearly independent* if for any linear combination of these such that

$$\lambda_1 v_1 + \lambda_2 v_2 + ... + \lambda_n v_n = \mathbf{0}$$

we have $\lambda_1 = \lambda_2 = ... = \lambda_n = 0$.

A set of vectors that fails to be linearly independent is said to be *linearly dependent.*

Examples (a) Consider $\begin{pmatrix} 1 \\ 2 \end{pmatrix}$ and $\begin{pmatrix} 3 \\ 1 \end{pmatrix}$ in $\boldsymbol{R}^2$.

Suppose that $\lambda_1 \begin{pmatrix} 1 \\ 2 \end{pmatrix} + \lambda_2 \begin{pmatrix} 3 \\ 1 \end{pmatrix} = \begin{pmatrix} 0 \\ 0 \end{pmatrix}$.

Solving, we find that $\lambda_1 = \lambda_2 = 0$, and hence the vectors are linearly independent.

(b) Consider $\begin{pmatrix} 1 \\ 2 \end{pmatrix}$ and $\begin{pmatrix} -3 \\ -6 \end{pmatrix}$ in $\boldsymbol{R}^2$.

Since $3\begin{pmatrix} 1 \\ 2 \end{pmatrix} + 1\begin{pmatrix} -3 \\ -6 \end{pmatrix} = \begin{pmatrix} 0 \\ 0 \end{pmatrix}$ the vectors are linearly dependent.

5.2 Proposition A set of non-zero vectors $\{v_1, v_2, ..., v_n\}$ is linearly dependent if and only if at least one of these vectors may be expressed as a linear combination of the others.

Proof: Suppose that $\{v_1, v_2, ..., v_n\}$ is linearly dependent. Then there are scalars $\lambda_1, \lambda_2, ..., \lambda_n$, not all zero, such that

$$\lambda_1 v_1 + \lambda_2 v_2 + ... + \lambda_n v_n = \mathbf{0}.$$

Suppose without loss of generality that $\lambda_1 \neq 0$. Then we may write

$$v_1 = \tfrac{-\lambda_2}{\lambda_1} v_2 + ... + \tfrac{-\lambda_n}{\lambda_1} v_n,$$

so that we have expressed v_1 as a linear combination of v_2 through v_n.

Conversely, suppose that $v_1 = \mu_2 v_2 + ... + \mu_n v_n$ for scalars $\mu_2, ...\mu_n$, not all zero since $v_1 \neq \mathbf{0}$, then we have $-v_1 + \mu_2 v_2 + ... + \mu_n v_n = \mathbf{0}$ and hence the vectors are linearly dependent.

5.3 Definition Suppose that $S = \{v_1, v_2, ..., v_n\}$ is a set of vectors in a vector space V. We say that S *spans* V if any $v \in V$ may be expressed as a linear combination of vectors in S.

Examples (a) Consider $\begin{pmatrix} 3 \\ 1 \end{pmatrix}$ and $\begin{pmatrix} 1 \\ 2 \end{pmatrix}$ in $\boldsymbol{R}^2$.

Solving $\begin{pmatrix} x \\ y \end{pmatrix} = \lambda_1 \begin{pmatrix} 3 \\ 1 \end{pmatrix} + \lambda_2 \begin{pmatrix} 1 \\ 2 \end{pmatrix}$ we obtain $\lambda_1 = \tfrac{2}{5}x - \tfrac{1}{5}y$ and $\lambda_2 = -\tfrac{1}{5}x + \tfrac{3}{5}y$ so that

$\begin{pmatrix} x \\ y \end{pmatrix} = (\tfrac{2}{5}x - \tfrac{1}{5}y)\begin{pmatrix} 3 \\ 1 \end{pmatrix} + (-\tfrac{1}{5}x + \tfrac{3}{5}y)\begin{pmatrix} 1 \\ 2 \end{pmatrix}$. Hence the set spans $\boldsymbol{R}^2$.

(b) Consider $\begin{pmatrix} 1 \\ 2 \end{pmatrix}$ and $\begin{pmatrix} -3 \\ -6 \end{pmatrix}$ in $\boldsymbol{R}^2$.

Suppose that we try to express $\begin{pmatrix} 1 \\ 1 \end{pmatrix}$ as a linear combination of these:

$$\lambda_1\begin{pmatrix}1\\2\end{pmatrix}+\lambda_2\begin{pmatrix}-3\\-6\end{pmatrix}=\begin{pmatrix}1\\1\end{pmatrix}$$

We have $\lambda_1-3\lambda_2=1$, and also $2\lambda_1-6\lambda_2=1$ so that $\lambda_1-3\lambda_2=\frac{1}{2}$. These equations are inconsistent and have no solution.

Hence $\begin{pmatrix}1\\1\end{pmatrix}$ cannot be expressed as a linear combination of these vectors, and so the set does not span $\boldsymbol{R}^2$.

5.4 Definition Suppose that $S=\{v_1,v_2,\ldots,v_n\}$ is a set of vectors in a vector space V. We say that S is a *basis* for V if

(i) S is linearly independent

(ii) S spans V.

Examples (a) The *standard basis* for $\boldsymbol{R}^3$ is

$$\left\{e_1=\begin{pmatrix}1\\0\\0\end{pmatrix},e_2=\begin{pmatrix}0\\1\\0\end{pmatrix},e_3=\begin{pmatrix}0\\0\\1\end{pmatrix}\right\}.$$

When we are using the standard basis for $\boldsymbol{R}^3$ then $\begin{pmatrix}x\\y\\z\end{pmatrix}$ is the linear combination

$$xe_1+ye_2+ze_3.$$

(b) Another basis for $\boldsymbol{R}^3$ is the set of vectors

$$\left\{\begin{pmatrix}1\\1\\0\end{pmatrix},\begin{pmatrix}0\\1\\1\end{pmatrix},\begin{pmatrix}1\\2\\3\end{pmatrix}\right\}$$

First we check that the vectors are linearly independent. Suppose that

$$\lambda_1 \begin{pmatrix} 1 \\ 1 \\ 0 \end{pmatrix} + \lambda_2 \begin{pmatrix} 0 \\ 1 \\ 1 \end{pmatrix} + \lambda_3 \begin{pmatrix} 1 \\ 2 \\ 3 \end{pmatrix} = \begin{pmatrix} 0 \\ 0 \\ 0 \end{pmatrix}$$

Notice that this can be written in matrix form as

$$\begin{pmatrix} 1 & 0 & 1 \\ 1 & 1 & 2 \\ 0 & 1 & 3 \end{pmatrix} \begin{pmatrix} \lambda_1 \\ \lambda_2 \\ \lambda_3 \end{pmatrix} = \begin{pmatrix} 0 \\ 0 \\ 0 \end{pmatrix}$$

The determinant of the matrix is 2 and so the matrix is invertible, hence we must have $\lambda_1 = \lambda_2 = \lambda_3 = 0$.

We now check that the vectors span $\boldsymbol{R}^3$. We wish to write

$$\begin{pmatrix} x \\ y \\ z \end{pmatrix} = \lambda_1 \begin{pmatrix} 1 \\ 1 \\ 0 \end{pmatrix} + \lambda_2 \begin{pmatrix} 0 \\ 1 \\ 1 \end{pmatrix} + \lambda_3 \begin{pmatrix} 1 \\ 2 \\ 3 \end{pmatrix}$$

Again notice that this can be written in matrix form as

$$\begin{pmatrix} x \\ y \\ z \end{pmatrix} = \begin{pmatrix} 1 & 0 & 1 \\ 1 & 1 & 2 \\ 0 & 1 & 3 \end{pmatrix} \begin{pmatrix} \lambda_1 \\ \lambda_2 \\ \lambda_3 \end{pmatrix}$$

We can find λ_1, λ_2 and λ_3 in terms of x, y and z provided that the coefficient matrix is invertible. This is the case, since as we have already seen it has non-zero determinant.

5.5 Proposition Suppose that a vector space V has a basis with n vectors.

(i) Any set of more than n vectors in V is linearly dependent, and so is not a basis.

(ii) Any set of fewer than n vectors in V fails to span V, and so is not a basis.

The proofs of these facts are not very interesting, and so are omitted. But the important conclusion that we draw is:

For each vector space, all bases have the same number of vectors.

5.6 Definition The number of vectors in each basis for a vector space V is called the *dimension* of V, written $\dim V$.

Examples As we might expect, $\boldsymbol{R}^3$ is of dimension 3, because each basis for $\boldsymbol{R}^3$ has three vectors. We write $\dim \boldsymbol{R}^3 = 3$.

A basis for $V = \{ax + b : a, b \in \boldsymbol{R}\}$ is $\{x, 1\}$ and so $\dim V = 2$..

The discussion above suggests a procedure to check that a set of vectors S is a basis for $\boldsymbol{R}^n$:

(i) Check that S has precisely n elements;

(ii) Form the square matrix that has the vectors of S as its columns and check that this matrix has non-zero determinant.

The following key result explains why bases are important:

5.7 Proposition Suppose that $S = \{v_1, v_2, ..., v_n\}$ is a basis for a vector space V.

Any vector in V may be written uniquely as a linear combination of vectors in S.

Proof: Since S spans V, any $v \in V$ may be written as

$$\lambda_1 v_1 + \lambda_2 v_2 + ... + \lambda_n v_n$$

To prove the result, it remains to establish uniqueness.

Suppose that v may also be written as

$$\mu_1 v_1 + \mu_2 v_2 + ... + \mu_n v_n.$$

Since these two linear combinations are equal we have

$$(\lambda_1 - \mu_1)v_1 + (\lambda_2 - \mu_2)v_2 + ... + (\lambda_n - \mu_n)v_n = \mathbf{0}.$$

Finally, since S is linearly independent we have

$$\lambda_1 - \mu_1 = \lambda_2 - \mu_2 = ... = \lambda_n - \mu_n = 0$$

and hence $\lambda_1 = \mu_1, \lambda_2 = \mu_2, ..., \lambda_n = \mu_n$, so there is only one way of writing v as a linear combination of basis vectors.

5.8 Proposition Suppose that $S = \{v_1, v_2, ..., v_n\}$ is a basis for V, and that $f: V \to W$ is a linear transformation of vector spaces. Then $f(v)$ is determined by $f(v_1), ..., f(v_n)$.

Proof: Since S is a basis, we can write any $v \in V$ uniquely as

$$v = \lambda_1 v_1 + \lambda_2 v_2 + ... + \lambda_n v_n$$

for scalars $\lambda_1, \lambda_2, ..., \lambda_n$. Then we have

$$\begin{aligned} f(v) &= f(\lambda_1 v_1 + \lambda_2 v_2 + ... + \lambda_n v_n) \\ &= \lambda_1 f(v_1) + \lambda_2 f(v_2) + ... + \lambda_n f(v_n). \end{aligned}$$

5.9 Corollary Suppose that V and W are vector spaces over F of the same finite dimension.

Then $V \cong W$.

Proof: Suppose that V has basis $\{v_1, v_2, ..., v_n\}$ and that W has basis $\{w_1, w_2, ..., w_n\}$. Define a linear transformation $f: V \to W$ by $f(v_i) = w_i$ for $i = 1, 2, ..., n$. Since f has an inverse given by $f^{-1}(w_i) = v_i$ for $i = 1, 2, ..., n$, it is also a bijection and hence an isomorphism.

It follows that for each natural number n there is essentially only one vector space of dimension n for a given field F.

We now present our deepest result so far:

5.10 The Rank Theorem Suppose that $f: V \to W$ is a linear transformation of vector spaces. Then we have

$$\dim V = \dim(\ker f) + \dim(\operatorname{im} f)$$

Proof: Let $S = \{v_1, v_2, ..., v_r\}$ be a basis for $\ker f$ and let $\{w_1, w_2, ..., w_s\}$ be a basis for $\operatorname{im} f$, so that

$$\dim(\ker f) = r \text{ and } \dim(\operatorname{im} f) = s.$$

Choose vectors $T = \{u_1, u_2, ..., u_s\}$ from V such that $f(u_i) = w_i$ for each i. We show that $S \cup T$ is a basis for V so that $\dim V = r + s$. This is sufficient to prove the result.

$S \cup T$ is linearly independent:

Suppose that $\lambda_1 v_1 + ... + \lambda_r v_r + \mu_1 u_1 + ... + \mu_s u_s = \mathbf{0}$. Then

$$f(\lambda_1 v_1 + ... + \lambda_r v_r + \mu_1 u_1 + ... + \mu_s u_s) = f(\mathbf{0}) = \mathbf{0}.$$

But since the v_i are in $\ker f$ we also have

$$\begin{aligned} &f(\lambda_1 v_1 + ... + \lambda_r v_r + \mu_1 u_1 + ... + \mu_s u_s) \\ &= f(\mu_1 u_1 + ... + \mu_s u_s) \\ &= \mu_1 f(u_1) + ... + \mu_s f(u_s) \text{ by linearity} \\ &= \mu_1 w_1 + ... + \mu_s w_s. \end{aligned}$$

Hence we have $\mu_1 w_1 + ... + \mu_s w_s = \mathbf{0}$, and since the w_i are linearly independent we have $\mu_i = 0$ for each i. It follows that $\lambda_1 v_1 + ... + \lambda_r v_r = \mathbf{0}$, and since the v_i are linearly independent we have $\lambda_i = 0$ for each i. This proves that the set $S \cup T$ is linearly independent.

$S \cup T$ spans V:

Consider a vector $v \in V$. Since $f(v) \in \text{im} f$ it may be expressed uniquely as a linear combination of the basis vectors w_i:

$$f(v) = \mu_1 w_1 + ... + \mu_s w_s.$$

Let $u = \mu_1 u_1 + ... + \mu_s u_s$ so that $f(v) = f(u)$. Then $f(v - u) = f(v) - f(u) = \mathbf{0}$, so that $v - u \in \ker f$, and so may be expressed as a linear combination of the basis vectors v_i:

$$v - u = \lambda_1 v_1 + ... + \lambda_r v_r$$

so that finally

$$v = \lambda_1 v_1 + ... + \lambda_r v_r + u = \lambda_1 v_1 + ... + \lambda_r v_r + \mu_1 u_1 + ... + \mu_s u_s.$$

Example Consider the linear transformation $f: \boldsymbol{R}^5 \to \boldsymbol{R}^4$ given by

$$f\begin{pmatrix} v \\ w \\ x \\ y \\ z \end{pmatrix} = \begin{pmatrix} 0 \\ x \\ y \\ z \end{pmatrix}.$$

Clearly $\operatorname{im} f$ is of dimension 3.

$$\ker f = \left\{ \begin{pmatrix} v \\ w \\ 0 \\ 0 \\ 0 \end{pmatrix} : v, w \in \boldsymbol{R} \right\}$$

and so $\ker f$ is of dimension 2. Then

$$\dim(\ker f) + \dim(\operatorname{im} f) = 2 + 3 = 5 = \dim \boldsymbol{R}^5.$$

5.11 Change of Basis

The set $E = \left\{ e_1 = \begin{pmatrix} 1 \\ 0 \end{pmatrix}, e_2 = \begin{pmatrix} 0 \\ 1 \end{pmatrix} \right\}$ is the *standard basis* for $\boldsymbol{R}^2$.

The set $E = \left\{ \begin{pmatrix} 1 \\ 0 \\ 0 \end{pmatrix}, \begin{pmatrix} 0 \\ 1 \\ 0 \end{pmatrix}, \begin{pmatrix} 0 \\ 0 \\ 1 \end{pmatrix} \right\}$ is the standard basis for $\boldsymbol{R}^3$.

When working with the standard basis for $\boldsymbol{R}^2$ we shall write $\begin{pmatrix} x \\ y \end{pmatrix}_E$ for $xe_1 + ye_2$.

Now suppose that we choose another basis $B = \{v_1, v_2\}$ for $\boldsymbol{R}^2$. We shall write $\begin{pmatrix} x \\ y \end{pmatrix}_B$ for $xv_1 + yv_2$.

For example, if we choose the basis $\left\{\begin{pmatrix}1\\3\end{pmatrix},\begin{pmatrix}1\\2\end{pmatrix}\right\}$ then $\begin{pmatrix}3\\4\end{pmatrix}_B$ means

$$3\begin{pmatrix}1\\3\end{pmatrix}+4\begin{pmatrix}1\\2\end{pmatrix}=\begin{pmatrix}1&1\\3&2\end{pmatrix}\begin{pmatrix}3\\4\end{pmatrix}_B=\begin{pmatrix}7\\17\end{pmatrix}_E$$

The example shows that writing the new basis vectors as the columns of a square matrix gives a matrix which multiplies a vector written with respect to the *new* basis to give a vector written with respect to the *standard* basis. Such a matrix is called a *transition matrix*. The inverse of this matrix multiplies a vector written with respect to the *standard* basis to give a vector written with respect to the *new* basis.

Example Consider the linear transformation $f: \boldsymbol{R}^3 \to \boldsymbol{R}^2$ given by

$$f\begin{pmatrix}x\\y\\z\end{pmatrix}=\begin{pmatrix}x+y\\y+z\end{pmatrix}$$

With respect to the standard bases f is represented by the matrix

$$M=\begin{pmatrix}1&1&0\\0&1&1\end{pmatrix}$$

Now suppose that the basis for $\boldsymbol{R}^3$ is changed to

$$B=\left\{\begin{pmatrix}1\\1\\1\end{pmatrix},\begin{pmatrix}1\\0\\-1\end{pmatrix},\begin{pmatrix}2\\0\\0\end{pmatrix}\right\}$$

so that the corresponding transition matrix is

$$P=\begin{pmatrix}1&1&2\\1&0&0\\1&-1&0\end{pmatrix}$$

and that the basis for $\boldsymbol{R}^2$ is changed to

$$C = \left\{ \begin{pmatrix} 1 \\ 3 \end{pmatrix}, \begin{pmatrix} 1 \\ 2 \end{pmatrix} \right\}$$

so that the corresponding transition matrix is

$$Q = \begin{pmatrix} 1 & 1 \\ 3 & 2 \end{pmatrix}$$

We wish to find the matrix N representing f with respect to the new bases. We need to do the following in the order given:

- change from the new to the standard basis in $\boldsymbol{R}^3$
- apply the matrix representing f with respect to the standard bases
- change from the standard to the new basis in $\boldsymbol{R}^2$

Hence we require

$$\begin{aligned} N &= Q^{-1}MP \\ &= \begin{pmatrix} -2 & 1 \\ 3 & -1 \end{pmatrix} \begin{pmatrix} 1 & 1 & 0 \\ 0 & 1 & 1 \end{pmatrix} \begin{pmatrix} 1 & 1 & 2 \\ 1 & 0 & 0 \\ 1 & -1 & 0 \end{pmatrix} \\ &= \begin{pmatrix} -2 & -3 & -4 \\ 4 & 4 & 6 \end{pmatrix} \end{aligned}$$

We summarise what we have discovered as follows:

5.12 Proposition Suppose that $f: V \to W$ is a linear transformation of vector spaces, and that f is represented with respect to the standard bases by the matrix M. Suppose that V and W are each provided with a new basis, having transition matrices P and Q respectively. Then f is represented with respect to the new bases by the matrix

$$N = Q^{-1}MP.$$

Sketch proof: Consider the following diagram:

$$\begin{array}{lccc} & & M & \\ \text{Standard bases:} & V & \rightarrow & W \\ & P\uparrow & & \uparrow Q \\ \text{New bases:} & V & \rightarrow & W \\ & & N & \end{array}$$

To apply f to a vector v expressed with respect to the new basis for V we do the following:

1. Pre multiply by P to transform v to a vector Pv with respect to the standard basis for V.

2. Pre multiply by M to effect the linear transformation to give a vector MPv with respect to the standard basis for W.

3. Pre multiply by the inverse of Q to transform MPv to $Q^{-1}MPv$ with respect to the new basis for W.

The combined effect of these three steps is to pre multiply by $Q^{-1}MP$, and hence the result.

Example Consider the linear transformation $f: \boldsymbol{R}^2 \to \boldsymbol{R}^3$ given by

$$f\begin{pmatrix} x \\ y \end{pmatrix} = \begin{pmatrix} x+y \\ 2x \\ -y \end{pmatrix}.$$

With respect to the usual bases f is represented by

$$M = \begin{pmatrix} 1 & 1 \\ 2 & 0 \\ 0 & -1 \end{pmatrix}.$$

Suppose $\boldsymbol{R}^2$ is given the new basis $\left\{ \begin{pmatrix} 1 \\ 1 \end{pmatrix}, \begin{pmatrix} 1 \\ -1 \end{pmatrix} \right\}$ so that

$$P = \begin{pmatrix} 1 & 1 \\ 1 & -1 \end{pmatrix}.$$

Suppose that $\boldsymbol{R}^3$ is given the new basis $\left\{ \begin{pmatrix} 1 \\ 1 \\ 0 \end{pmatrix}, \begin{pmatrix} 1 \\ 0 \\ 1 \end{pmatrix}, \begin{pmatrix} 0 \\ 1 \\ 1 \end{pmatrix} \right\}$ so that

$$Q = \begin{pmatrix} 1 & 1 & 0 \\ 1 & 0 & 1 \\ 0 & 1 & 1 \end{pmatrix}.$$

Then with respect to the new bases f is represented by

$$N = Q^{-1}MP = \begin{pmatrix} 1 & 1 & 0 \\ 1 & 0 & 1 \\ 0 & 1 & 1 \end{pmatrix}^{-1} \begin{pmatrix} 1 & 1 \\ 2 & 0 \\ 0 & -1 \end{pmatrix} \begin{pmatrix} 1 & 1 \\ 1 & -1 \end{pmatrix} = \frac{1}{2} \begin{pmatrix} 5 & 1 \\ -1 & -1 \\ -1 & 3 \end{pmatrix}.$$

Why make a change of basis? We shall see in the next chapter that a well chosen basis allows a linear transformation to be represented by a matrix of a very simple form.

Exercises 5

1. Consider the following vectors in $\boldsymbol{R}^2$:

$$\begin{pmatrix} 2 \\ 1 \end{pmatrix}, \begin{pmatrix} 1 \\ -1 \end{pmatrix}.$$

Show that (i) the vectors are linearly independent
(ii) the vectors span $\boldsymbol{R}^2$.

2. By finding the determinants of suitable matrices, decide whether or not each of the following sets of vectors is a basis for $\boldsymbol{R}^3$:

(i) $$\left\{ \begin{pmatrix} 1 \\ 3 \\ 0 \end{pmatrix}, \begin{pmatrix} 1 \\ -2 \\ -1 \end{pmatrix}, \begin{pmatrix} 2 \\ 1 \\ 1 \end{pmatrix} \right\}$$

(ii) $$\left\{ \begin{pmatrix} 1 \\ 3 \\ 0 \end{pmatrix}, \begin{pmatrix} 1 \\ -2 \\ -1 \end{pmatrix}, \begin{pmatrix} 2 \\ 1 \\ -1 \end{pmatrix} \right\}$$

3. State *why* each of the following sets of vectors *fails* to be a basis for $\boldsymbol{R}^3$:

(i) $$\left\{ \begin{pmatrix} 1 \\ 0 \\ 0 \end{pmatrix}, \begin{pmatrix} 1 \\ 1 \\ 1 \end{pmatrix} \right\}$$

(ii) $$\left\{ \begin{pmatrix} 1 \\ 0 \\ 1 \end{pmatrix}, \begin{pmatrix} 1 \\ 1 \\ 0 \end{pmatrix}, \begin{pmatrix} 0 \\ 1 \\ 1 \end{pmatrix}, \begin{pmatrix} 1 \\ 1 \\ 1 \end{pmatrix} \right\}$$

(iii) $$\left\{ \begin{pmatrix} 1 \\ 0 \\ -1 \end{pmatrix}, \begin{pmatrix} -1 \\ 1 \\ 0 \end{pmatrix}, \begin{pmatrix} 0 \\ 1 \\ -1 \end{pmatrix} \right\}.$$

4. Consider the vector space of polynomials with real coefficients of degree not exceeding 2:

$$P = \{ax^2 + bx + c : a, b, c \in \boldsymbol{R}\}$$

By defining a suitable linear transformation show that $P \cong \boldsymbol{R}^3$.

5. Let E be the standard basis for $\boldsymbol{R}^2$ and let B be the basis for $\boldsymbol{R}^2$ given by

$$\left\{ \begin{pmatrix} 7 \\ 2 \end{pmatrix}, \begin{pmatrix} 3 \\ 1 \end{pmatrix} \right\}$$

(i) Write down the transition matrix for change of basis from B to E.

(ii) If $v = \begin{pmatrix} 2 \\ 3 \end{pmatrix}$ with respect to B, find v with respect to E.

(iii) If $v = \begin{pmatrix} 5 \\ 1 \end{pmatrix}$ with respect to E, find v with respect to B.

6. Consider the linear transformation $f : \boldsymbol{R}^2 \to \boldsymbol{R}^3$ given by

$$f\begin{pmatrix} x \\ y \end{pmatrix} = \begin{pmatrix} x + 3y \\ 2x - y \\ 4x + y \end{pmatrix}$$

Write down the matrix M representing f with respect to the standard bases.

Now suppose that the basis for $\boldsymbol{R}^2$ is changed to

$$\left\{ \begin{pmatrix} 1 \\ 1 \end{pmatrix}, \begin{pmatrix} 1 \\ 3 \end{pmatrix} \right\}$$

and that the basis for $\boldsymbol{R}^3$ is changed to

$$\left\{ \begin{pmatrix} 1 \\ 1 \\ 0 \end{pmatrix}, \begin{pmatrix} 1 \\ 0 \\ 1 \end{pmatrix}, \begin{pmatrix} 0 \\ 1 \\ 1 \end{pmatrix} \right\}.$$

Write down transition matrices for the new bases, and find the matrix N representing f with respect to the new bases.

7. Decide whether or not each of the following is a linearly independent set of vectors in $\boldsymbol{R}^3$:

(i) $\left\{ \begin{pmatrix} 1 \\ 2 \\ 3 \end{pmatrix} \right\}$ (ii) $\left\{ \begin{pmatrix} 0 \\ 0 \\ 0 \end{pmatrix} \right\}$

(iii) $\left\{ \begin{pmatrix} 1 \\ 2 \\ 3 \end{pmatrix}, \begin{pmatrix} 0 \\ 0 \\ 0 \end{pmatrix} \right\}.$

Chapter 6: Eigenvalues and Eigenvectors

In this chapter we are concerned with linear transformations which have the same vector space as both domain and codomain. These are known as *endomorphisms*. As a first example, consider the linear transformation $f: \boldsymbol{R}^2 \to \boldsymbol{R}^2$ represented with respect to the standard basis by the matrix

$$\begin{pmatrix} -1 & 3 \\ -4 & 6 \end{pmatrix}$$

We may notice that for certain vectors in $\boldsymbol{R}^2$, when we apply f we obtain a scalar multiple of the original vector:

$$\begin{pmatrix} -1 & 3 \\ -4 & 6 \end{pmatrix}\begin{pmatrix} 1 \\ 1 \end{pmatrix} = \begin{pmatrix} 2 \\ 2 \end{pmatrix} = 2\begin{pmatrix} 1 \\ 1 \end{pmatrix}$$

$$\begin{pmatrix} -1 & 3 \\ -4 & 6 \end{pmatrix}\begin{pmatrix} 3 \\ 4 \end{pmatrix} = \begin{pmatrix} 9 \\ 12 \end{pmatrix} = 3\begin{pmatrix} 3 \\ 4 \end{pmatrix}$$

The non-zero vectors that have this property are called *eigenvectors*, and the scalars giving the multiples of the original vectors are called *eigenvalues*.

In our example, $\begin{pmatrix} 1 \\ 1 \end{pmatrix}$ is an eigenvector with eigenvalue 2 and $\begin{pmatrix} 3 \\ 4 \end{pmatrix}$ is an eigenvector with eigenvalue 3.

This leads us to the following definition:

6.1 Definition Suppose that $f: \boldsymbol{R}^n \to \boldsymbol{R}^n$ is an endomorphism represented by a matrix M. A non-zero vector v is called an *eigenvector* if

$$Mv = \lambda v$$

for some scalar $\lambda \in F$. The scalar is called an *eigenvalue*.

We present a method of finding the eigenvectors and eigenvalues of a linear transformation $f: \boldsymbol{R}^n \to \boldsymbol{R}^n$ represented by a matrix M. If v is an eigenvector with eigenvalue λ then we have

$$Mv = \lambda v = \lambda I v$$

where I is the $n \times n$ identity matrix. Rearranging and factorising gives

$$(M - \lambda I)v = \mathbf{0}.$$

Since we are only interested in non-zero vectors v, the kernel of $M - \lambda I$ is non-trivial so that $M - \lambda I$ is not an isomorphism and hence not invertible. It follows that

$$\det(M - \lambda I) = 0.$$

This is called the *characteristic equation*, and its roots are the eigenvalues.

Notice that if v is an eigenvector with eigenvalue λ, then any non-zero scalar multiple av of v is also an eigenvector with the same eigenvalue, since

$$Mav = aMv = a(\lambda v) = \lambda(av).$$

Hence for each eigenvalue we may choose an eigenvector of any non-zero magnitude.

Examples (a)

$$M = \begin{pmatrix} -1 & -2 \\ 10 & 8 \end{pmatrix}$$

$$M - \lambda I = \begin{pmatrix} -1-\lambda & -2 \\ 10 & 8-\lambda \end{pmatrix}$$

Calculating the determinant we obtain the characteristic equation

$$(-1-\lambda)(8-\lambda) - (-2) \times 10 = \lambda^2 - 7\lambda + 12 = 0$$

which factorises as $(\lambda - 3)(\lambda - 4) = 0$

Hence the eigenvalues are $\lambda = 3, \lambda = 4$.

Next we find the eigenvectors. Recall that for an eigenvector v we require

$$(M-\lambda I)v=\mathbf{0}$$

In our example, for $\lambda=3$ we have $M-3I=\begin{pmatrix}-4 & -2\\ 10 & 5\end{pmatrix}$ so we require

$$\begin{pmatrix}-4 & -2\\ 10 & 5\end{pmatrix}\begin{pmatrix}x\\ y\end{pmatrix}=\begin{pmatrix}0\\ 0\end{pmatrix}.$$

Hence x and y must satisfy $2x+y=0$, and the eigenvector is any non-zero multiple of

$$\begin{pmatrix}1\\ -2\end{pmatrix}.$$

For $\lambda=4$ we have $M-4I=\begin{pmatrix}-5 & -2\\ 10 & 4\end{pmatrix}$ so we require

$$\begin{pmatrix}-5 & -2\\ 10 & 4\end{pmatrix}\begin{pmatrix}x\\ y\end{pmatrix}=\begin{pmatrix}0\\ 0\end{pmatrix}.$$

Hence x and y must satisfy $5x+2y=0$ and the eigenvector is any non-zero multiple of

$$\begin{pmatrix}2\\ -5\end{pmatrix}.$$

(b)

$$M=\begin{pmatrix}1 & 2 & -1\\ 1 & 0 & 1\\ 4 & -4 & 5\end{pmatrix}$$

$$M-\lambda I=\begin{pmatrix}1-\lambda & 2 & -1\\ 1 & -\lambda & 1\\ 4 & -4 & 5-\lambda\end{pmatrix}$$

Calculating the determinant we obtain

$$(1-\lambda)\det\begin{pmatrix} -\lambda & 1 \\ -4 & 5-\lambda \end{pmatrix} - 2\det\begin{pmatrix} 1 & 1 \\ 4 & 5-\lambda \end{pmatrix} + (-1)\det\begin{pmatrix} 1 & -\lambda \\ 4 & -4 \end{pmatrix}$$

$$= (1-\lambda)(\lambda^2 - 5\lambda + 4) - 2(-\lambda + 1) - (-4 + 4\lambda)$$

$$= (1-\lambda)(\lambda^2 - 5\lambda + 6) = -(\lambda - 1)(\lambda - 2)(\lambda - 3).$$

Hence the characteristic equation is

$$(\lambda - 1)(\lambda - 2)(\lambda - 3) = 0$$

and so the eigenvalues are $\lambda = 1, \lambda = 2$ and $\lambda = 3$.

Next we find the eigenvectors. For $\lambda = 1$ we require

$$\begin{pmatrix} 0 & 2 & -1 \\ 1 & -1 & 1 \\ 4 & -4 & 4 \end{pmatrix}\begin{pmatrix} x \\ y \\ z \end{pmatrix} = \begin{pmatrix} 0 \\ 0 \\ 0 \end{pmatrix}$$

The coefficient matrix reduces by row operations to $\begin{pmatrix} 1 & 1 & 0 \\ 0 & 1 & -\frac{1}{2} \\ 0 & 0 & 0 \end{pmatrix}$.

Hence we obtain the eigenvector $\begin{pmatrix} -1 \\ 1 \\ 2 \end{pmatrix}$.

For $\lambda = 2$ we require

$$\begin{pmatrix} -1 & 2 & -1 \\ 1 & -2 & 1 \\ 4 & -4 & 3 \end{pmatrix}\begin{pmatrix} x \\ y \\ z \end{pmatrix} = \begin{pmatrix} 0 \\ 0 \\ 0 \end{pmatrix}$$

The coefficient matrix reduces by row operations to $\begin{pmatrix} 1 & 0 & \frac{1}{2} \\ 0 & 1 & -\frac{1}{4} \\ 0 & 0 & 0 \end{pmatrix}$.

Hence we obtain the eigenvector $\begin{pmatrix} -2 \\ 1 \\ 4 \end{pmatrix}$.

For $\lambda = 3$ we require

$$\begin{pmatrix} -2 & 2 & -1 \\ 1 & -3 & 1 \\ 4 & -4 & 2 \end{pmatrix}\begin{pmatrix} x \\ y \\ z \end{pmatrix} = \begin{pmatrix} 0 \\ 0 \\ 0 \end{pmatrix}$$

The coefficient matrix reduces by row operations to $\begin{pmatrix} 1 & 1 & 0 \\ 0 & 1 & -\frac{1}{4} \\ 0 & 0 & 0 \end{pmatrix}$.

Hence we obtain the eigenvector $\begin{pmatrix} -1 \\ 1 \\ 4 \end{pmatrix}$.

(c) Note that not every real square matrix has real eigenvalues.

For example the matrix

$$\begin{pmatrix} 2 & 1 \\ -7 & -3 \end{pmatrix}$$

has characteristic equation $\det\begin{pmatrix} 2-\lambda & 1 \\ -7 & -3-\lambda \end{pmatrix} = \lambda^2 + \lambda + 1 = 0$,

which has no real roots.

6.2 Basis of Eigenvectors

Suppose that an endomorphism $f : \boldsymbol{R}^n \to \boldsymbol{R}^n$ is represented with respect to the standard basis by the matrix M. It turns out that if the eigenvectors of M are a basis for $\boldsymbol{R}^n$ then the matrix N representing f with respect to this basis is particularly simple.

Examples (a) Consider the linear transformation $f : \boldsymbol{R}^2 \to \boldsymbol{R}^2$ represented with respect to the standard basis by the matrix

$$M = \begin{pmatrix} -1 & -2 \\ 10 & 8 \end{pmatrix}$$

We found previously that M has eigenvalues $\lambda = 3$ and $\lambda = 4$ with respective eigenvectors

$$\begin{pmatrix} 1 \\ -2 \end{pmatrix} \text{and} \begin{pmatrix} 2 \\ -5 \end{pmatrix}.$$

It is easily checked that these are a basis for $\boldsymbol{R}^2$.

For our new basis $\left\{ \begin{pmatrix} 1 \\ -2 \end{pmatrix}, \begin{pmatrix} 2 \\ -5 \end{pmatrix} \right\}$ the transition matrix is

$$P = \begin{pmatrix} 1 & 2 \\ -2 & -5 \end{pmatrix}$$

With respect to the basis of eigenvectors, f is represented by

$$\begin{aligned} N &= P^{-1} M P \\ &= \begin{pmatrix} 5 & 2 \\ -2 & -1 \end{pmatrix} \begin{pmatrix} -1 & -2 \\ 10 & 8 \end{pmatrix} \begin{pmatrix} 1 & 2 \\ -2 & -5 \end{pmatrix} \\ &= \begin{pmatrix} 5 & 2 \\ -2 & -1 \end{pmatrix} \begin{pmatrix} 3 & 8 \\ -6 & -20 \end{pmatrix} = \begin{pmatrix} 3 & 0 \\ 0 & 4 \end{pmatrix}. \end{aligned}$$

The matrix N is an example of a *diagonal matrix*, having entries of zero everywhere except the leading diagonal. The entries on the leading diagonal are the eigenvalues. These appear in the order that the corresponding eigenvectors appear as columns of the transition matrix P.

(b) Consider the linear transformation $f: \boldsymbol{R}^3 \to \boldsymbol{R}^3$ represented with respect to the standard basis by the matrix

$$M = \begin{pmatrix} 1 & 2 & -1 \\ 1 & 0 & 1 \\ 4 & -4 & 5 \end{pmatrix}$$

We found that M has eigenvectors

$$\begin{pmatrix} -1 \\ 1 \\ 2 \end{pmatrix}, \begin{pmatrix} -2 \\ 1 \\ 4 \end{pmatrix} \text{ and } \begin{pmatrix} -1 \\ 1 \\ 4 \end{pmatrix}.$$

Once again, these are a basis for $\boldsymbol{R}^3$ with transition matrix

$$P = \begin{pmatrix} -1 & -2 & -1 \\ 1 & 1 & 1 \\ 2 & 4 & 4 \end{pmatrix}.$$

The matrix representing f with respect to the basis of eigenvectors is

$$P^{-1}MP = \begin{pmatrix} 1 & 0 & 0 \\ 0 & 2 & 0 \\ 0 & 0 & 3 \end{pmatrix}.$$

Once again, we obtain a diagonal matrix featuring the eigenvalues on the leading diagonal. This suggests the following result:

6.3 Proposition Suppose that $f : \boldsymbol{R}^n \to \boldsymbol{R}^n$ is an endomorphism, and that we have a basis for $\boldsymbol{R}^n$ of eigenvectors. Then f is represented with respect to this basis by a diagonal matrix featuring the eigenvalues on the leading diagonal.

Proof: If $B = \{v_1, v_2, ..., v_n\}$ is a basis of eigenvectors, then any v can be expressed uniquely as a linear combination of these:

$$v = a_1 v_1 + a_2 v_2 + ... + a_n v_n$$

which is equivalent to the column vector $v = \begin{pmatrix} a_1 \\ a_2 \\ . \\ . \\ a_n \end{pmatrix}_B$ with respect to the eigenvector basis. Applying f we obtain

$$
\begin{aligned}
f(v) &= a_1 f(v_1) + a_2 f(v_2) + \ldots + a_n f(v_n) \\
&= \lambda_1 a_1 v_1 + \lambda_2 a_2 v_2 + \ldots + \lambda_n a_n v_n \\
&= \begin{pmatrix} \lambda_1 a_1 \\ \lambda_2 a_2 \\ \cdot \\ \cdot \\ \lambda_n a_n \end{pmatrix}_B = \begin{pmatrix} \lambda_1 & & & & \\ & \lambda_2 & & 0 & \\ & & \cdot & & \\ & 0 & & \cdot & \\ & & & & \lambda_n \end{pmatrix} \begin{pmatrix} a_1 \\ a_2 \\ \cdot \\ \cdot \\ a_n \end{pmatrix}.
\end{aligned}
$$

A matrix representing a linear transformation for which we can carry out this process is said to be *diagonalisable.*

6.4 Corollary Suppose that M is a diagonalisable square matrix. The product of the eigenvalues of M is equal to the determinant of M.

Proof: Suppose M is diagonalisable, and that $\Lambda = P^{-1}MP$ is the diagonal matrix of eigenvalues.

We have $\det \Lambda = \lambda_1 \lambda_2 \ldots \lambda_n$.

But $\det \Lambda = \det P^{-1}MP = (\det P)^{-1} \det M \det P = \det M$.

We have calculated eigenvalues for a matrix representing a linear transformation with respect to a particular basis. However, it turns out that the eigenvalues do not depend upon the basis we choose but are intrinsic to the linear transformation itself, as the following result shows.

6.5 Theorem Suppose that M is a square matrix and P is an invertible matrix of the same size. Then M and $P^{-1}MP$ have the same characteristic equation and hence the same eigenvalues.

Proof:

$$
\begin{aligned}
\det(P^{-1}MP - \lambda I) &= \det(P^{-1}MP - \lambda P^{-1}IP) = \det(P^{-1}(MP - \lambda IP)) \\
&= \det(P^{-1}(M - \lambda I)P) = \det P^{-1} \det(M - \lambda I) \det P \\
&= \det(M - \lambda I).
\end{aligned}
$$

In each of the previous examples the eigenvalues were distinct, and the set of eigenvectors was a basis. In fact, this will always be the case where we have distinct eigenvalues:

6.6 Theorem Suppose that a linear transformation f has *distinct* eigenvalues $\lambda_1, \lambda_2, ..., \lambda_n$.

The set of corresponding eigenvectors $\{v_1, v_2, ..., v_n\}$ is linearly independent.

Proof: We prove the result by induction on the number of eigenvalues n.

For $n = 1$ there is just a single eigenvalue and clearly $\{v_1\}$ is linearly independent.

Now suppose that the result is true for $n = k$:

If $\lambda_1, ..., \lambda_k$ are distinct then $\{v_1, v_2, ..., v_k\}$ is linearly independent. Suppose that λ_{k+1} is an eigenvalue that is distinct from $\lambda_1, ..., \lambda_k$.

We need to show that $\{v_1, ..., v_k, v_{k+1}\}$ is linearly independent. Suppose

$$\mu_1 v_1 + ... + \mu_k v_k + \mu_{k+1} v_{k+1} = \mathbf{0}.$$

Then $f(\mu_1 v_1 + \mu_2 v_2 + ... + \mu_k v_k + \mu_{k+1} v_{k+1}) = f(\mathbf{0}) = \mathbf{0}.$

Hence $\lambda_1 \mu_1 v_1 + \lambda_2 \mu_2 v_2 + ... + \lambda_k \mu_k v_k + \lambda_{k+1} \mu_{k+1} v_{k+1} = \mathbf{0}.$

But we also have

$$\lambda_{k+1} \mu_1 v_1 + \lambda_{k+1} \mu_2 v_2 + ... + \lambda_{k+1} \mu_k v_k + \lambda_{k+1} \mu_{k+1} v_{k+1} = \mathbf{0}.$$

Subtracting,

$$(\lambda_1 - \lambda_{k+1})\mu_1 v_1 + (\lambda_2 - \lambda_{k+1})\mu_2 v_2 + ... + (\lambda_k - \lambda_{k+1})\mu_k v_k = \mathbf{0}.$$

Hence $(\lambda_1 - \lambda_{k+1})\mu_1 = 0, (\lambda_2 - \lambda_{k+1})\mu_2 = 0, ..., (\lambda_k - \lambda_{k+1})\mu_k = 0.$

But since $\lambda_{k+1} \neq \lambda_i$ for each i we have $\mu_i = 0$ for $i = 1, 2, ..., k.$

Clearly we must have $\mu_{k+1} = 0$ also.

The set of eigenvectors for a given scalar together with the zero vector form a subspace, as the next result shows.

6.7 Proposition Let $f: V \to V$ be an endomorphism of a vector space V over F, and let $\lambda \in F$.

The set $U = \{v : f(v) = \lambda v\}$ is a subspace of V.

Proof: We apply the test for a subspace 4.2:

(i) Suppose $v_1, v_2 \in U$, so that $f(v_1) = \lambda v_1$ and $f(v_2) = \lambda v_2$.

Then $f(v_1 + v_2) = f(v_1) + f(v_2) = \lambda v_1 + \lambda v_2 = \lambda(v_1 + v_2)$ and hence $v_1 + v_2 \in U$.

(ii) If $v \in U$, so that $f(v) = \lambda v$ then $f(av) = af(v) = a\lambda v = \lambda(av)$, and hence $av \in U$.

Such a subspace is called the *eigenspace* of λ.

For most scalars λ the corresponding eigenspace will be the trivial subspace $\{\mathbf{0}\}$. Non-trivial eigenspaces occur for values of λ that are eigenvalues, and are spanned by the corresponding eigenvectors.

In our examples so far, non-trivial eigenspaces have always been of dimension 1. This is not always the case, as we shall see in the next section.

6.8 Algebraic and Geometric Multiplicity

In our previous examples the eigenvalues of the matrices were distinct. Each eigenvalue appeared precisely once as a root of the characteristic equation. It is possible for the characteristic equation of a matrix to have repeated roots. Where an eigenvalue λ appears n times as the root of the characteristic equation, we say that it has *algebraic multiplicity n*.

The dimension of the eigenspace, which is the number of linearly independent eigenvectors of λ, is called its *geometric multiplicity*. It turns out that the geometric multiplicity can never exceed the algebraic multiplicity. However, the geometric multiplicity may sometimes be less than the algebraic multiplicity. It is precisely under these circumstances that the matrix fails to be diagonalisable.

Example

$$M = \begin{pmatrix} -2 & 18 & -3 \\ -1 & 7 & -1 \\ -2 & 12 & -1 \end{pmatrix}$$

has characteristic equation $(\lambda - 1)(\lambda - 1)(\lambda - 2) = 0$.

$\lambda = 1$ has algebraic multiplicity 2;

$\lambda = 2$ has algebraic multiplicity 1.

For $\lambda = 1$ we require

$$\begin{pmatrix} -3 & 18 & -3 \\ -1 & 6 & -1 \\ -2 & 12 & -2 \end{pmatrix} \begin{pmatrix} x \\ y \\ z \end{pmatrix} = \begin{pmatrix} 0 \\ 0 \\ 0 \end{pmatrix}$$

By row operations this reduces to

$$\begin{pmatrix} 1 & -6 & 1 \\ 0 & 0 & 0 \\ 0 & 0 & 0 \end{pmatrix} \begin{pmatrix} x \\ y \\ z \end{pmatrix} = \begin{pmatrix} 0 \\ 0 \\ 0 \end{pmatrix}$$

The eigenspace is the plane defined by $x - 6y + z = 0$. We may choose a pair of linearly independent eigenvectors, for example

$$\begin{pmatrix} 6 \\ 1 \\ 0 \end{pmatrix} \text{and} \begin{pmatrix} 1 \\ 0 \\ -1 \end{pmatrix}$$

There are other choices.

For $\lambda = 2$ we find the single eigenvector

$$\begin{pmatrix} 3 \\ 1 \\ 2 \end{pmatrix}$$

So $\lambda = 1$ has geometric multiplicity 2 and $\lambda = 2$ has geometric multiplicity 1. These coincide with the algebraic multiplicities.

We have now obtained the basis of eigenvectors

$$\left\{ \begin{pmatrix} 6 \\ 1 \\ 0 \end{pmatrix}, \begin{pmatrix} 1 \\ 0 \\ -1 \end{pmatrix}, \begin{pmatrix} 3 \\ 1 \\ 2 \end{pmatrix} \right\}.$$

With respect to the basis of eigenvectors the linear transformation is represented by

$$N = P^{-1}MP = \begin{pmatrix} 1 & 0 & 0 \\ 0 & 1 & 0 \\ 0 & 0 & 2 \end{pmatrix}$$

We might hope that for each eigenvalue the geometric multiplicity was always equal to the algebraic multiplicity. Unfortunately this is not the case, as the following example illustrates:

Example

$$M = \begin{pmatrix} 3 & 4 \\ 0 & 3 \end{pmatrix}$$

has characteristic equation $(\lambda - 3)(\lambda - 3) = 0$. Hence the eigenvalue $\lambda = 3$ has algebraic multiplicity 2. The corresponding eigenvectors must satisfy

$$\begin{pmatrix} 0 & 4 \\ 0 & 0 \end{pmatrix} \begin{pmatrix} x \\ y \end{pmatrix} = \begin{pmatrix} 0 \\ 0 \end{pmatrix}$$

and we find that there is only a single eigenvector, namely

$$\begin{pmatrix} 1 \\ 0 \end{pmatrix}.$$

Hence the eigenvalue $\lambda = 3$ has geometric multiplicity 1. Since we cannot obtain a basis of eigenvectors, the matrix cannot be diagonalised.

We have shown in theorem 6.6 that we can be sure of finding a basis of eigenvectors provided that our matrix has distinct eigenvalues. However, it is clear that we can improve upon this, since we previously found a basis of eigenvectors for a matrix with an eigenvalue of algebraic multiplicity 2. We would like to be able to state necessary and sufficient conditions for a matrix to be diagonalisable.

We first present an important theorem, stated without proof:

6.9 The Cayley-Hamilton Theorem A square matrix satisfies its own characteristic equation. That is, if M has characteristic equation

$$a_n\lambda^n + a_{n-1}\lambda^{n-1} + ... + a_1\lambda + a_0 = 0$$

then

$$a_nM^n + a_{n-1}M^{n-1} + ... + a_1M + a_0I = 0$$

Example The matrix

$$M = \begin{pmatrix} -1 & -2 \\ 10 & 8 \end{pmatrix}$$

has characteristic equation $\lambda^2 - 7\lambda + 12 = 0$. It may be verified that $M^2 - 7M + 12I = 0$:

$$\begin{pmatrix} -1 & -2 \\ 10 & 8 \end{pmatrix}\begin{pmatrix} -1 & -2 \\ 10 & 8 \end{pmatrix} - 7\begin{pmatrix} -1 & -2 \\ 10 & 8 \end{pmatrix} + 12\begin{pmatrix} 1 & 0 \\ 0 & 1 \end{pmatrix} = \begin{pmatrix} 0 & 0 \\ 0 & 0 \end{pmatrix}.$$

As well as satisfying its own characteristic equation, a matrix M may satisfy other polynomial equations. The unique polynomial $m(\lambda)$ of lowest degree for which

$$m(M) = 0$$

is called the *minimal polynomial*. Clearly the degree of the minimal polynomial cannot exceed that of the characteristic polynomial. We have the following theorems, again stated without proof:

6.10 Theorem Each irreducible factor of the characteristic polynomial is also an irreducible factor of the minimal polynomial.

6.11 Theorem A matrix is diagonalisable if and only if its minimal polynomial can be factorised into distinct linear factors.

6.12 Corollary Suppose that the characteristic polynomial of a matrix M is

$$(\lambda-\lambda_1)^{m_1}(\lambda-\lambda_2)^{m_2}...(\lambda-\lambda_n)^{m_n}$$

and let $f(\lambda)=(\lambda-\lambda_1)(\lambda-\lambda_2)...(\lambda-\lambda_n)$.

Then M is diagonalisable if and only if $f(M)=0$.

Examples (a)

$$M=\begin{pmatrix}-2 & 18 & -3\\ -1 & 7 & -1\\ -2 & 12 & -1\end{pmatrix}$$

has characteristic equation $(\lambda-1)^2(\lambda-2)=0$, and so

$$f(\lambda)=(\lambda-1)(\lambda-2).$$

It may be checked that

$$(M-I)(M-2I)=\begin{pmatrix}-3 & 18 & -3\\ -1 & 6 & -1\\ -2 & 12 & -2\end{pmatrix}\begin{pmatrix}-4 & 18 & -3\\ -1 & 5 & -1\\ -2 & 12 & -3\end{pmatrix}=\begin{pmatrix}0 & 0 & 0\\ 0 & 0 & 0\\ 0 & 0 & 0\end{pmatrix},$$

and hence M is diagonalisable.

(b)

$$M=\begin{pmatrix}1 & 1 & -1\\ 1 & 1 & 0\\ 2 & -2 & 3\end{pmatrix}$$

has characteristic equation $(\lambda-2)^2(\lambda-1)=0$, and so

$$f(\lambda)=(\lambda-2)(\lambda-1)$$

It may be checked that

$$(M-2I)(M-I)=\begin{pmatrix}-1 & 1 & -1\\ 1 & -1 & 0\\ 2 & -2 & 1\end{pmatrix}\begin{pmatrix}0 & 1 & -1\\ 1 & 0 & 0\\ 2 & -2 & 2\end{pmatrix}=\begin{pmatrix}-1 & 1 & -1\\ -1 & 1 & -1\\ 0 & 0 & 0\end{pmatrix},$$

and hence M is *not* diagonalisable.

Exercises 6

1. For each of the following matrices find the characteristic equation, and hence determine the eigenvalues. Find the eigenvector(s) associated with each eigenvalue.

(i) $\begin{pmatrix} 3 & 7 \\ 6 & 4 \end{pmatrix}$ (ii) $\begin{pmatrix} 15 & 20 \\ -6 & -7 \end{pmatrix}$

(iii) $\begin{pmatrix} -2 & 12 & 6 \\ -4 & 14 & 8 \\ 6 & -18 & -11 \end{pmatrix}$ (iv) $\begin{pmatrix} 3 & 1 & 1 \\ 2 & 4 & 2 \\ 1 & 1 & 3 \end{pmatrix}$

2. Consider the linear transformation $f: \boldsymbol{R}^2 \to \boldsymbol{R}^2$ which is represented with respect to the standard basis by the matrix

$$M = \begin{pmatrix} 9 & -7 \\ 14 & -12 \end{pmatrix}$$

Find the eigenvalues and the eigenvectors for M. Calculate (or write down!) the matrix N representing f with respect to the basis of eigenvectors.

3. Repeat question 2 for the linear transformation $f: \boldsymbol{R}^3 \to \boldsymbol{R}^3$ which is represented with respect to the standard basis by the matrix

$$M = \begin{pmatrix} 4 & 0 & 1 \\ 2 & 3 & 2 \\ 1 & 0 & 4 \end{pmatrix}$$

4. Let V be the vector space of real-valued functions defined on the unit interval that are differentiable an arbitrary number of times. Consider the differential operator $D: V \to V$ that maps a function to its derivative.

Show that e^{ax} is an eigenvector for D and state the corresponding eigenvalue. Show that $\sin ax$ and $\cos ax$ are eigenvectors for D^2 and state the corresponding eigenvalue.

Chapter 7: Inner Product Spaces

In this chapter we give a vector space extra structure, in the form of a generalisation of the scalar product known as an *inner product.* We define a special type of linear transformation called a *symmetric* linear transformation, for which the basis of eigenvectors will also be special.

Recall from chapter 3 that the *scalar product* on $\boldsymbol{R}^2$ is the mapping $\boldsymbol{R}^2 \times \boldsymbol{R}^2 \to \boldsymbol{R}$ calculated as follows:

If $u = \begin{pmatrix} x_1 \\ y_1 \end{pmatrix}$ and $v = \begin{pmatrix} x_2 \\ y_2 \end{pmatrix}$ then $u.v = x_1x_2 + y_1y_2$.

7.1 Proposition The scalar product on $\boldsymbol{R}^2$ satisfies the following:

(i) $u.v = v.u$

(ii) $(u+v).w = u.w + v.w$

(iii) $(\lambda u).v = \lambda(u.v) = u.(\lambda v)$

(iv) $v.v \geq 0$ and $v.v = 0$ iff $v = \mathbf{0}$

for all $u, v, w \in \boldsymbol{R}^2$ and all $\lambda \in \boldsymbol{R}$.

Proof: The proofs are left as an exercise.

Recall also that the *magnitude* of a vector $v = \begin{pmatrix} x \\ y \end{pmatrix}$ in $\boldsymbol{R}^2$ was defined by

$$|v| = \sqrt{x^2 + y^2}\,.$$

It may easily be verified that

$$|v| = \sqrt{v.v}\,.$$

Using the scalar product on $\boldsymbol{R}^2$ as a motivating example, we make the following definition:

7.2 Definition Suppose that V is a vector space over $\boldsymbol{R}$. A mapping $V \times V \to \boldsymbol{R}$, written $\langle u, v \rangle$, is called a *real inner product* if it satisfies:

(i) $\langle u, v \rangle = \langle v, u \rangle$

(ii) $\langle u+v, w \rangle = \langle u, w \rangle + \langle v, w \rangle$

(iii) $\langle \lambda u, v \rangle = \lambda \langle u, v \rangle$

(iv) $\langle v, v \rangle \geq 0$ and $\langle v, v \rangle = 0$ iff $v = \mathbf{0}$

for all $u, v, w \in V$ and all $\lambda \in \boldsymbol{R}$.

A real vector space equipped with a real inner product is called a *real inner product space.*

Examples (a) For $\boldsymbol{R}^n$ the product defined by

$$\begin{pmatrix} u_1 \\ \cdot \\ \cdot \\ \cdot \\ u_n \end{pmatrix} \cdot \begin{pmatrix} v_1 \\ \cdot \\ \cdot \\ \cdot \\ v_n \end{pmatrix} = u_1 v_1 + u_2 v_2 + \ldots + u_n v_n$$

is an inner product.

(b) Another inner product on $\boldsymbol{R}^2$ that is distinct from the standard scalar product is given by

$$\left\langle \begin{pmatrix} x_1 \\ y_1 \end{pmatrix}, \begin{pmatrix} x_2 \\ y_2 \end{pmatrix} \right\rangle = 2x_1x_2 + 3y_1y_2.$$

(c) Let V be the vector space of sequences $x_1, x_2, x_3 \ldots$ of real numbers such that the series

$$\sum_{n=1}^{\infty} x_n^2$$

is convergent. We define an inner product on V by

$$\langle (x_n), (y_n) \rangle = x_1 y_1 + x_2 y_2 + x_3 y_3 + \ldots$$

(d) Let V be the vector space of real valued continuous functions of one variable with domain the unit interval $[0, 1] = \{x : 0 \leq x \leq 1\}$. We define an inner product by

$$\langle f, g \rangle = \int_0^1 f(t)g(t)dt$$

for functions f and g in V.

7.3 Proposition In addition to properties (i) to (iv) of 7.2, any real inner product will satisfy:

(ii') $\langle u, v + w \rangle = \langle u, v \rangle + \langle u, w \rangle$

(iii') $\langle u, \lambda v \rangle = \lambda \langle u, v \rangle$

(v) $\langle v, \mathbf{0} \rangle = \langle \mathbf{0}, v \rangle = 0$

for all $u, v, w \in V$ and all $\lambda \in \boldsymbol{R}$.

Proof: (ii')

$$\begin{aligned} \langle u, v + w \rangle &= \langle v + w, u \rangle \text{ by 7.2(i)} \\ &= \langle v, u \rangle + \langle w, u \rangle \text{ by 7.2(ii)} \\ &= \langle u, v \rangle + \langle u, w \rangle \text{ by 7.2(i) again} \end{aligned}$$

The proofs of (iii') and (v) are left as exercises.

We may use an inner product to define a *norm* on a vector space as follows:

$$\| v \| = \sqrt{\langle v, v \rangle}.$$

Example For the inner product of example (d) above we may define a norm by

$$\| f \| = \sqrt{\int_0^1 f(t)^2 dt}.$$

It is possible to recover the inner product from the norm, as the following proposition shows:

7.4 Proposition Suppose that V is a real inner product space. Then

$$\langle u, v\rangle = \frac{1}{4}\|u+v\|^2 - \frac{1}{4}\|u-v\|^2$$

for all $u, v \in V$.

Proof: The proof is left as an exercise.

7.5 Cauchy-Schwartz Inequality $|\langle u, v\rangle| \leq \|u\| \cdot \|v\|$

for all $u, v \in V$.

Proof: Let λ be a real number. Then $\langle \lambda u + v, \lambda u + v\rangle \geq 0$ and hence

$$\lambda^2\langle u, u\rangle + 2\lambda\langle u, v\rangle + \langle v, v\rangle \geq 0.$$

Now consider the quadratic equation in λ given by

$$\lambda^2\langle u, u\rangle + 2\lambda\langle u, v\rangle + \langle v, v\rangle = 0.$$

By the previous inequality this has at most one root, and hence the discriminant $(2\langle u, v\rangle)^2 - 4\langle u, u\rangle\langle v, v\rangle$ is not positive. From this we obtain $\langle u, v\rangle^2 \leq \langle u, u\rangle\langle v, v\rangle$ and hence the result.

7.6 The Triangle Inequality $\|u+v\| \leq \|u\| + \|v\|$

for all $u, v \in V$.

Proof:

$$\begin{aligned}
\|u+v\|^2 &= \langle u+v, u+v\rangle \\
&= \langle u, u\rangle + 2\langle u, v\rangle + \langle v, v\rangle \\
&\leq \langle u, u\rangle + 2\|u\| \cdot \|v\| + \langle v, v\rangle \text{ by 7.5} \\
&= \|u\|^2 + 2\|u\| \cdot \|v\| + \|v\|^2 \\
&= (\|u\| + \|v\|)^2.
\end{aligned}$$

Hence the result.

7.7 Definition Suppose that V is a real inner product space. A pair of non-zero vectors u and v are said to be *orthogonal* if $\langle u, v\rangle = 0$. A set of non-zero vectors is called an *orthogonal set* if any pair of these are orthogonal.

A basis $\{v_1, v_2, ..., v_n\}$ is called an *orthogonal basis* if

$$\langle v_i, v_j \rangle = 0 \text{ for all } i \neq j.$$

Examples (a) Consider $\boldsymbol{R}^3$ with the scalar product. The vectors

$\begin{pmatrix} 1 \\ 2 \\ 3 \end{pmatrix}$ and $\begin{pmatrix} -14 \\ 4 \\ 2 \end{pmatrix}$ are orthogonal since $1\times -14 + 2\times 4 + 3\times 2 = 0$.

For a vector space $\boldsymbol{R}^n$ with the scalar product, a pair of vectors are orthogonal if they are at right angles to one another.

(b) Under the scalar product on $\boldsymbol{R}^2$ the vectors $\begin{pmatrix} 1 \\ 1 \end{pmatrix}$ and $\begin{pmatrix} 1 \\ -1 \end{pmatrix}$ are orthogonal.

However, for the alternative inner product of example (b) above we have

$$\left\langle \begin{pmatrix} 1 \\ 1 \end{pmatrix}, \begin{pmatrix} 1 \\ -1 \end{pmatrix} \right\rangle = 2 - 3 = -1,$$

so the vectors are not orthogonal. The point of this example is that orthogonality depends not only on the vectors, but also on which inner product in being used.

(c) Consider the real inner product space of continuous functions defined on the unit interval described above. Although we have no intuition of what it might mean for a pair of functions to be "at right angles" to one another, we can still use the concept of orthogonality. For example, the functions x and $3x-2$ are orthogonal since

$$\langle x, 3x-2\rangle = \int_0^1 x(3x-2)dx = \int_0^1 3x^2 - 2x.dx = [x^3 - x^2]_0^1 = 0.$$

7.8 Proposition An orthogonal set of vectors is linearly independent.

Proof: Suppose that $\{v_1, v_2, ..., v_n\}$ is an orthogonal set of vectors and that

$$\lambda_1 v_1 + \lambda_2 v_2 + ... + \lambda_n v_n = \mathbf{0}.$$

We need to show that all of the scalars of this linear combination are zero.

For $i = 1, 2, ..., n$ we have $\langle \lambda_1 v_1 + \lambda_2 v_2 + ... + \lambda_n v_n, v_i \rangle = \langle \mathbf{0}, v_i \rangle = 0.$

Hence $\lambda_1 \langle v_1, v_i \rangle + \lambda_2 \langle v_2, v_i \rangle + ... + \lambda_n \langle v_n, v_i \rangle = 0.$

Since the vectors are orthogonal we have $\lambda_i \langle v_i, v_i \rangle = 0$ and hence $\lambda_i = 0$ for each i.

For what follows we need an orthogonal basis for an inner product space. We describe a method of obtaining an orthogonal basis of eigenvectors from any given basis of eigenvectors.

7.9 The Gram-Schmidt Process

Suppose that $\{v_1, v_2, ..., v_n\}$ is a basis for a real inner product space V. An orthogonal basis $\{w_1, w_2, ..., w_n\}$ may be constructed as follows:

- Let $w_1 = v_1$
- Let $w_2 = v_2 - \alpha_1 w_1$ where the scalar α_1 is such that $\langle w_1, w_2 \rangle = 0.$
- Let $w_3 = v_3 - \beta_1 w_1 - \beta_2 w_2$ where the scalars β_1 and β_2 are such that $\langle w_1, w_3 \rangle = \langle w_2, w_3 \rangle = 0.$

and so on.

Notice that if v_1, v_2 and v_3 are eigenvectors for an eigenvalue λ then w_1, w_2 and w_3 are also eigenvectors for λ.

Example Consider $\boldsymbol{R}^3$ with the scalar product, and suppose that we start with the basis:

$$v_1 = \begin{pmatrix} 1 \\ 1 \\ 1 \end{pmatrix} \quad v_2 = \begin{pmatrix} 1 \\ 1 \\ 0 \end{pmatrix} \quad v_3 = \begin{pmatrix} 1 \\ 0 \\ 0 \end{pmatrix}$$

Let $w_1 = v_1 = \begin{pmatrix} 1 \\ 1 \\ 1 \end{pmatrix}$.

Let $w_2 = v_2 - a_1 w_1 = \begin{pmatrix} 1 \\ 1 \\ 0 \end{pmatrix} - a_1 \begin{pmatrix} 1 \\ 1 \\ 1 \end{pmatrix} = \begin{pmatrix} 1-a_1 \\ 1-a_1 \\ -a_1 \end{pmatrix}$.

We require $\begin{pmatrix} 1 \\ 1 \\ 1 \end{pmatrix} \cdot \begin{pmatrix} 1-a_1 \\ 1-a_1 \\ -a_1 \end{pmatrix} = 0$, which is satisfied for $a_1 = \frac{2}{3}$.

Hence $w_2 = \begin{pmatrix} \frac{1}{3} \\ \frac{1}{3} \\ -\frac{2}{3} \end{pmatrix}$.

Let $w_3 = v_3 - \beta_1 w_1 - \beta_2 w_2 = \begin{pmatrix} 1 \\ 0 \\ 0 \end{pmatrix} - \beta_1 \begin{pmatrix} 1 \\ 1 \\ 1 \end{pmatrix} - \beta_2 \begin{pmatrix} \frac{1}{3} \\ \frac{1}{3} \\ -\frac{2}{3} \end{pmatrix}$

$$= \begin{pmatrix} 1 - \beta_1 - \frac{1}{3}\beta_2 \\ -\beta_1 - \frac{1}{3}\beta_2 \\ -\beta_1 + \frac{2}{3}\beta_2 \end{pmatrix}.$$

We require $\begin{pmatrix} 1 \\ 1 \\ 1 \end{pmatrix} \cdot \begin{pmatrix} 1 - \beta_1 - \frac{1}{3}\beta_2 \\ -\beta_1 - \frac{1}{3}\beta_2 \\ -\beta_1 + \frac{2}{3}\beta_2 \end{pmatrix} = 0$, which is satisfied for

$\beta_1 = \frac{1}{3}$,

and $\begin{pmatrix} \frac{1}{3} \\ \frac{1}{3} \\ -\frac{2}{3} \end{pmatrix} \cdot \begin{pmatrix} 1-\beta_1-\frac{1}{3}\beta_2 \\ -\beta_1-\frac{1}{3}\beta_2 \\ -\beta_1+\frac{2}{3}\beta_2 \end{pmatrix} = 0$, which is satisfied for $\beta_2 = \frac{1}{2}$.

Hence $w_3 = \begin{pmatrix} \frac{1}{2} \\ -\frac{1}{2} \\ 0 \end{pmatrix}$.

Thus we have obtained the orthogonal basis:

$$w_1 = \begin{pmatrix} 1 \\ 1 \\ 1 \end{pmatrix} \qquad w_2 = \begin{pmatrix} \frac{1}{3} \\ \frac{1}{3} \\ -\frac{2}{3} \end{pmatrix} \qquad w_3 = \begin{pmatrix} \frac{1}{2} \\ -\frac{1}{2} \\ 0 \end{pmatrix}$$

For a given vector v, we may obtain a vector with magnitude 1 that has the same direction as v by taking

$$\frac{1}{|v|}v.$$

We say that we have *normalised* the vector v. By normalising each vector in our orthogonal basis we obtain the basis

$$\left\{ \begin{pmatrix} \frac{1}{\sqrt{3}} \\ \frac{1}{\sqrt{3}} \\ \frac{1}{\sqrt{3}} \end{pmatrix}, \begin{pmatrix} \frac{1}{\sqrt{6}} \\ \frac{1}{\sqrt{6}} \\ -\frac{2}{\sqrt{6}} \end{pmatrix}, \begin{pmatrix} \frac{1}{\sqrt{2}} \\ -\frac{1}{\sqrt{2}} \\ 0 \end{pmatrix} \right\}$$

Such a basis is called an *orthonormal* basis, and will be important in what follows.

7.10 Definition A basis $\{v_1, v_2, ..., v_n\}$ for an inner product space V is an *orthonormal basis* if

$$\langle v_i, v_j \rangle = \begin{cases} 1 \text{ if } i = j \\ 0 \text{ if } i \neq j \end{cases}.$$

That is, a basis is orthonormal if the inner product of a basis vector with itself is 1, and the inner product of each pair of distinct basis vectors is 0.

Notice that in particular the standard basis for $\boldsymbol{R}^n$ is orthonormal.

7.11 Orthogonal Matrices and Symmetric Transformations

The *transpose* of an $m \times n$ matrix M is the $n \times m$ matrix M^T obtained by interchanging rows and columns, so that the first row of M becomes the first column of M^T, the second row of M becomes the second column of M^T, and so on.

Example

$$\text{If } M = \begin{pmatrix} 1 & 2 & 3 \\ 4 & 5 & 6 \end{pmatrix} \text{ then } M^T = \begin{pmatrix} 1 & 4 \\ 2 & 5 \\ 3 & 6 \end{pmatrix}.$$

It may easily be verified that $(M+N)^T = M^T + N^T$ and that $(MN)^T = N^T M^T$ for matrices of compatible size.

We use the transpose to define two important types of matrix.

A square matrix that is equal to its own transpose is called a *symmetric matrix*.

Example

$$M = \begin{pmatrix} 1 & 2 & 3 \\ 2 & 4 & 5 \\ 3 & 5 & 6 \end{pmatrix} \text{ is symmetric since } M = M^T.$$

A square matrix M that has transpose equal to its inverse, so that $M^T = M^{-1}$, is called an *orthogonal matrix*. This is equivalent to

$$MM^T = I = M^T M$$

where I is the identity matrix.

Example (a) $M = \begin{pmatrix} \frac{1}{\sqrt{2}} & -\frac{1}{\sqrt{2}} \\ \frac{1}{\sqrt{2}} & \frac{1}{\sqrt{2}} \end{pmatrix}$ is orthogonal, since we have

$$MM^T = \begin{pmatrix} \frac{1}{\sqrt{2}} & -\frac{1}{\sqrt{2}} \\ \frac{1}{\sqrt{2}} & \frac{1}{\sqrt{2}} \end{pmatrix} \begin{pmatrix} \frac{1}{\sqrt{2}} & \frac{1}{\sqrt{2}} \\ -\frac{1}{\sqrt{2}} & \frac{1}{\sqrt{2}} \end{pmatrix} = \begin{pmatrix} 1 & 0 \\ 0 & 1 \end{pmatrix} \text{ and } M^T M = I.$$

(b) $$M = \begin{pmatrix} \frac{1}{\sqrt{3}} & \frac{1}{\sqrt{6}} & \frac{1}{\sqrt{2}} \\ \frac{1}{\sqrt{3}} & \frac{1}{\sqrt{6}} & -\frac{1}{\sqrt{2}} \\ \frac{1}{\sqrt{3}} & -\frac{2}{\sqrt{6}} & 0 \end{pmatrix}$$ is an orthogonal matrix.

To verify this, calculate MM^T and M^TM and observe that each is equal to the identity.

7.12 Theorem Suppose that S is an orthonormal basis for $\boldsymbol{R}^n$ with the scalar product. Then the transition matrix for S is an orthogonal matrix.

Proof: Let $S = \{v_1, v_2, ..., v_n\}$ be an orthonormal basis for $\boldsymbol{R}^n$. The transition matrix P for the basis S has the vectors $v_1, v_2, ..., v_n$ as its columns, and P^T has the vectors $v_1, v_2, ..., v_n$ as its rows. Consider the product of matrices P^TP.

The entry at the intersection of the *i*th row and the *j*th column of this product will be $v_i.v_j$, so that the matrix has 1's on the diagonal and 0's elsewhere. Hence $P^TP = I$. Similarly, $PP^T = I$. Hence P is orthogonal.

Example Alert readers will have already noticed that the orthogonal matrix of example (b) above is the transition matrix for the orthonormal basis obtained on page 100.

We shall now define a special type of linear transformation:

7.13 Definition Suppose that V is a real inner product space and that $f: V \to V$ is a linear transformation. We say that f is *symmetric* if

$$\langle f(u), v\rangle = \langle u, f(v)\rangle$$

for all $u, v \in V$.

Example Consider $f: \boldsymbol{R}^2 \to \boldsymbol{R}^2$ defined by

$$f\begin{pmatrix} x \\ y \end{pmatrix} = \begin{pmatrix} 2x + y \\ x + 3y \end{pmatrix}$$

where the inner product on $\boldsymbol{R}^2$ is the scalar product. Then we have

$$\left\langle f\begin{pmatrix} x_1 \\ y_1 \end{pmatrix}, \begin{pmatrix} x_2 \\ y_2 \end{pmatrix} \right\rangle = \left\langle \begin{pmatrix} 2x_1 + y_1 \\ x_1 + 3y_1 \end{pmatrix}, \begin{pmatrix} x_2 \\ y_2 \end{pmatrix} \right\rangle$$

$$= 2x_1x_2 + x_2y_1 + x_1y_2 + 3y_1y_2$$

$$= \left\langle \begin{pmatrix} x_1 \\ y_1 \end{pmatrix}, \begin{pmatrix} 2x_2 + y_2 \\ x_2 + 3y_2 \end{pmatrix} \right\rangle$$

$$= \left\langle \begin{pmatrix} x_1 \\ y_1 \end{pmatrix}, f\begin{pmatrix} x_2 \\ y_2 \end{pmatrix} \right\rangle.$$

We observe that, with respect to the standard basis, this linear transformation f is represented by the matrix

$$\begin{pmatrix} 2 & 1 \\ 1 & 3 \end{pmatrix}$$

and that this is a symmetric matrix. This suggests the following theorem:

7.14 Theorem Suppose that V is a real inner product space and that $f: V \to V$ is a linear transformation. Then f is symmetric if and only if the matrix representing f with respect to an orthonormal basis is symmetric.

Proof: Suppose that f is symmetric and that $B = \{v_1, v_2, ..., v_n\}$ is an orthonormal basis for V, so that

$$\langle v_i, v_j \rangle = \begin{cases} 1 \text{ if } i = j \\ 0 \text{ if } i \neq j. \end{cases}$$

Suppose also that, with respect to the orthonormal basis, f is represented by the matrix

$$M = \begin{pmatrix} m_{11} & m_{12} & . & . & m_{1n} \\ m_{21} & m_{22} & . & . & m_{2n} \\ . & & & & . \\ . & & & & . \\ m_{n1} & m_{n2} & . & . & m_{nn} \end{pmatrix}$$

Then

$$f(v_i) = M\begin{pmatrix} 0 \\ . \\ 1 \\ . \\ 0 \end{pmatrix}_B \text{, where the 1 appears in the } i^{th} \text{component}$$

$$= \begin{pmatrix} m_{1i} \\ m_{2i} \\ . \\ . \\ m_{ni} \end{pmatrix}_B = m_{1i}v_1 + m_{2i}v_2 + \ldots + m_{ni}v_n$$

Hence $\langle f(v_i), v_j \rangle = \langle m_{1i}v_1 + m_{2i}v_2 + \ldots + m_{ni}v_n, v_j \rangle = m_{ji}$

and $\quad \langle v_i, f(v_j) \rangle = \langle v_i, m_{1j}v_1 + m_{2j}v_2 + \ldots + m_{nj}v_n \rangle = m_{ij}$

Now if f is symmetric then $\langle f(v_i), v_j \rangle = \langle v_i, f(v_j) \rangle$ and it follows that $m_{ji} = m_{ij}$, so that the matrix M is symmetric.

Conversely, if the matrix is symmetric then for any pair of basis vectors we have

$$\langle f(v_i), v_j \rangle = \langle v_i, f(v_j) \rangle .$$

The fact that the symmetry condition holds for basis vectors is enough to establish that f is symmetric.

If a symmetric linear transformation has distinct eigenvalues, then the basis of eigenvectors that we obtain will be orthogonal, as the following result shows.

7.15 Proposition Let $f: V \to V$ be a symmetric linear transformation of a real inner product space V. If λ_i and λ_j are distinct eigenvalues of f with corresponding eigenvectors v_i and v_j, then

$$\langle v_i, v_j \rangle = 0.$$

Proof:

$$\lambda_i\langle v_i, v_j\rangle = \langle \lambda_i v_i, v_j\rangle = \langle f(v_i), v_j\rangle$$
$$= \langle v_i, f(v_j)\rangle = \langle v_i, \lambda_j v_j\rangle = \lambda_j\langle v_i, v_j\rangle.$$

Since $\lambda_i \neq \lambda_j$ we must have $\langle v_i, v_j\rangle = 0$.

It follows that for a symmetric linear transformation with distinct eigenvalues we can normalise the eigenvectors to obtain an orthonormal basis, which has an orthogonal transition matrix.

Example Consider the linear transformation $f: \boldsymbol{R}^2 \to \boldsymbol{R}^2$ given by

$$f\begin{pmatrix} x \\ y \end{pmatrix} = \begin{pmatrix} 7x - 6y \\ -6x - 2y \end{pmatrix}$$

With respect to the standard basis this is represented by

$$M = \begin{pmatrix} 7 & -6 \\ -6 & -2 \end{pmatrix}$$

which is symmetric. We may obtain the eigenvalues $\lambda = -5$ and $\lambda = 10$.

The eigenvectors are $\begin{pmatrix} 1 \\ 2 \end{pmatrix}$ and $\begin{pmatrix} -2 \\ 1 \end{pmatrix}$.

Notice that these are orthogonal.

Normalising the eigenvectors we obtain $\begin{pmatrix} \frac{1}{\sqrt{5}} \\ \frac{2}{\sqrt{5}} \end{pmatrix}$ and $\begin{pmatrix} -\frac{2}{\sqrt{5}} \\ \frac{1}{\sqrt{5}} \end{pmatrix}$, so that the transition matrix is

$$P = \begin{pmatrix} \frac{1}{\sqrt{5}} & -\frac{2}{\sqrt{5}} \\ \frac{2}{\sqrt{5}} & \frac{1}{\sqrt{5}} \end{pmatrix}.$$

It may be checked that

$$P^T M P = \begin{pmatrix} -5 & 0 \\ 0 & 10 \end{pmatrix}.$$

If a symmetric matrix has repeated eigenvalues, then provided that we have a basis of eigenvectors we can use the Gram-Schmidt process to obtain an orthogonal basis for the eigenspace of each repeated eigenvalue. This allows us to obtain an orthonormal basis of eigenvectors as before.

Example Consider the linear transformation $f: \boldsymbol{R}^3 \to \boldsymbol{R}^3$ given by

$$f\begin{pmatrix} x \\ y \\ z \end{pmatrix} = \begin{pmatrix} y+z \\ x+z \\ x+y \end{pmatrix}$$

With respect to the standard basis this is represented by

$$M = \begin{pmatrix} 0 & 1 & 1 \\ 1 & 0 & 1 \\ 1 & 1 & 0 \end{pmatrix}$$

which is symmetric. We may obtain the eigenvalues $\lambda = 2$, $\lambda = -1$ and $\lambda = -1$.

For $\lambda = 2$ the eigenvector is $\begin{pmatrix} 1 \\ 1 \\ 1 \end{pmatrix}$. The situation for $\lambda = -1$ is a little more complicated, as the eigenvalue has algebraic multiplicity 2. Vectors in the eigenspace must satisfy

$$x + y + z = 0.$$

Hence the eigenspace is of dimension 2. We choose an arbitrary basis for the eigenspace, for example

$$\begin{pmatrix} -1 \\ 1 \\ 0 \end{pmatrix} \text{and} \begin{pmatrix} -1 \\ 0 \\ 1 \end{pmatrix}$$

By applying the Gram-Schmidt process we obtain an orthogonal basis for the eigenspace:

$$\begin{pmatrix} -1 \\ 1 \\ 0 \end{pmatrix} \text{ and } \begin{pmatrix} -\frac{1}{2} \\ -\frac{1}{2} \\ 1 \end{pmatrix}$$

Finally, we normalise each of the three eigenvectors to obtain an orthonormal basis of eigenvectors:

$$\left\{ \begin{pmatrix} \frac{1}{\sqrt{3}} \\ \frac{1}{\sqrt{3}} \\ \frac{1}{\sqrt{3}} \end{pmatrix}, \begin{pmatrix} -\frac{1}{\sqrt{2}} \\ \frac{1}{\sqrt{2}} \\ 0 \end{pmatrix}, \begin{pmatrix} -\frac{1}{\sqrt{6}} \\ -\frac{1}{\sqrt{6}} \\ \frac{2}{\sqrt{6}} \end{pmatrix} \right\}.$$

Hence the transition matrix is

$$P = \begin{pmatrix} \frac{1}{\sqrt{3}} & -\frac{1}{\sqrt{2}} & -\frac{1}{\sqrt{6}} \\ \frac{1}{\sqrt{3}} & \frac{1}{\sqrt{2}} & -\frac{1}{\sqrt{6}} \\ \frac{1}{\sqrt{3}} & 0 & \frac{2}{\sqrt{6}} \end{pmatrix}$$

and f may be represented with respect to the orthonormal basis of eigenvectors by

$$N = P^T MP = \begin{pmatrix} 2 & 0 & 0 \\ 0 & -1 & 0 \\ 0 & 0 & -1 \end{pmatrix}.$$

Having considered real inner product spaces we now turn our attention to the complex case. Recall that for a complex number $z = x + yi$ the *complex conjugate* is given by $\overline{z} = x - yi$. It is easily verified that $z\overline{z} = x^2 + y^2$, and so is real. Notice also that if $z = \overline{z}$ then we have $y = 0$ so that z is real.

It is also easy to verify that for complex numbers z_1 and z_2 we have

$$\overline{z_1 + z_2} = \overline{z_1} + \overline{z_2}$$

$$\overline{z_1 z_2} = \overline{z_1}\,\overline{z_2}$$

We now give the definition of a complex inner product space:

7.16 Definition Suppose that V is a vector space over $\boldsymbol{C}$. A mapping $V \times V \to \boldsymbol{C}$, written $\langle u, v\rangle$, is called a *complex inner product* if it satisfies:

(i) $\langle u, v\rangle = \overline{\langle v, u\rangle}$

(ii) $\langle u+v, w\rangle = \langle u, w\rangle + \langle v, w\rangle$

(iii) $\langle \lambda u, v\rangle = \lambda\langle u, v\rangle$

(iv) $\langle v, v\rangle \geq 0$ and $\langle v, v\rangle = 0$ iff $v = \mathbf{0}$

for all $u, v, w \in V$ and all $\lambda \in \boldsymbol{C}$. A complex vector space equipped with a complex inner product is called a *complex inner product space.*

Condition (iv) may at first appear puzzling, as the complex numbers are not ordered. However, by (i) we have $\langle v, v\rangle = \overline{\langle v, v\rangle}$ and so $\langle v, v\rangle$ is real for all v, making the inequality meaningful.

Examples (a) The complex scalar product on $\boldsymbol{C}^n$ is defined by:

$$\begin{pmatrix} z_1 \\ z_2 \\ \vdots \\ z_n \end{pmatrix} \cdot \begin{pmatrix} w_1 \\ w_2 \\ \vdots \\ w_n \end{pmatrix} = z_1\overline{w_1} + z_2\overline{w_2} + \ldots + z_n\overline{w_n}.$$

(b) Let V be the vector space of continuous functions $[0, 1] \to \boldsymbol{C}$.

For $f, g \in V$ we may define a complex inner product by

$$\langle f, g\rangle = \int_0^1 f(t)\overline{g(t)}dt.$$

7.17 Proposition In addition to properties (i) to (iv) any complex inner product satisfies:

(ii') $\langle u, v+w\rangle = \langle u, v\rangle + \langle u, w\rangle$

(iii') $\langle u, \lambda v\rangle = \overline{\lambda}\langle u, v\rangle$

(v) $\langle v, \mathbf{0}\rangle = \langle \mathbf{0}, v\rangle = 0$

Proof: (iii')

$$\langle u, \lambda v\rangle = \overline{\langle \lambda v, u\rangle} \text{ by (i)}$$
$$= \overline{\lambda\langle v, u\rangle} \text{ by (iii)}$$
$$= \overline{\lambda}\ \overline{\langle v, u\rangle}$$
$$= \overline{\lambda}\langle u, v\rangle \text{ by (i) again.}$$

The proof of (ii') is left as an exercise. (v) is the same as the real case.

We define orthogonality in the same way as for a real inner product space.

Example Consider $\boldsymbol{C}^2$ with the scalar product. We have

$$\begin{pmatrix} 1+i \\ 1+2i \end{pmatrix} \cdot \begin{pmatrix} 2+i \\ -1-i \end{pmatrix} = (1+i)(2-i) + (1+2i)(-1+i) = 0.$$

Hence $\begin{pmatrix} 1+i \\ 1+2i \end{pmatrix}$ and $\begin{pmatrix} 2+i \\ -1-i \end{pmatrix}$ are orthogonal.

Next we seek the appropriate analogues of symmetric and orthogonal matrices. For a matrix M with complex entries, the complex conjugate $\overline{M}$ is obtained by taking the complex conjugate of each entry in M.

Example

For $M = \begin{pmatrix} 2+i & 3-2i \\ -1+4i & 5 \end{pmatrix}$ we have $\overline{M} = \begin{pmatrix} 2-i & 3+2i \\ -1-4i & 5 \end{pmatrix}$.

The *conjugate transpose* of a matrix M is given by

$$M^* = \left(\overline{M}\right)^T = \overline{(M^T)}.$$

Example For the previous example we have

$$M^* = \begin{pmatrix} 2-i & -1-4i \\ 3+2i & 5 \end{pmatrix}.$$

A complex square matrix M such that $M^* = M$ is said to be *Hermitian*.

Examples

$$\begin{pmatrix} 1 & 3+i \\ 3-i & 2 \end{pmatrix} \text{ and } \begin{pmatrix} 2 & 1+2i & 4-i \\ 1-2i & -1 & -3+2i \\ 4+i & -3-2i & 0 \end{pmatrix} \text{ are Hermitian.}$$

Notice that the diagonal entries of a Hermitian matrix are real.

A complex square matrix M such that $M^* = M^{-1}$ is said to be *unitary*.

Examples $\frac{1}{\sqrt{2}}\begin{pmatrix} 1 & i \\ i & 1 \end{pmatrix}$ and $\frac{1}{3}\begin{pmatrix} \frac{4}{\sqrt{5}}+\frac{2}{\sqrt{5}}i & 2+i \\ -\sqrt{5} & 2 \end{pmatrix}$ are unitary.

7.18 Definition Suppose that V is a complex inner product space and that $f\colon V \to V$ is a linear transformation. We say that f is *Hermitian* if

$$\langle f(u), v\rangle = \langle u, f(v)\rangle$$

for all $u, v \in V$.

Example Consider the linear transformation $f\colon \boldsymbol{C}^2 \to \boldsymbol{C}^2$ given by

$$f\begin{pmatrix} z_1 \\ z_2 \end{pmatrix} = \begin{pmatrix} z_1 + (3+i)z_2 \\ (3-i)z_1 + 2z_2 \end{pmatrix}$$

We have

$$\left\langle f\begin{pmatrix} z_1 \\ z_2 \end{pmatrix}, \begin{pmatrix} w_1 \\ w_2 \end{pmatrix} \right\rangle = \left\langle \begin{pmatrix} z_1 + (3+i)z_2 \\ (3-i)z_1 + 2z_2 \end{pmatrix}, \begin{pmatrix} w_1 \\ w_2 \end{pmatrix} \right\rangle$$

$$= z_1\overline{w_1} + (3+i)z_2\overline{w_1} + (3-i)z_1\overline{w_2} + 2z_2\overline{w_2}$$

$$= z_1\overline{w_1} + z_2\overline{(3-i)w_1} + z_1\overline{(3+i)w_2} + 2z_2\overline{w_2}$$

$$= \left\langle \begin{pmatrix} z_1 \\ z_2 \end{pmatrix}, \begin{pmatrix} w_1 + (3+i)w_2 \\ (3-i)w_1 + 2w_2 \end{pmatrix} \right\rangle$$

$$= \left\langle \begin{pmatrix} z_1 \\ z_2 \end{pmatrix}, f\begin{pmatrix} w_1 \\ w_2 \end{pmatrix} \right\rangle.$$

Hence f is Hermitian. We also observe that the matrix representing f with respect to the standard basis is the Hermitian matrix of the previous example. This leads us to our next theorem:

7.19 Theorem Suppose that V is a complex inner product space and that $f: V \to V$ is a linear transformation. Then f is Hermitian if and only if the matrix representing f with respect to an orthonormal basis is Hermitian.

Proof: The proof is very similar to theorem 7.14 for symmetric linear transformations, and is left as an exercise.

In general it is reasonable to expect an endomorphism of a complex vector space to have complex eigenvalues, and indeed this is usually the case. However, Hermitian endomorphisms are a special case - their eigenvalues are always *real*.

7.20 Theorem Suppose that V is a complex inner product space and that $f: V \to V$ is a Hermitian linear transformation. The eigenvalues of f are real.

Proof: Suppose that λ is an eigenvalue of f with eigenvector v. Then we have

$$\lambda\langle v, v\rangle = \langle \lambda v, v\rangle = \langle f(v), v\rangle = \langle v, f(v)\rangle = \langle v, \lambda v\rangle = \bar{\lambda}\langle v, v\rangle.$$

Now since $\langle v, v\rangle \neq 0$ we have $\lambda = \bar{\lambda}$, and hence λ is real.

Example The matrix $\begin{pmatrix} 2 & 4+2i \\ 4-2i & 1 \end{pmatrix}$ is Hermitian.

The characteristic equation is $(2-\lambda)(1-\lambda)-(4+2i)(4-2i)=0$.

This simplifies to $\lambda^2 - 3\lambda - 18 = 0$ which factorises as $(\lambda+3)(\lambda-6)=0$, giving the real eigenvalues $\lambda = -3$ and $\lambda = 6$.

As in the real case, distinct eigenvalues of a Hermitian linear transformation have orthogonal eigenvectors. Continuing with the same example, for $\lambda = -3$ we have

$\begin{pmatrix} 5 & 4+2i \\ 4-2i & 4 \end{pmatrix}\begin{pmatrix} w \\ z \end{pmatrix} = \begin{pmatrix} 0 \\ 0 \end{pmatrix}$ giving the eigenvector $\begin{pmatrix} 4+2i \\ -5 \end{pmatrix}$.

Similarly, for $\lambda = 6$ we have

$\begin{pmatrix} -4 & 4+2i \\ 4-2i & -5 \end{pmatrix}\begin{pmatrix} w \\ z \end{pmatrix} = \begin{pmatrix} 0 \\ 0 \end{pmatrix}$ giving the eigenvector $\begin{pmatrix} 2+i \\ 2 \end{pmatrix}$.

It is readily checked that these are orthogonal:

$$\begin{pmatrix} 4+2i \\ -5 \end{pmatrix} \cdot \begin{pmatrix} 2+i \\ 2 \end{pmatrix} = (4+2i)(2-i) + (-5)\times 2 = 0.$$

We may normalise the eigenvectors to obtain an orthonormal basis:

$$\left\{ \frac{1}{3\sqrt{5}}\begin{pmatrix} 4+2i \\ -5 \end{pmatrix}, \frac{1}{3}\begin{pmatrix} 2+i \\ 2 \end{pmatrix} \right\}.$$

With respect to this orthonormal basis f is represented by the diagonal matrix $\begin{pmatrix} -3 & 0 \\ 0 & 6 \end{pmatrix}$.

Exercise 7

1. Show that the scalar product on $\boldsymbol{R}^2$ is a real inner product.

2. Show that for a real inner product we have

$$\langle u, \lambda v\rangle = \lambda\langle u, v\rangle$$

$$\langle v, \mathbf{0}\rangle = \langle \mathbf{0}, v\rangle = 0$$

3. For $u = \begin{pmatrix} x_1 \\ y_1 \end{pmatrix}$ and $v = \begin{pmatrix} x_2 \\ y_2 \end{pmatrix}$ in $\boldsymbol{R}^2$ define

$$\langle u, v\rangle = 5x_1x_2 + 3y_1y_2.$$

Show that this is a real inner product.

4. State, with reasons, whether or not each of the following is a real inner product on $\boldsymbol{R}^2$, with u and v as in Question 3:

(i) $\langle u, v\rangle = x_1x_2 + 2y_1y_2.$

(ii) $\langle u, v\rangle = x_1x_2 - y_1y_2.$

(iii) $\langle u, v\rangle = x_1 + x_2 + y_1 + y_2.$

(iv) $\langle u, v\rangle = \begin{pmatrix} x_1x_2 \\ y_1y_2 \end{pmatrix}.$

5. Let V be a real inner product space. Show that for all $u, v \in V$ we have:

(i) $\|u + v\|^2 = \|u\|^2 + 2\langle u, v\rangle + \|v\|^2$

(ii) $\|u - v\|^2 = \|u\|^2 - 2\langle u, v\rangle + \|v\|^2$

(iii) $\langle u, v\rangle = \frac{1}{4}\{\|u + v\|^2 - \|u - v\|^2\}.$

6 Let V be the vector space of continuous real-valued functions defined on the interval $[0, 1]$ described in chapter 3. Show that

$$\langle f, g\rangle = \int_0^1 f(t)g(t)dt$$

is a real inner product on V. Show that in V the functions x^2 and $4x - 3$ are orthogonal.

Show also that the functions $\sin \pi x$ and $\cos \pi x$ are orthogonal

More difficult, show that the functions $\sin \pi x$ and $\sin 2\pi x$ are orthogonal

7. For each of the following symmetric linear transformations find an orthonormal basis of eigenvectors and write down the transition matrix P. Find the matrix N representing the linear transformation with respect to this basis:

(i) $f: \boldsymbol{R}^2 \to \boldsymbol{R}^2$ defined by

$$f\begin{pmatrix} x \\ y \end{pmatrix} = \begin{pmatrix} x + 2y \\ 2x + y \end{pmatrix}$$

(ii) $f: \boldsymbol{R}^3 \to \boldsymbol{R}^3$ defined by

$$f\begin{pmatrix} x \\ y \\ z \end{pmatrix} = \begin{pmatrix} x + z \\ y \\ x + z \end{pmatrix}$$

8. Normalise each of the following vectors:

$$\begin{pmatrix} 1 \\ 2 \end{pmatrix} \quad \begin{pmatrix} 1 \\ 2 \\ 3 \end{pmatrix} \quad \begin{pmatrix} 1 \\ -1 \\ 1 \end{pmatrix} \quad \begin{pmatrix} 1 \\ -2 \\ 3 \\ -4 \end{pmatrix}$$

9. Let V be a real inner product space and $v \in V$.

Let $v^{\perp} = \{u \in V : \langle u, v \rangle = 0\}$.

Show that $v^{\perp}$ is a subspace of V.

This is known as the *orthogonal complement* of v.

10. Define a mapping $\boldsymbol{C}^2 \times \boldsymbol{C}^2 \to \boldsymbol{C}$ by

$$\left\langle \begin{pmatrix} w_1 \\ w_2 \end{pmatrix}, \begin{pmatrix} z_1 \\ z_2 \end{pmatrix} \right\rangle = w_1\overline{z_1} + w_2\overline{z_2}.$$

Verify that this is a complex inner product.

11. Prove 7.17 (ii'): For a complex inner product

$\langle u, v + w \rangle = \langle u, v \rangle + \langle u, w \rangle$

Chapter 8: Cosets and Quotient Groups

We begin by reviewing the concepts of *group* and *subgroup* and show how a subgroup gives rise to a partition of the elements of a group into *cosets*. We then describe how the set of cosets may be given a group structure to form a *quotient group*.

Recall from chapter 1 that a *group* is a set G with a binary operation $*$ on G satisfying:

Associativity: $(a * b) * c = a * (b * c)$ for all $a, b, c \in G$

Identity: There is an element $e \in G$ such that

$e * a = a = a * e$ for all $a \in G$

Inverses: For each $a \in G$ there is an element $a^{-1} \in G$ such that $a * a^{-1} = e = a^{-1} * a$

These three statements are known as the *group axioms*. If we also have

Commutativity: $a * b = b * a$ for all $a, b \in G$

then we say that the group is *abelian*.

A group that has a finite number of elements is called a *finite group*. The number of elements in a finite group G is called the *order* of the group, and is written $|G|$.

Recall from chapter 2 that if G is a group, then a subset $H \subseteq G$ is a *subgroup* if H is itself a group under the same operation. There is an easy test to determine whether or not a given non-empty subset is a subgroup: H is a subgroup if and only if

(i) if $a, b \in H$ then $ab \in H$

(ii) if $a \in H$ then $a^{-1} \in H$.

Where H is a subgroup of G we may write $H \leq G$. For any group G the trivial group $\{e\}$ and G itself are subgroups. Where a subgroup H is not equal to G we say it is a *proper subgroup* and write $H < G$.

Examples (a) For each natural number n the additive group $n\boldsymbol{Z}$ is a subgroup of $\boldsymbol{Z}$.

(b) The proper subgroups of $\boldsymbol{Z}_6$ are $\{0\},\{0,3\},\{0,2,4\}$.

(c) The proper subgroups of D_3 are

$$\{I\},\{I,R,R^2\},\{I,M_1\},\{I,M_2\},\{I,M_3\}.$$

Notice that the subgroups of an abelian group are abelian.

Let G be a group and $a \in G$. Recall that the subgroup generated by a is

$$\langle a \rangle = \{\ldots, a^{-1}, e, a, a*a, a*a*a, \ldots\}$$

We have already remarked that the number of elements in a finite group is called the *order* of the group. We may also speak of the order of an individual element: for $a \in G$ the *order* of a is smallest natural number n such that

$$\underbrace{a*a*\ldots*a}_{n} = e$$

This dual use of the word "order" makes sense because the order of an element a in a group is the order of the subgroup generated by a.

Example In $\boldsymbol{Z}_6$ we have $\langle 2 \rangle = \{0,2,4\}$ so the element 2 has order 3, and $\langle 1 \rangle = \boldsymbol{Z}_6$ so the element 1 has order 6.

8.2 Definition Let G be a group and suppose that H is a subgroup of G and $a \in G$.

A *left coset* is a set of the form $aH = \{ah : h \in H\}$.

A *right coset* is a set of the form $Ha = \{ha : h \in H\}$.

Notice that if G is abelian then the left and right cosets are the same. In an additive abelian group we prefer to write a coset as $a + H$.

Examples (a) Consider the subgroup $5\mathbf{Z}$ of $\mathbf{Z}$. Left cosets include

$$3+5\mathbf{Z}=\{\ldots,-2,3,8,13,18,23,\ldots\}$$

and
$$7+5\mathbf{Z}=\{\ldots,-3,2,7,12,17,22,\ldots\}$$

(b) Consider the subgroup $\{I,R,R^2\}$ of D_3. Left cosets include

$$R\{I,R,R^2\}=\{R,R^2,I\}$$

$$M_1\{I,R,R^2\}=\{M_1,M_2,M_3\}$$

and
$$M_2\{I,R,R^2\}=\{M_2,M_3,M_1\}.$$

Notice that the last two of these are the same.

Now consider the subgroup $\{I,M_1\}$ of D_3. Left cosets include

$$R\{I,M_1\}=\{R,M_3\} \text{ and } M_2\{I,M_1\}=\{M_2,R^2\}.$$

Right cosets include

$$\{I,M_1\}R=\{R,M_2\} \text{ and } \{I,M_1\}M_2=\{M_2,R\}.$$

Notice that left and right cosets do not coincide, for example

$$R\{I,M_1\}\neq\{I,M_1\}R.$$

We saw in the first part of example (b) that a pair of cosets aH and bH can turn out to be the same. The next result tells us precisely when this happens.

8.3 Proposition Suppose H is a subgroup of G and that $a,b\in G$.

Then $aH=bH$ if and only if $b^{-1}a\in H$.

Proof: Clearly $a\in aH$.

If $aH=bH$ then $a\in bH$ so that $a=bh$ for some $h\in H$

Hence $b^{-1}a=h\in H$, as required.

Conversely, suppose $b^{-1}a\in H$.

If $x\in aH$ so that $x=ah$ for some $h\in H$, then

$$x=(bb^{-1})ah=b(b^{-1}a)h\in bH.$$

If $x \in bH$ so that $x = bh$ for some $h \in H$, then

$$x = (aa^{-1})bh = a(a^{-1}b)h = a(b^{-1}a)^{-1}h \in aH.$$

Hence $x \in aH$ iff $x \in bH$ so that $aH = bH$.

We also observe from the examples above that for a given subgroup H, each coset aH has the same number of elements as H. This is easily proved by observing that the mapping $f : H \to aH$ given by $f(h) = ah$ has an inverse $f^{-1} : aH \to H$ given by $f^{-1}(ah) = h$ and so is a bijection. It follows that all cosets have the same number of elements.

Finally, we show that distinct cosets are disjoint:

8.4 Proposition

Suppose H is a subgroup of a group G and that $a, b \in G$.

If $aH \neq bH$ then $aH \cap bH = \varnothing$.

Proof: Suppose that cosets have a common element, $x \in aH \cap bH$.

Then $x = ah_1$ and $x = bh_2$ for some $h_1, h_2 \in H$.

It follows that $ah_1 = bh_2$ so that $b^{-1}ah_1 = h_2$ and $b^{-1}a = h_2h_1^{-1} \in H$.

By 8.3 we have $aH = bH$.

The number of left cosets of H in G is called the *index of H in G*, and is denoted $[G : H]$. Since each coset has $|H|$ elements, and distinct cosets are disjoint we have the following:

8.5 Lagrange's Theorem Let H be a subgroup of a finite group G. Then

$$[G : H] \cdot |H| = |G|$$

Notice that an immediate corollary to Lagrange's Theorem is that the order of a subgroup is a divisor of the order of the group to which it belongs. This is a useful fact to have at hand when finding the subgroups of a finite group.

Example Consider the subgroup $H = \{0, 5\}$ of $\boldsymbol{Z}_{10}$ with $|H| = 2$.

There are five cosets: $\{0,5\}, \{1,6\}, \{2,7\}, \{3,8\}, \{4,9\}$.

Hence the index of H in $\boldsymbol{Z}_{10}$ is $[\boldsymbol{Z}_{10} : H] = 5$. We have $5 \times 2 = 10$, the order of $\boldsymbol{Z}_{10}$.

8.6 Corollary

For each prime number p the only group of order p is $\boldsymbol{Z}_p$.

Proof: Let G be a group of order p, where p is prime, and suppose $a \in G$.

Consider the subgroup $\langle a \rangle$ generated by a. If a is the identity then $\langle a \rangle$ is the trivial group. Otherwise, by Lagrange's theorem $\langle a \rangle$ must be of order p. Hence $G \cong \langle a \rangle$, and this group is isomorphic to $\boldsymbol{Z}_p$.

8.7 Definition Suppose that $a, b \in G$. We say that b is a *conjugate* of a if $b = gag^{-1}$ for some $g \in G$. More particularly, gag^{-1} is the *conjugation* of a by g.

Example In D_3 the conjugation of M_1 by R is

$$RM_1R^{-1} = M_3R^2 = M_2.$$

8.8 Proposition Conjugacy is an equivalence relation.

Proof:

Reflexive: $a = eae^{-1}$ and so a is a conjugate of itself for each $a \in G$.

Symmetric: Suppose that b is a conjugate of a, so that $b = gag^{-1}$ for some $g \in G$.

Then $a = g^{-1}bg$ so that a is a conjugate of b.

Transitive: Suppose that c is a conjugate of b, and b is a conjugate of a, so that $c = g_1bg_1^{-1}$ and $b = g_2ag_2^{-1}$ for some $g_1, g_2 \in G$.

Then $c = g_1(g_2 a g_2^{-1})g_1^{-1} = (g_1 g_2)a(g_1 g_2)^{-1}$ so that c is a conjugate of a.

The equivalence classes under conjugation are called the *conjugacy classes*. Notice that for the identity we have $geg^{-1} = gg^{-1} = e$, so that in any group e is always in a singleton conjugacy class of its own.

Example The conjugacy classes of D_3 are

$$\{I\}, \{R, R^2\}, \{M_1, M_2, M_3\}.$$

We shall now define a special type of subgroup known as a *normal* subgroup. Normal subgroups play a role in group theory analogous to that of *ideals* in the theory of rings, studied in chapter 15.

8.9 Definition Let H be a subgroup of a group G. We say that H is a *normal* subgroup if it also satisfies the following:

$$\text{if } h \in H \text{ then } ghg^{-1} \in H \text{ for any } g \in G.$$

Thus a normal subgroup is a subgroup that is closed under conjugation. Where H is a normal subgroup of G we may write $H \lhd G$.

Notice that in any group G we have $G \lhd G$ and $\{e\} \lhd G$.

Examples (a) Suppose G is an abelian group. Then for any subgroup H of G and $h \in H$ we have $ghg^{-1} = gg^{-1}h = h$ for all $g \in G$.

Thus in an abelian group *all* subgroups are normal.

(b) Consider again the non-abelian group D_3. The subgroup $H = \{I, R, R^2\}$ is normal. To see this, think about an element of the form ghg^{-1} where $h \in H, g \in D_3$. If g is a rotation, then ghg^{-1} is of the form

$$(\text{rotation})(\text{rotation})(\text{rotation}) = (\text{rotation})$$

and so belongs to H. If g is a reflection, then ghg^{-1} is of the form

$$(\text{reflection})(\text{rotation})(\text{reflection}) = (\text{reflection})(\text{reflection}) = (\text{rotation})$$

and so again belongs to H.

By contrast, consider the subgroup $H = \{I, M_1\}$ of D_3.

We have $RM_1R^{-1} = R(M_1R^2) = RM_3 = M_2 \notin H$. Hence H is *not* a normal subgroup.

8.10 Proposition If H is a normal subgroup of G and $a \in G$, then

$$aH = Ha.$$

Proof: Suppose that $x \in aH$ so that $x = ah$ for some $h \in H$.

Then $x = ah(a^{-1}a) = (aha^{-1})a \in Ha$.

Similarly, suppose that $x \in Ha$ so that $x = ha$ for some $h \in H$.

Then $x = (aa^{-1})ha = a(a^{-1}ha) \in aH$.

Hence $x \in aH$ iff $x \in Ha$ so that $aH = Ha$ as required.

Thus when considering normal subgroups we need not trouble with the distinction between left and right cosets. From now on we shall only consider left cosets. We have already seen that the cosets of a given subgroup form a partition of the elements of the group. We show how the set of cosets may be made into a group by defining a suitable operation.

8.11 Proposition Suppose that H is a normal subgroup of a group G and that $a, b \in G$.

The set of cosets forms a group under the binary operation given by

$$aH \cdot bH = abH.$$

Proof: We need first to show that this binary operation is well-defined, in that it is independent of the choice of different ways of writing the same coset.

Suppose that $a_1H = a_2H$ and that $b_1H = b_2H$.

Then by proposition 8.3 we have $a_2^{-1}a_1 \in H$ and $b_2^{-1}b_1 \in H$.

Now

$$
\begin{aligned}
(a_2 b_2)^{-1}(a_1 b_1) &= b_2^{-1} a_2^{-1} a_1 b_1 \\
&= b_2^{-1}(a_2^{-1} a_1)(b_2 b_2^{-1}) b_1 \\
&= (b_2^{-1}(a_2^{-1} a_1) b_2)(b_2^{-1} b_1) \in H
\end{aligned}
$$

Hence by 8.3 again $a_1 b_1 H = a_2 b_2 H$ so that

$$a_1 H \cdot b_1 H = a_2 H \cdot b_2 H$$

as required. Next we verify the groups axioms:

Associativity:

$$
\begin{aligned}
(aH \cdot bH) \cdot cH = abH \cdot cH = (ab)cH &= a(bc)H \\
&= aH \cdot bcH = aH \cdot (bH \cdot cH).
\end{aligned}
$$

The *identity* is given by eH, since

$$aH \cdot eH = aeH = aH \text{ and } eH \cdot aH = eaH = aH.$$

The *inverse* of a coset aH is $a^{-1}H$, since

$$aH \cdot a^{-1}H = aa^{-1}H = eH = a^{-1}aH = a^{-1}H \cdot aH.$$

A group of cosets obtained in this way is called a *quotient group*, and denoted G/H. Since the elements of G/H are precisely the cosets of H in G, we have $|G/H| = [G : H]$. Hence by Lagrange's Theorem 8.5 we have

8.12 Corollary If H is a normal subgroup of a finite group G then

$$|G/H| = \frac{|G|}{|H|}.$$

Examples (a) Consider the subgroup $5\boldsymbol{Z}$ of the group of integers $\boldsymbol{Z}$. The cosets are

$$
\begin{aligned}
0 + 5\boldsymbol{Z} &= \{..., -5, 0, 5, 10, 15, ...\} \\
1 + 5\boldsymbol{Z} &= \{..., -4, 1, 6, 11, 16, ...\} \\
2 + 5\boldsymbol{Z} &= \{..., -3, 2, 7, 12, 17, ...\} \\
3 + 5\boldsymbol{Z} &= \{..., -2, 3, 8, 13, 18, ...\} \\
4 + 5\boldsymbol{Z} &= \{..., -1, 4, 9, 14, 19, ...\}
\end{aligned}
$$

The quotient group $\boldsymbol{Z}/5\boldsymbol{Z}$ is the group of integers modulo 5, written $\boldsymbol{Z}_5$.

(b) Consider the normal subgroup $H = \{I, R, R^2\}$ in D_3. There are two cosets, namely $IH = \{I, R, R^2\}$ and $M_1H = \{M_1, M_2, M_3\}$.

Hence the quotient group D_3/H has two elements. We shall see presently that this group is isomorphic to $\boldsymbol{Z}_2$.

Recall from chapter 2 that for a pair of groups G and H a mapping $f: G \rightarrow H$ is called a *homomorphism* if

$$f(ab) = f(a)f(b) \text{ for all } a, b \in G.$$

Example Let $(\boldsymbol{R}, +)$ be the group of real numbers under addition and $(\boldsymbol{R}^*, \cdot)$ be the group of non-zero real numbers under multiplication. The mapping $f: (\boldsymbol{R}, +) \rightarrow (\boldsymbol{R}^*, \cdot)$ given by $f(a) = 2^a$ is a homomorphism. To see this observe that

$$f(a+b) = 2^{a+b} = 2^a \cdot 2^b = f(a) \cdot f(b) \text{ for all } a, b \in \boldsymbol{R}$$

Recall also that if $f: G \rightarrow H$ is a homomorphism we have $f(e) = e$ and $f(a^{-1}) = f(a)^{-1}$ for all $a \in G$.

A homomorphism that is bijective is called an *isomorphism*. If for a pair of groups G and H there is an isomorphism $f: G \rightarrow H$ then we say that the groups are *isomorphic* and we write $G \cong H$.

Example For the quotient group D_3/H of example (b) above we may define a mapping $f: D_3/H \rightarrow \boldsymbol{Z}_2$ by

$$f(IH) = f(\{I, R, R^2\}) = 0$$

$$f(M_1H) = f(\{M_1, M_2, M_3\}) = 1.$$

It is left to the reader to check that this is a homomorphism and a bijection. Hence we have $D_3/H \cong \boldsymbol{Z}_2$.

Recall that the *kernel* of a group homomorphism $f: G \to H$ is

$$\ker f = \{a \in G : f(a) = e\}.$$

We now present our main result on quotient groups:

8.13 First Isomorphism Theorem Suppose that $f: G \to H$ is a group homomorphism and is surjective. Then $\ker f$ is a normal subgroup of G, and the mapping $\bar{f}: G/\ker f \to H$ given by

$$\bar{f}(a(\ker f)) = f(a)$$

is an isomorphism, so that $G/\ker f \cong H$.

Some authors call this result the *homomorphism theorem.*

Proof: The proof is broken down into four parts.

(I) $\ker f$ is a normal subgroup of G

Suppose $a, b \in \ker f$, so that $f(a) = f(b) = e$.

Then $f(ab) = f(a)f(b) = ee = e$ and $f(a^{-1}) = f(a)^{-1} = e^{-1} = e$, so that $ab \in \ker f$ and $a^{-1} \in \ker f$, showing that $\ker f$ is a subgroup.

Now suppose $a \in \ker f$ and $g \in G$.

Then $f(gag^{-1}) = f(g)f(a)f(g)^{-1} = f(g)ef(g)^{-1} = f(g)f(g)^{-1} = e$, and hence $gag^{-1} \in \ker f$, showing that $\ker f$ is a normal subgroup.

(II) The mapping $\bar{f}$ is well-defined

Suppose that $a_1(\ker f) = a_2(\ker f)$. By 8.3 we have $a_2^{-1}a_1 \in \ker f$.

Hence $f(a_2^{-1}a_1) = f(a_2)^{-1}f(a_1) = e$ so that $f(a_1) = f(a_2)$.

It follows that $\bar{f}(a_1(\ker f)) = \bar{f}(a_2(\ker f))$.

(III) The mapping $\bar{f}$ is a homomorphism

$$\begin{aligned}\bar{f}(a(\ker f)\cdot b(\ker f)) &= \bar{f}(ab(\ker f))\\ &= f(ab)\\ &= f(a)f(b) \text{ since } f \text{ is a homomorphism}\\ &= \bar{f}(a(\ker f))\bar{f}(b(\ker f)).\end{aligned}$$

(IV) The mapping $\bar{f}$ is a bijection

Since f is surjective, for any $h \in H$ there is $a \in G$ such that $f(a) = h$ and hence $\bar{f}(a(\ker f)) = f(a) = h$. It follows that $\bar{f}$ is also surjective.

Finally, if $\bar{f}(a(\ker f)) = \bar{f}(b(\ker f))$ so that $f(a) = f(b)$ then we have $f(b^{-1}a) = f(b)^{-1}f(a) = e$, so that $b^{-1}a \in \ker f$ and by 8.3 we have $a(\ker f) = b(\ker f)$. Hence $\bar{f}$ is injective and so is an isomorphism.

Examples

(a) Define $f: \boldsymbol{Z} \to \boldsymbol{Z}_6$ by $f(x) = [x]$, the congruence class of x modulo 6.

We have $\operatorname{im} f = \boldsymbol{Z}_6$ and $\ker f = 6\boldsymbol{Z}$ and so by the first isomorphism theorem $\boldsymbol{Z}/6\boldsymbol{Z} \cong \boldsymbol{Z}_6$.

(b) Define $f: D_3 \to \boldsymbol{Z}_2$ by

$f(I) = f(R) = f(R^2) = 0$ and $f(M_1) = f(M_2) = f(M_3) = 1$.

Then f is a homomorphism (you should check this) with kernel $H = \{I, R, R^2\}$ and is clearly surjective.

Hence by the first isomorphism theorem $D_3/H \cong \boldsymbol{Z}_2$.

Exercises 8

1. Write down all of the left cosets of $3\boldsymbol{Z}$ in $\boldsymbol{Z}$.

2. Write down all of the left cosets in D_4 of:

   (i) $\{I, R^2\}$

   (ii) $\{I, R, R^2, R^3\}$

   (iii) $\{I, M_1\}$.

   Also write down all of the right cosets for each of these three.

   What do you notice?

3. For a group G and an element $g \in G$ show that the mapping $\phi_g : G \to G$ given by $\phi_g(a) = gag^{-1}$ is an isomorphism.

4. Find all of the non-trivial proper subgroups of

   (i) $\boldsymbol{Z}_{12}$ (ii) D_4 (iii) V (iv) Q_8

   Which of these are normal subgroups?

5. Using the first isomorphism theorem 8.3 show that

   (i) $\boldsymbol{Z}/3\boldsymbol{Z} \cong \boldsymbol{Z}_3$

   (ii) $D_4/\{I, R, R^2, R^3\} \cong \boldsymbol{Z}_2$

   (iii) $D_4/\{I, R^2\} \cong V$

6. Let N be a normal subgroup of G. Show that if G is abelian then G/N is also abelian.

   Is the converse true?

Chapter 9: Group Actions

9.1 Definition Suppose that G is a group and that X is a set. A *group action* is a mapping $G \times X \to X$ that maps the pair $g \in G$ and $x \in X$ to $g \cdot x \in X$ and satisfies

(i) $e \cdot x = x$ for all $x \in X$, where e is the identity of G

(ii) $g_1 g_2 \cdot x = g_1 \cdot (g_2 \cdot x)$ for all $g_1, g_2 \in G$ and all $x \in X$.

We say that the group G *acts* on the set X.

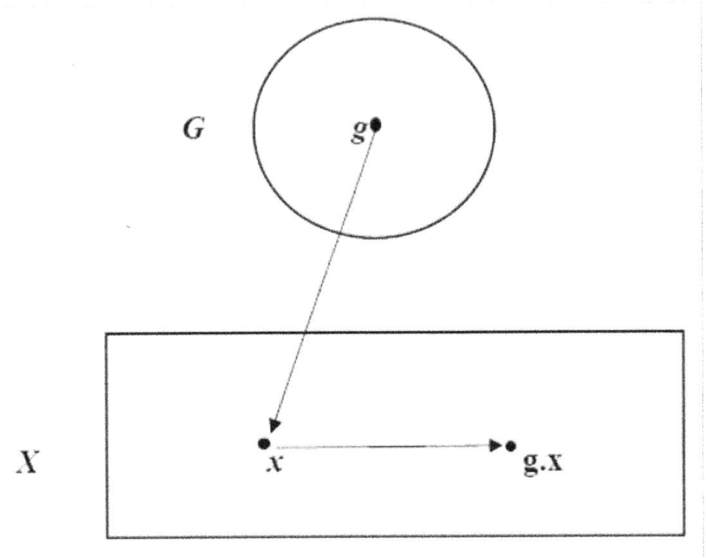

Examples (a) Let $G = D_3$ and $X = \mathbf{Z}[x, y, z]$, the set of polynomials with integer coefficients in three variables x, y and z. Suppose that the three corners 1, 2 and 3 of the equilateral triangle are labelled with the three variables x, y and z.

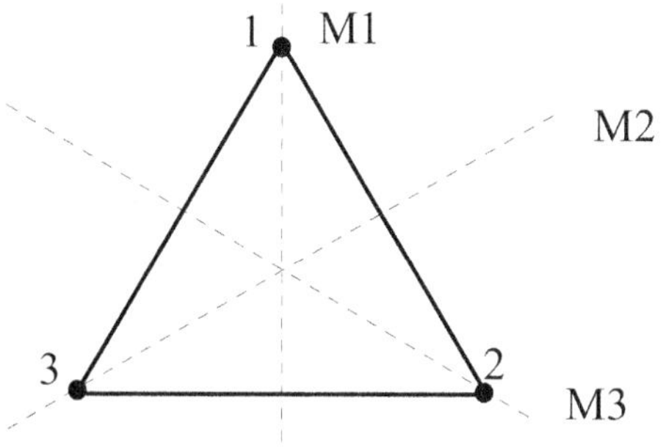

A symmetry from the dihedral group acts on a polynomial by permuting the variables. For example

$R \cdot (5x + xz + y^2) = 5y + yx + z^2$ and $M_1 \cdot (xy + 2z) = xz + 2y.$

(b) Consider a square subdivided into four equal subsquares. We assign one of two colours, say black and white, to each of the four squares. There are $2^4 = 16$ different colourings indicated below:

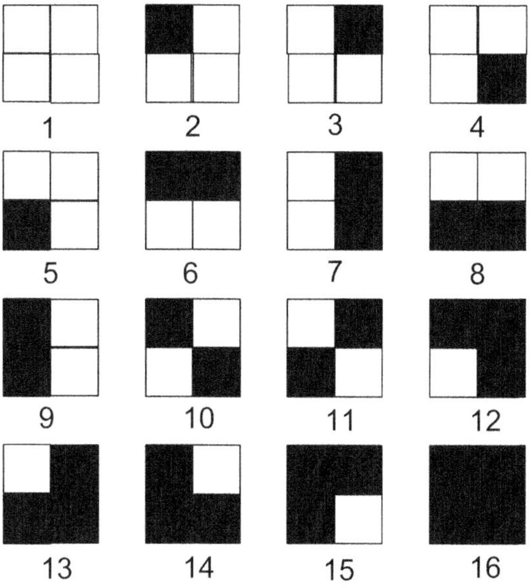

We let X be the set of sixteen colourings and let $G = \{I, R, R^2, R^3\}$ be the group of rotational symmetries of the square, where R is a rotation of 90 degrees clockwise. Then we may define an action of G on X by allowing each rotation in G to rotate a colouring in X to give a new colouring.

For example $R \cdot 2 = 3,\ R^2 \cdot 2 = 4,\ R \cdot 10 = 11$ and so on.

(c) We may take X to be G, the set of elements of the group, and define an action by conjugation:

$$g \cdot x = gxg^{-1} \text{ for all } g \in G \text{ and all } x \in X.$$

It is left as an exercise to show that this is a group action.

(d) Let $G = GL(2, \boldsymbol{R})$ and $X = \boldsymbol{R}^2$.

We may define the action of matrix M on a vector v by matrix multiplication Mv.

For example $\begin{pmatrix} 1 & 2 \\ 3 & 4 \end{pmatrix}$ acts on the vector $\begin{pmatrix} 1 \\ 1 \end{pmatrix}$ to give $\begin{pmatrix} 3 \\ 7 \end{pmatrix}$.

9.2 Definition Suppose that a group G acts on a set X.

For $x \in X$ the *orbit* of x is

$$orb(x) = \{g \cdot x : g \in G\} \subseteq X,$$

and the *stabiliser* of x is

$$G_x = \{g \in G : g \cdot x = x\}.$$

Notice that we always have $x = e \cdot x \in orb(x)$ and $e \in G_x$.

Examples (a) For example (a) above the orbit of $x + yz$ is

$$\{x + yz, y + zx, z + xy\}$$

and the stabiliser of is $\{I, M_1\}$. The orbit of xyz is $\{xyz\}$ and the stabiliser is the whole of D_3.

(b) For example (b) above, the orbit of colouring 2 is $\{2,3,4,5\}$ and its stabiliser is $\{I\}$.

The orbit of colouring 10 is $\{10,11\}$ and its stabiliser is $\{I,R^2\}$.

The orbit of colouring 16 is $\{16\}$ and its stabiliser is $\{I,R,R^2,R^3\}$.

(c) For example (c) above, the orbit of x is the conjugacy class of x. The stabiliser of x is the centraliser of x, introduced in 2.4:

$$G_x = \{g \in G : g \cdot x = x\} = \{g \in G : gxg^{-1} = x\}$$
$$= \{g \in G : gx = xg\} = C(x).$$

Orbits have rather similar properties to those of cosets. Notice that two elements may determine the same orbit. For example, in (b) the orbit of 3 is the same as the orbit of 2. The next proposition makes it clear when this happens.

9.3 Proposition Suppose that a group G acts on a set X, and that $x_1, x_2 \in X$.

Then $orb(x_1) = orb(x_2)$ if and only if $x_2 = g \cdot x_1$ for some $g \in G$.

Proof: Suppose that $orb(x_1) = orb(x_2)$.

Then $x_2 \in orb(x_1)$ and hence $x_2 = g \cdot x_1$ for some $g \in G$.

Conversely, suppose that $x_2 = g \cdot x_1$ for some $g \in G$.

If $y \in orb(x_2)$ then $y = h \cdot x_2$ for some $h \in G$.

Then $y = h \cdot (g \cdot x_1) = hg \cdot x_1$ so that $y \in orb(x_1)$.

Similarly if $y \in orb(x_1)$ then $y = h \cdot x_1$ for some $h \in G$.

Then $y = h \cdot (g^{-1} \cdot x_2) = hg^{-1} \cdot x_2$ so that $y \in orb(x_2)$.

Hence $orb(x_1) = orb(x_2)$ as required.

Next, we show that distinct orbits are disjoint:

9.4 Proposition

If $orb(x_1) \neq orb(x_2)$ then $orb(x_1) \cap orb(x_2) = \varnothing$.

Proof: Suppose that the orbits $orb(x_1)$ and $orb(x_2)$ have a common element y.

Then $y = g_1 \cdot x_1$ and $y = g_2 \cdot x_2$ for some $g_1, g_2 \in G$.

It follows that

$$x_2 = e \cdot x_2 = g_2^{-1} g_2 \cdot x_2 = g_2^{-1} \cdot (g_2 \cdot x_2) = g_2^{-1} \cdot (g_1 \cdot x_1) = g_2^{-1} g_1 \cdot x_1$$

and hence by 9.3 we have $orb(x_1) = orb(x_2)$.

9.5 Proposition

Suppose that a group G acts on a set X and that $x \in X$.

The stabiliser G_x is a subgroup of G.

Proof: The proof is left as an exercise. The strategy is to apply the test for a subgroup 2.2.

We now present our main result about orbits and stabilisers.

9.6 The Orbit-Stabiliser Theorem Suppose that a group G acts on a set X. For $x \in X$ the number of elements in the orbit of x is equal to the index of the stabiliser in G, that is

$$|orb(x)| = [G : G_x].$$

Proof: Let S be the set of distinct left cosets of G_x in G.

Notice that $[G : G_x] = |S|$. Define a mapping $f: orb(x) \to S$ by

$$f(g \cdot x) = gG_x.$$

We first show that f is well-defined in that it does not depend upon how we choose to specify a particular element of an orbit. Suppose that $g_1 \cdot x = g_2 \cdot x$.

Then $g_2^{-1} \cdot (g_1 \cdot x) = g_2^{-1} \cdot (g_2 \cdot x)$ so $g_2^{-1} g_1 \cdot x = g_2{}^{1} g_2 \cdot x = e \cdot x = x$

Hence $g_2^{-1}g_1 \in G_x$ and by 8.3 we have $g_1 G_x = g_2 G_x$ and so $f(g_1 \cdot x) = f(g_2 \cdot x)$.

Next we show that f is injective. Suppose that $f(g_1 \cdot x) = f(g_2 \cdot x)$, then $g_1 G_x = g_2 G_x$.

By 8.3 we have $g_2^{-1}g_1 \in G_x$ so that $g_2^{-1}g_1 \cdot x = x$, and consequently $g_1 \cdot x = g_2 \cdot x$.

Clearly f is surjective: for any coset gG_x we have $f(g \cdot x) = gG_x$.

Hence f is a bijection between $orb(x)$ and S, so these have the same number of elements.

Example Consider the action of D_3 on $\mathbf{Z}[x, y, z]$ above.

We saw that the orbit of $x + yz$ has three elements. The stabiliser of $x + yz$ is $\{I, M_1\}$ which gives three cosets in D_3.

9.7 Corollary Suppose that a finite group G acts on a set X. Then

$$|orb(x)| \cdot |G_x| = |G|$$

Proof: By Lagrange's theorem 8.5 we have $[G : G_x] \cdot |G_x| = |G|$

The result follows from the orbit-stabiliser theorem 9.6.

Examples (a) For the action of $\{I, R, R^2, R^3\}$ on the set of sixteen colourings the orbit of colouring 10 has 2 elements.

Since the stabiliser is $\{I, R^2\}$ we have $\dfrac{|G|}{|G_x|} = \dfrac{4}{2} = 2.$

The orbit of colouring 2 has 4 elements.

Since the stabiliser is $\{I\}$ we have $\dfrac{|G|}{|G_x|} = \dfrac{4}{1} = 4.$

The orbit of colouring 1 has 1 element.

Since the stabiliser is $\{I, R, R^2, R^3\}$ we have $\dfrac{|G|}{|G_x|} = \dfrac{4}{4} = 1.$

(b) Suppose a group G acts on G by conjugation. As described above, the orbits are the conjugacy classes, and for an element $x \in G$ the stabiliser is the centraliser $C(x)$.

Hence by the orbit-stabiliser theorem we have $|orb(x)| = [G : C(x)]$.

By applying Lagrange's theorem we obtain a formula for the size of the conjugacy class of x:

$$|orb(x)| = \frac{|G|}{|C(x)|}.$$

This formula will be useful to us in chapter 10.

9.8 Burnside's Counting Lemma Suppose that a finite group G acts on a finite set X.

For each $g \in G$ let $X_g = \{x \in X : g \cdot x = x\}$ and let m be the number of orbits in X. Then we have

$$\sum_{g \in G} |X_g| = m \cdot |G|.$$

Although this result is often named after Burnside, it was in fact first proved by Frobenius.

Proof: Let n be the number of ordered pairs $(g, x) \in G \times X$ such that $g \cdot x = x$.

For each $g \in G$ there are $|X_g|$ such pairs, so $n = \sum_{g \in G} |X_g|$.

For each $x \in X$ there are $|G_x|$ such pairs, so $n = \sum_{x \in X} |G_x|$.

Hence we have

$$\sum_{g \in G} |X_g| = \sum_{x \in X} |G_x| = \sum_{x \in X} \frac{|G|}{|orb(x)|} = |G| \sum_{x \in X} \frac{1}{|orb(x)|}$$

$$= |G| \sum_{\text{orbits } \Omega} \sum_{x \in \Omega} \frac{1}{|orb(x)|} = |G| \sum_{\text{orbits } \Omega} 1 = m \cdot |G|$$

Examples (a) Consider the group $G = \{I, R, R^2, R^3\}$ of rotational symmetries of a square, and let X be the set of sixteen colourings from above. Then X_I is the whole of X, and

$$X_R = X_{R^3} = \{1, 16\} \qquad \text{and} \qquad X_{R^2} = \{1, 10, 11, 16\}$$

Hence $\sum_{g \in G} |X_g| = 16 + 2 + 4 + 2 = 24$.

By lemma 9.8 we have $24 = 4m$ so there are $24 \div 4 = 6$ orbits.

(b) Consider a square subdivided into nine squares of equal size:

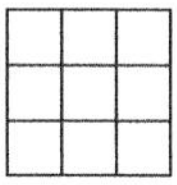

Each of the nine squares is coloured either black or white. Let X be the set of such colourings. Then $|X| = 2^9 = 512$. We allow the group of rotations $\{I, R, R^2, R^3\}$ to act on X as before.

How many orbits are there?

Since the identity rotation I fixes any colouring we have $|X_I| = 512$.

The rotation R fixes each of the following eight colourings:

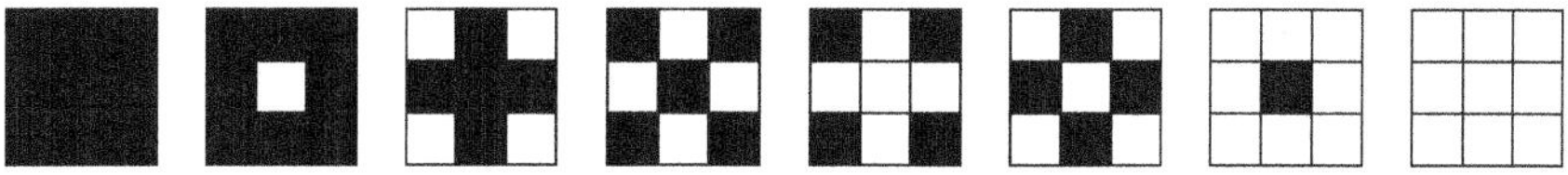

These are also fixed by R^3. Hence $|X_R| = |X_{R^3}| = 8$. The rotation R^2 fixes all of those colourings fixed by R. It also fixes the following twelve and their complements obtained by swapping white squares for black and vice versa:

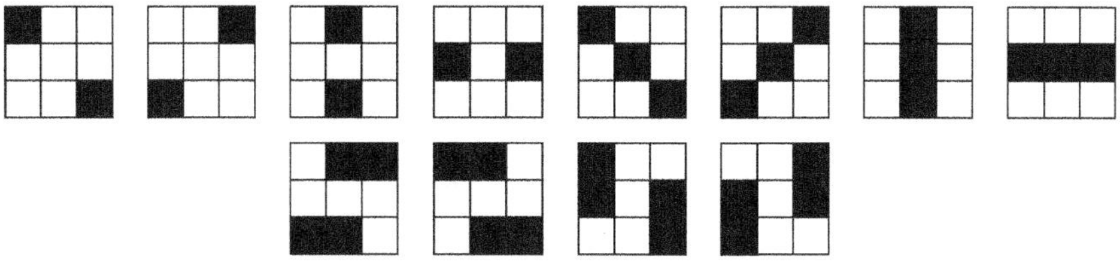

Hence $|X_{R^2}| = 32$ and so $\sum_{g \in G} |X_g| = 512 + 8 + 32 + 8 = 560$.

By lemma 9.8 there are $560 \div 4 = 140$ orbits.

We have already learnt from Lagrange's theorem that if H is a subgroup of a finite group G then $|H|$ divides $|G|$. However it is *not always* the case that if d is a divisor of $|G|$ then G has a subgroup of order d. Nevertheless, we do have a partial converse to Lagrange's theorem that guarantees the existence of a subgroup whose order is the highest power of a prime that divides $|G|$. To prove this we need the following lemma:

9.9 Lemma Suppose that p is a prime and does not divide k. Then

$$\binom{kp^n}{p^n} \equiv k(\text{mod}\, p).$$

Proof: Recall the binomial theorem:

$$(1+x)^n = 1 + nx + \frac{n(n-1)}{2!}x^2 + \frac{n(n-1)(n-2)}{3!}x^3 + \ldots + x^n.$$

For p a prime we have

$$p, \frac{p(p-1)}{2!}, \frac{p(p-1)(p-2)}{3!}, \ldots, \frac{p(p-1)\ldots2}{(p-1)!} \equiv 0(\text{mod}\, p).$$

Hence $(1+x)^p \equiv 1 + x^p(\text{mod}\, p)$.

Next we show by induction that

$$(1+x)^{p^n} \equiv 1 + x^{p^n}(\text{mod}\, p) \text{ for } n = 1, 2, 3, \ldots$$

We have already proved the case $n = 1$. Assume the result holds for n. Then

$$\begin{aligned}(1+x)^{p^{n+1}} &= (1+x)^{p^n p} = ((1+x)^{p^n})^p \\ &\equiv (1+x^{p^n})^p \equiv 1 + (x^{p^n})^p = 1 + x^{p^{n+1}}(\text{mod}\, p).\end{aligned}$$

Finally, $\binom{kp^n}{p^n}$ is the coefficient of x^{p^n} in

$$(1+x)^{kp^n} = ((1+x)^{p^n})^k \equiv (1+x^{p^n})^k = 1 + kx^{p^n} + \ldots + x^{kp^n}(\text{mod}\, p).$$

9.10 The First Sylow Theorem (1872) Let G be a group of order kp^n where p is a prime that does not divide k. Then G has a subgroup of order p^n.

Proof: Let X be the set of all those subsets S of G that have p^n elements. By lemma 9.9 the number of ways of choosing a p^n element subset from the total kp^n elements is

$$\binom{kp^n}{p^n} \equiv k(\text{mod}\, p),$$

so we have $|X| \equiv k(\text{mod}\, p)$. Define an action of G on X by

$$g \cdot S = gS = \{gx : x \in S\},$$

so that X is the disjoint union of orbits, say

$$X = orb(S_1) \cup orb(S_2) \cup \ldots \cup orb(S_r).$$

We have $|X| = |orb(S_1)| + |orb(S_2)| + \ldots + |orb(S_r)|$.

It cannot be that $|orb(S_i)| \equiv 0(\text{mod}\, p)$ for each i, since this would imply $|X| \equiv 0(\text{mod}\, p)$.

Suppose that $|orb(S)| = m$ for some m such that $m \not\equiv 0(\text{mod}\, p)$.

We claim that the stabiliser G_S is a subgroup of G of order p^n.

By 9.7 we have $|G_S| = \dfrac{|G|}{|orb(S)|} = \dfrac{kp^n}{m}$.

Now $|G_S|$ is a positive integer, and p does not divide m, so $\dfrac{k}{m}$ is a positive integer t and we have $|G_S| = tp^n$, so that $|G_S| \geq p^n$.

For $g \in G_S$ we have $g \cdot S = S$.

Hence for $g \in G_S$ and $x \in S$ we have $gx \in S$, and so the right coset $G_S\, x$ is a subset of S. It follows that $|G_S| = |G_S\, x| \leq |S| = p^n$.

Hence $|G_S| = p^n$, as required.

Examples (a) The dihedral group D_5 is of order $10 = 2 \times 5$. The First Sylow Theorem predicts subgroups of order 2 and 5. The identity together with a reflection is a subgroup of order 2 and the set of all rotational symmetries is a subgroup of order 5.

(b) The symmetric group S_5 is of order $5! = 120 = 2^3 \times 3 \times 5$.

The First Sylow Theorem predicts subgroups of order 8, 3 and 5.

Exercises 9

1. Let G be a group and let $X = G$. We may define an action of G on X by conjugation:

$$g \cdot x = gxg^{-1} \text{ for } g \in G \text{ and } x \in X.$$

Verify that this satisfies the definition of a group action.

2. (a) Consider the group action of D_3 on $\boldsymbol{Z}[x, y, z]$ described on page 128.

Find the orbit and the stabiliser of

(i) $x + y + z$ (ii) $x^2 + yz$

(iii) $x + 2y + 3z$ (iv) $(x - y)(y - z)(z - x)$

(b) Consider the group action of $\{I, R, R^2, R^3\}$ on the set of sixteen colourings on page 129.

Find the orbit and the stabiliser of

(i) colouring 6 (ii) colouring 12

(iii) colouring 1.

(c) Consider the group action of $GL(2, \boldsymbol{R})$ on $\boldsymbol{R}^2$.

Find the orbit and the stabiliser of

(i) $\begin{pmatrix} 0 \\ 0 \end{pmatrix}$ (ii) $\begin{pmatrix} 1 \\ 0 \end{pmatrix}$ (iii) $\begin{pmatrix} 2 \\ 3 \end{pmatrix}$.

3. Suppose that a group G acts on a set X and that $x \in X$.

Show that the stabiliser G_x is a subgroup of G.

4. Consider a square subdivided into four subsquares. Each subsquare is coloured one of three colours, red, green or blue.

How many colourings are there?

Let the group of rotations of $\{I, R, R^2, R^3\}$ act on the set of colourings. How many orbits of colourings are there under this action?

Let the dihedral group D_4 act on the set of colourings. How many orbits of colourings are there under this action?

5. Suppose that the dihedral group D_4 acts on itself by conjugation.

What are the orbits?

What is the stabiliser for each orbit?

Repeat this question for the quaternion group Q_8.

6. List the orders of subgroups predicted by Sylow's First Theorem for the following groups:

(i) $\mathbf{Z}_{200}$ (ii) D_6 (iii) S_6.

Can you find subgroups of the predicted orders?

Chapter 10: Simple Groups

Suppose that G is a non-trivial group. The trivial group $\{e\}$ and the whole group G are normal subgroups of G. If these are the *only* normal subgroups of G then we say that G is a *simple* group. Simple groups play a role in group theory analogous to that of prime numbers in arithmetic. The trivial group $\{e\}$ is not counted as a simple group in much the same way that 1 is not counted as a prime number. We will look at three families of finite simple group. The first family comprises the additive modular arithmetic groups modulo a prime. These are the only finite simple groups that are abelian. Next we look at a family of simple groups that arise as groups of permutations. Finally we turn our attention to groups of matrices.

10.1 Proposition For p a prime number the additive modular arithmetic group $\boldsymbol{Z}_p$ is simple.

Proof: We saw from 8.5 that the order of a subgroup divides the order of the group. Since the order of $\boldsymbol{Z}_p$ is the prime number p, its only divisors are 1 and p. The only subgroup of order 1 is $\{e\}$ and the only subgroup of order p is $\boldsymbol{Z}_p$ itself. Hence $\boldsymbol{Z}_p$ is simple.

Examples $\boldsymbol{Z}_2, \boldsymbol{Z}_3$ and $\boldsymbol{Z}_5$ are simple groups

$\boldsymbol{Z}_4$ is *not* a simple group since it has $\{0,2\}$ as a proper normal subgroup.

In fact, the groups $\boldsymbol{Z}_p$ are the only *abelian* finite simple groups. However, there are a number of interesting families of non-abelian finite simple groups.

Recall from 1.10 that a *permutation* of a set S is a bijection $S \to S$. The permutations of S form a group under composition. Where S is a finite set having n elements, this is the *symmetric group* S_n of order $n!$

Example The six permutations of S_3 are written as

$$\begin{pmatrix} 1 & 2 & 3 \\ 1 & 2 & 3 \end{pmatrix}, \begin{pmatrix} 1 & 2 & 3 \\ 2 & 1 & 3 \end{pmatrix}, \begin{pmatrix} 1 & 2 & 3 \\ 3 & 2 & 1 \end{pmatrix}, \begin{pmatrix} 1 & 2 & 3 \\ 1 & 3 & 2 \end{pmatrix}, \begin{pmatrix} 1 & 2 & 3 \\ 3 & 1 & 2 \end{pmatrix}, \begin{pmatrix} 1 & 2 & 3 \\ 2 & 3 & 1 \end{pmatrix}.$$

Recall that every permutation may be uniquely factorised as a product of disjoint cycles.

Example In S_{10} we have

$$\begin{pmatrix} 1 & 2 & 3 & 4 & 5 & 6 & 7 & 8 & 9 & 10 \\ 3 & 4 & 5 & 6 & 1 & 8 & 7 & 10 & 9 & 2 \end{pmatrix} = (2\ 4\ 6\ 8\ 10)(1\ 3\ 5).$$

Notice the convention whereby cycles of length one are omitted.

A *transposition* is a cycle of length two. Any cycle may be written as a product of transpositions, for example

$$(1\ 2\ 3\ 4\ 5) = (4\ 5)(3\ 4)(2\ 3)(1\ 2).$$

It follows that every permutation may be written as a product of transpositions. However, such a product is not unique; a permutation may be written as a product of transpositions in more than one way. For example in S_5 we have

$$(1\ 2\ 3\ 4\ 5) = (1\ 2)(1\ 3)(1\ 4)(1\ 5) = (4\ 5)(3\ 4)(2\ 3)(1\ 2).$$

Nor is it necessary that two ways of writing a permutation have the same number of transpositions.

For example in S_3 we have

$$\begin{pmatrix} 1 & 2 & 3 \\ 2 & 1 & 3 \end{pmatrix} = (1\ 2) = (1\ 3)(2\ 3)(1\ 3).$$

However, we do have the following:

10.2 Proposition For a permutation σ in S_n, *either* each way of expressing σ as a product of transpositions has an even number of transpositions *or* each way of expressing σ is a product of transpositions has an odd number of transpositions.

Before we can prove this result we need the following lemma:

10.3 Lemma Suppose that $\sigma \in S_n$ and that $(a\ b)$ is a transposition in S_n. Suppose also that σ and $\sigma \circ (a\ b)$ are each written as a product of disjoint cycles. Then the number of cycles in σ and the number of cycles in $\sigma \circ (a\ b)$ will always differ by one.

Proof: *Case 1:* Suppose a and b belong to the same cycle in σ. Then

$$(i\,a\,...\,j\,b\,...)(a\,b) = (i\,b\,...)(j\,a\,...).$$

Hence $\sigma \circ (a\,b)$ has one more cycle than σ.

Case 2: Suppose a and b belong to different cycles in σ. Then

$$(i\,a\,...\,)(j\,b\,...)(a\,b) = (i\,b\,..j\,a\,...).$$

Hence $\sigma \circ (a\,b)$ has one less cycle than σ.

Proof of 10.2: Suppose that $\sigma \in S_n$, where n is even. The identity permutation in S_n is a product of evenly many cycles of length 1: $e = (1)(2)...(n)$.

By the lemma $(a_1 b_1)$ is a product of an odd number of cycles,

$(a_1 b_1)(a_2 b_2)$ is a product of an even number of cycles,

$\vdots$

$(a_1 b_1)(a_2 b_2)...(a_r b_r)$ is a product of an odd number of cycles if r is odd, and an even number of cycles if r is even.

Since σ can be written *uniquely* as a product of disjoint cycles there cannot be both cases where r is odd and cases where r is even.

Now suppose that $\sigma \in S_n$, where n is odd. The identity permutation in S_n is a product of oddly many cycles $e = (1)(2)...(n)$.

By the lemma $(a_1 b_1)$ is a product of an even number of cycles,

$(a_1 b_1)(a_2 b_2)$ is a product of an odd number of cycles,

$\vdots$

$(a_1 b_1)(a_2 b_2)...(a_r b_r)$ is a product of an even number of cycles if r is odd, and an odd number of cycles if r is even.

As before, the result follows from the uniqueness of the factorisation of σ into disjoint cycles.

We call a permutation *odd* or *even* where it is expressible as respectively an odd or an even number of transpositions. For example, in S_3

$\begin{pmatrix} 1 & 2 & 3 \\ 2 & 1 & 3 \end{pmatrix} = (1\ 2)$ is odd, whereas $\begin{pmatrix} 1 & 2 & 3 \\ 2 & 3 & 1 \end{pmatrix} = (1\ 2\ 3) = (2\ 3)(1\ 2)$ is even.

10.4 Definition The set of even permutations of an n element set forms a group A_n, called an *alternating group*.

Note that $A_1 = A_2 = \{e\}$ and $A_3 \cong \boldsymbol{Z}_3$.

10.5 Proposition A_n is a normal subgroup of S_n.

The proof is left as an exercise.

10.6 Proposition For $n \geq 2$ we have $S_n/A_n \cong \boldsymbol{Z}_2$.

Consequently, the groups A_n are of order $\frac{n!}{2}$.

Proof: Define a mapping $f: S_n \to \boldsymbol{Z}_2$ by

$$f(\sigma) = \begin{cases} 0 \text{ if } \sigma \text{ is even} \\ 1 \text{ if } \sigma \text{ is odd.} \end{cases}$$

This is a surjective homomorphism with kernel A_n. The result follows by the first isomorphism theorem, 8.13.

Examples (a) A_4 has $\frac{4!}{2} = 12$ elements:

$$\{e, (123), (132), (124), (142), (134), (143), (234), (243), (12)(34), (13)(24), (14)(23)\}$$

(b) A_5 has $\frac{5!}{2} = 60$ elements.

The group A_4 has a normal subgroup comprising the permutations

$$\{e, (12)(34), (13)(24), (14)(23)\}.$$

With this exception, all of the other alternating groups A_n are either simple or trivial.

We show that A_5 is simple. To do this, we need a couple of preliminary results.

10.7 Lemma Suppose that $a \in G$ is an element of order *n*.

Then any conjugate gag^{-1} of a is also of order *n*.

Proof: If $a \in G$ is of order *n* then $a^n = e$ and $a^k \neq e$ for $1 \leq k \leq n-1$.

We have $(gag^{-1})^n = (gag^{-1})(gag^{-1})...(gag^{-1}) = ga^n g^{-1} = geg^{-1} = e$.

Also, if $(gag^{-1})^k = e$ for any value of *k* such that $1 \leq k \leq n-1$ then $ga^k g^{-1} = e$ so that $ga^k = g$ and $a^k = e$. Since this is not the case we have $(gag^{-1})^k \neq e$ for $1 \leq k \leq n-1$.

Hence gag^{-1} is of order *n*.

10.8 Lemma If *H* is a normal subgroup of *G* then *H* is a union of conjugacy classes of *G*, including the singleton conjugacy class of the identity.

Proof: Recall that for a normal subgroup, if $h \in H$ then for any conjugate of h we have $ghg^{-1} \in H$. Hence if *H* contains *one* element of a conjugacy class then it contains *all* of the elements of that class. Since any subgroup contains the identity element, the singleton conjugacy class $\{e\}$ must be included.

10.9 Proposition A_5 is a simple group.

Proof: A_5 has $\frac{5!}{2} = 60$ elements. By Lagrange's theorem 8.5 the order of any subgroup divides the order of the group, and so feasible orders for non-trivial proper subgroups are 2, 3, 4, 5, 6, 10, 12, 15, 20 and 30.

We shall find the sizes of the conjugacy classes. We first tabulate the numbers and orders of elements of different types.

Type	*Number of elements*	*Order*
e	1	1
$(ab)(cd)$	$\frac{5 \times 4 \times 3 \times 2}{2 \times 2 \times 2} = 15$	2

(abc)	$\frac{5\times4\times3}{3}=20$	3
$(abcde)$	$\frac{5\times4\times3\times2\times1}{5}=24$	5

The identity e belongs to the singleton conjugacy class $\{e\}$.

We find the size of the conjugacy class of $(12)(34)$. Since by 10.7 conjugation does not change the order of an element, this class has at most fifteen elements. To find the exact size of this class we find the centraliser of $(12)(34)$. Suppose that

$$\pi=\begin{pmatrix}1&2&3&4&5\\p&q&r&s&t\end{pmatrix}$$

Then if $(12)(34)\pi=\pi(12)(34)$ we have $\pi^{-1}(12)(34)\pi=(12)(34)$, that is

$$\begin{pmatrix}p&q&r&s&t\\1&2&3&4&5\end{pmatrix}(12)(34)\begin{pmatrix}1&2&3&4&5\\p&q&r&s&t\end{pmatrix}=(12)(34).$$

Hence $(pq)(rs)=(12)(34)$. This yields four permutations, namely the identity, $(12)(34)$ itself,

$$\begin{pmatrix}1&2&3&4&5\\3&4&1&2&5\end{pmatrix}=(13)(24)\text{ and}\begin{pmatrix}1&2&3&4&5\\4&3&2&1&5\end{pmatrix}=(14)(23).$$

Hence the conjugacy class of $(12)(34)$ is of size $\frac{60}{4}=15$.

Next we find the size of the conjugacy class of (123) Since conjugation does not change the order of an element, this class has at most twenty elements. To find the exact size of this class we find the centraliser of (123). Suppose that π is as above.

Then if $(123)\pi=\pi(123)$ we have $\pi^{-1}(123)\pi=(123)$, that is

$$\begin{pmatrix}p&q&r&s&t\\1&2&3&4&5\end{pmatrix}(123)\begin{pmatrix}1&2&3&4&5\\p&q&r&s&t\end{pmatrix}=(123)$$

hence $(pqr)=(123)$. This yields three permutations, namely the identity, (123) itself, and

$$\begin{pmatrix} 1 & 2 & 3 & 4 & 5 \\ 3 & 1 & 2 & 4 & 5 \end{pmatrix} = (132).$$

Hence the conjugacy class of (123) is of size $\frac{60}{3} = 20$.

Finally we find the size of the conjugacy class of (12345) Since conjugation does not change the order of an element, this class has at most 24 elements. To find the exact size of this class we find the centraliser of (12345). Suppose that π is as above.

Then if $(12345)\pi = \pi(12345)$ we have $\pi^{-1}(12345)\pi = (12345)$, that is

$$\begin{pmatrix} p & q & r & s & t \\ 1 & 2 & 3 & 4 & 5 \end{pmatrix}(12345)\begin{pmatrix} 1 & 2 & 3 & 4 & 5 \\ p & q & r & s & t \end{pmatrix} = (12345)$$

hence $(pqrst) = (12345)$. This yields five permutations, namely the identity, (12345) itself,

$$\begin{pmatrix} 1 & 2 & 3 & 4 & 5 \\ 5 & 1 & 2 & 3 & 4 \end{pmatrix} = (15432), \begin{pmatrix} 1 & 2 & 3 & 4 & 5 \\ 4 & 5 & 1 & 2 & 3 \end{pmatrix} = (14253),$$

and $\begin{pmatrix} 1 & 2 & 3 & 4 & 5 \\ 3 & 4 & 5 & 1 & 2 \end{pmatrix} = (13524)$

Hence the conjugacy class of (12345) is of size $\frac{60}{5} = 12$. The remaining twelve permutations of order 5 form another conjugacy class.

Thus we have conjugacy classes of size 1, 15, 20, 12 and 12. By inspection we see that no sum of these, including 1, is a divisor of 60. Hence A_5 has no proper normal subgroup, and so is a simple group.

10.10 Theorem The alternating groups A_n are non-abelian and simple for all $n \geq 5$.

The proof is omitted, although you are invited to prove that A_6 is simple in the exercises.

The following result about S_5 is used at the end of chapter 13:

10.11 Proposition The cycle $\sigma = (12345)$ and the transposition (12) together generate the symmetric group S_5.

Proof: Since any permutation may be written as a product of transpositions, it suffices to show that we can obtain all ten of the transpositions in S_5 using just (12) and σ.

(12) is given. By repeatedly conjugating we obtain

$$\sigma(12)\sigma^{-1} = (15) \qquad \sigma(15)\sigma^{-1} = (45)$$

$$\sigma(45)\sigma^{-1} = (34) \qquad \sigma(34)\sigma^{-1} = (23)$$

$$(12)(23)(12) = (13) \quad (12)(15)(12) = (25)$$

$$(15)(45)(15) = (14) \quad (12)(14)(12) = (24)$$

$$(13)(15)(13) = (35)$$

More generally, any transposition together with any cycle of length 5 generate the group S_5.

For the remainder of this chapter we turn our attention to groups of matrices. These provide several more families of finite non-abelian simple group. We shall examine one such family, known as the *projective special linear groups*, in detail.

Let F be a field. Recall that the *general linear group* $GL(n, F)$ is the group of all $n \times n$ invertible matrices with entries in F. A matrix is invertible if and only if its determinant is non-zero. Suppose that we restrict attention to those matrices in $GL(n, F)$ which have determinant 1. If $\det M = 1$ and $\det N = 1$ then

$$\det MN = \det M \det N = 1$$

and also $\det M^{-1} = (\det M)^{-1} = 1$. Hence such matrices form a subgroup of $GL(n, F)$. This is known as the *special linear group* and is denoted $SL(n, F)$.

Previously we have studied the cases in which F has been one of the infinite fields $\boldsymbol{Q}$, $\boldsymbol{R}$ or $\boldsymbol{C}$. Now suppose that F is a finite field $\boldsymbol{Z}_p$ where p is a prime number. We shall write $GL(n, \boldsymbol{Z}_p)$ as $GL(n, p)$ and $SL(n, \boldsymbol{Z}_p)$ as $SL(n, p)$

Examples (a) $GL(2,2)$ consists of these six matrices:

$$\begin{pmatrix} 1 & 0 \\ 0 & 1 \end{pmatrix}, \begin{pmatrix} 0 & 1 \\ 1 & 0 \end{pmatrix}, \begin{pmatrix} 1 & 1 \\ 1 & 0 \end{pmatrix}, \begin{pmatrix} 1 & 0 \\ 1 & 1 \end{pmatrix}, \begin{pmatrix} 0 & 1 \\ 1 & 1 \end{pmatrix}, \begin{pmatrix} 1 & 1 \\ 0 & 1 \end{pmatrix}.$$

Since in $\boldsymbol{Z}_2$ the only non-zero element is 1 we have

$$SL(2,2) = GL(2,2).$$

(b) $SL(2,3)$ consists of these 24 matrices:

$$\begin{pmatrix} 1 & 0 \\ 0 & 1 \end{pmatrix}, \begin{pmatrix} 2 & 0 \\ 0 & 2 \end{pmatrix}, \begin{pmatrix} 1 & 1 \\ 0 & 1 \end{pmatrix}, \begin{pmatrix} 2 & 2 \\ 0 & 2 \end{pmatrix}, \begin{pmatrix} 1 & 2 \\ 0 & 1 \end{pmatrix}, \begin{pmatrix} 2 & 1 \\ 0 & 2 \end{pmatrix}, \begin{pmatrix} 1 & 0 \\ 1 & 1 \end{pmatrix}, \begin{pmatrix} 2 & 0 \\ 2 & 2 \end{pmatrix},$$

$$\begin{pmatrix} 1 & 0 \\ 2 & 1 \end{pmatrix}, \begin{pmatrix} 2 & 0 \\ 1 & 2 \end{pmatrix}, \begin{pmatrix} 2 & 1 \\ 1 & 1 \end{pmatrix}, \begin{pmatrix} 1 & 2 \\ 2 & 2 \end{pmatrix}, \begin{pmatrix} 1 & 1 \\ 1 & 2 \end{pmatrix}, \begin{pmatrix} 2 & 2 \\ 2 & 1 \end{pmatrix}, \begin{pmatrix} 0 & 1 \\ 2 & 0 \end{pmatrix}, \begin{pmatrix} 0 & 2 \\ 1 & 0 \end{pmatrix},$$

$$\begin{pmatrix} 1 & 1 \\ 2 & 0 \end{pmatrix}, \begin{pmatrix} 2 & 2 \\ 1 & 0 \end{pmatrix}, \begin{pmatrix} 2 & 1 \\ 2 & 0 \end{pmatrix}, \begin{pmatrix} 1 & 2 \\ 1 & 0 \end{pmatrix}, \begin{pmatrix} 0 & 1 \\ 2 & 1 \end{pmatrix}, \begin{pmatrix} 0 & 2 \\ 1 & 2 \end{pmatrix}, \begin{pmatrix} 0 & 1 \\ 2 & 2 \end{pmatrix}, \begin{pmatrix} 0 & 2 \\ 1 & 1 \end{pmatrix}.$$

10.12 Proposition The order of $GL(2,p)$ is $(p^2-1)(p^2-p)$.

Proof: Suppose that M is a matrix in $GL(2,p)$. Consider the first column of M. For the determinant to be non-zero this column must not be the zero vector. There are p^2-1 such columns.

Now consider the second column of M. For the determinant to be non-zero this column must not be a multiple of the first column. There are p^2-p such columns.

Examples (a) $GL(2,2)$ has $(2^2-1)(2^2-2)=6$ elements.

We saw these six matrices in the example (a) above.

(b) $GL(2,3)$ has $(3^2-1)(3^2-3)=8\times 6=48$ elements.

10.13 Proposition The order of $SL(2,p)$ is $(p^2-1)p$.

Proof: Let $(\boldsymbol{Z}_p^*,\cdot)$ denote the group of non-zero integers modulo p under multiplication.

Consider the homomorphism $\phi : GL(2,p) \to (\boldsymbol{Z}_p^*,\cdot)$ that maps each matrix to its determinant. Clearly ϕ is surjective and has kernel $SL(2,p)$.

Hence by the first isomorphism theorem 8.13 we have

$$GL(2,p)/SL(2,p) \cong (\boldsymbol{Z}_p^*,\cdot).$$

By the corollary 8.12 to Lagrange's Theorem we have

$$\frac{|GL(2,p)|}{|SL(2,p)|} = p-1$$

$$\text{and so } |SL(2,p)| = \frac{|GL(2,p)|}{p-1} = \frac{(p^2-1)(p^2-p)}{p-1}$$
$$= \frac{(p^2-1)p(p-1)}{p-1} = (p^2-1)p.$$

Examples (a) $|SL(2,3)| = (3^2-1)\times 3 = 24.$

We exhibited the 24 elements of $SL(2,3)$ above.

(b) $|SL(2,5)| = (5^2-1)\times 5 = 120.$

10.14 Definition A *scalar matrix* in $GL(n,F)$ is a matrix of the form

$$\Lambda = \begin{pmatrix} \lambda & 0 & . & . & . & 0 \\ 0 & \lambda & 0 & . & . & . \\ . & 0 & . & & & . \\ . & & & . & & . \\ . & & & & . & 0 \\ 0 & . & . & . & 0 & \lambda \end{pmatrix}$$

having the same non-zero $\lambda \in F$ as each entry of the leading diagonal and zeros elsewhere.

10.15 Proposition The set of scalar matrices Z in $GL(n, F)$ is a normal subgroup.

Proof: The proof is left as an exercise.

Hint: Observe that $M\Lambda = \Lambda M = \lambda M$ for any $M \in GL(n, F)$.

10.16 Definition Let Z be the normal subgroup of scalar matrices in $GL(n, F)$.

The *projective general linear group*, denoted $PGL(n, F)$, is the quotient group $GL(n, F)/Z$.

Similarly, if Z is the normal subgroup of scalar matrices in $SL(n, F)$ then the *projective special linear group*, denoted $PSL(n, F)$, is the quotient group $SL(n, F)/Z$.

Examples (a) In $GL(2, 2)$ the only scalar matrix is $\begin{pmatrix} 1 & 0 \\ 0 & 1 \end{pmatrix}$ and hence Z is trivial.

It follows that $PGL(2, 2)$ and $PSL(2, 2)$ are the same as $GL(2, 2)$.

(b) In $SL(2, 3)$ there are a pair of scalar matrices, namely

$$\begin{pmatrix} 1 & 0 \\ 0 & 1 \end{pmatrix} \text{and} \begin{pmatrix} 2 & 0 \\ 0 & 2 \end{pmatrix}.$$

The list of the 24 matrices of $SL(2, 3)$ above is written in pairs: in each case the second matrix of the pair is two times the first. These twelve pairs of matrices are the cosets of $PSL(2, 3)$.

It turns out that $PSL(2, 3) \cong A_4$.

10.17 Proposition

For p an odd prime $PSL(2, p)$ has $\dfrac{(p^2 - 1)p}{2}$ elements.

Proof: In $SL(2,p)$ with p odd the scalar matrices are $\begin{pmatrix} 1 & 0 \\ 0 & 1 \end{pmatrix}$ and $\begin{pmatrix} p-1 & 0 \\ 0 & p-1 \end{pmatrix}$, so Z has 2 elements.

Hence $|PSL(2,p)| = \dfrac{|SL(2,p)|}{2} = \dfrac{(p^2-1)p}{2}$.

Examples (a) $PSL(2,5)$ has $\dfrac{(5^2-1)\times 5}{2} = 60$ elements.

(b) $PSL(2,7)$ has $\dfrac{(7^2-1)\times 7}{2} = 168$ elements.

10.18 Theorem The groups $PSL(2,p)$ are simple for primes $p \geq 5$.

It turns out that the simple group $PSL(2,5)$ is isomorphic to the alternating group A_5. Hence our first new finite simple group is $PSL(2,7)$ of order 168. Curiously, this group is isomorphic to the group $GL(3,2)$.

10.19 Feit-Thompson Theorem (1963) Every non-abelian finite simple group is of even order.

The proof occupies 255 pages of a journal. We shall see another version of this theorem in chapter 11.

To summarise, we have seen three infinite families of simple group:

- Modular arithmetic groups $\boldsymbol{Z}_p$ for p a prime.
- Alternating groups A_n for $n \geq 5$.
- The projective special linear groups $PSL(2,p)$ for primes $p \geq 5$ which are examples of what are called *simple groups of Lie type.*

The projective special linear groups were already known in the 1870's, and the groups $\boldsymbol{Z}_p$ and A_n considerably earlier. In addition to these three families, Mathieu had discovered another five finite simple groups in 1861 and 1873 that do not belong to these families, the smallest of which is of order $2^4 \times 3^2 \times 5 \times 11 = 7920$. These became known as *sporadic groups*. After the other groups of Lie type (almost all of which are matrix groups) were classified during the 1950's it was hoped that the full classification of the finite simple groups was close to completion. However in the late 1960's further sporadic groups were discovered, which again belonged to none of infinite families of finite simple group. In total the number of sporadic groups rose to 26 with the last, known as *The Monster*, of order

$$\begin{aligned} &2^{46} \times 3^{20} \times 5^9 \times 7^6 \times 11^2 \times 13^3 \times 17 \times 19 \\ &\times 23 \times 29 \times 31 \times 41 \times 47 \times 59 \times 71 \\ \approx\ &8 \times 10^{53} \end{aligned}$$

discovered in 1981. After this it was believed that the classification was complete. However, the announcement of the completion of the classification in 1983 turned out to be premature. The difficulty was how to be sure there were no further sporadic groups to be discovered. The full details of the last part of the classification were not published until 2004.

It is estimated that the entire proof of the classification of finite simple groups occupies around 5000 pages of research journals.

Exercises 10

1. Show that the groups $\boldsymbol{Z}_6$ and $\boldsymbol{Z}_{10}$ are not simple.

2. Explain why A_1 and A_2 are trivial groups.

   Show that $A_3 \cong \boldsymbol{Z}_3$.

3. What is the order of S_4? Write each of the elements of S_4 as a product of disjoint cycles. Which of these permutations are odd and which are even?

4. Find subgroups of S_4 that are isomorphic to (i) $\boldsymbol{Z}_4$ (ii) D_4.

5. Show that A_n is a normal subgroup of S_n for each n.

   Show that the subgroup

   $$V = \{e, (12)(34), (13)(24), (14)(23)\}$$

   of S_4 is isomorphic to the Klein four group.

   Convince yourself that V is a normal subgroup of A_4.

6. Find the sizes of the conjugacy classes in each of the following groups:

   (i) S_3 (ii) A_4 (iii) S_4 (iv) S_5.

7. What is the order of A_6? What are the feasible orders of subgroups allowed by Lagrange's Theorem? What types of element does A_6 have? How many elements are there of each type, and what are their orders? Determine the sizes of the conjugacy classes of A_6 and hence show that A_6 is a simple group.

8. Show that $SL(n,F)$ is a normal subgroup of $GL(n,F)$ for each n and each field F.

9. Let Z be the set of scalar matrices in $GL(n,F)$. Show that Z is a normal subgroup of $GL(n,F)$.

10. Find the number of elements in each of the following groups:

(i)	$GL(2,11)$	(ii)	$SL(2,11)$	(iii)	$PSL(2,11)$
(iv)	$GL(2,13)$	(v)	$SL(2,13)$	(vi)	$PSL(2,13)$

11. Show that $GL(3,p)$ has $(p^3-1)(p^3-p)(p^3-p^2)$ elements for each prime p.

Hence show that $SL(3,p)$ has $(p^3-1)(p^3-p)p^2$ elements for each prime p.

Can you find similar formulae for the orders of $GL(4,p)$ and $SL(4,p)$?

12. Draw the Cayley table for S_3 with the six elements listed as

$$e,(12),(13),(23),(123),(132)$$

Find the order of each element of $GL(2,2)$. Show that

$$GL(2,2) \cong S_3.$$

Find the order of each element of $SL(2,3)$. Is $SL(2,3) \cong S_4$?

Chapter 11: Soluble Groups

The concept of a *soluble group* will be particularly important in chapter 13 when we study the solvability of polynomial equations. Unfortunately, the concept is not an easy one to grasp, so we break it down into three stages.

11.1 Definition A *subnormal series* for a group G is a finite set of groups such that

$$G = G_0 \rhd G_1 \rhd G_2 \rhd \ldots \rhd G_n = \{e\}.$$

We say that such a subnormal series has *length n.*

Examples (a) Subnormal series for $\boldsymbol{Z}$ include

$\boldsymbol{Z} \rhd 4\boldsymbol{Z} \rhd 20\boldsymbol{Z} \rhd 80\boldsymbol{Z} \rhd \{0\}$

and $\boldsymbol{Z} \rhd 2\boldsymbol{Z} \rhd 4\boldsymbol{Z} \rhd 8\boldsymbol{Z} \rhd 16\boldsymbol{Z} \rhd \{0\}$.

(b) A subnormal series for $\boldsymbol{Z}_{24}$ is

$\boldsymbol{Z}_{24} \rhd \{0, 3, 6, 9, 12, 15, 18, 21\} \rhd \{0, 12\} \rhd \{0\}$.

(c) A subnormal series for S_4 is $S_4 \rhd A_4 \rhd V \rhd \{e\}$,

where $V = \{e, (12)(34), (13)(24), (14)(23)\}$

Recall that a *simple group* has no proper non-trivial normal subgroups.

11.2 Definition A *composition series* for a finite group G is a subnormal series such that each quotient G_i/G_{i+1} for $i = 0, 1, 2, \ldots, n-1$ is a simple group. The quotient groups G_i/G_{i+1} are known as the *composition factors.*

Examples Consider the following subnormal series for $\boldsymbol{Z}_{24}$:

$$\boldsymbol{Z}_{24} \rhd \{0, 3, 6, 9, 12, 15, 18, 21\} \rhd \{0, 6, 12, 18\} \rhd \{0, 12\} \rhd \{0\}.$$

The composition factors are $\boldsymbol{Z}_{24}/\{0, 3, 6, 9, 12, 15, 18, 21\} \cong \boldsymbol{Z}_3$,

$\{0, 3, 6, 9, 12, 15, 18, 21\}/\{0, 6, 12, 18\} \cong \boldsymbol{Z}_2$,

$\{0, 6, 12, 18\}/\{0, 12\} \cong \boldsymbol{Z}_2$ and $\{0, 12\}/\{0\} \cong \boldsymbol{Z}_2$.

Since each of these is a simple group the series is a composition series.

Notice that the product of the orders of the composition factors is equal to the order of the group: $3 \times 2 \times 2 \times 2 = 24$. A composition series gives us, in a sense, a factorisation of a group into simple groups analogous to factorising a natural number into primes.

11.3 Definition A finite group G is *soluble* if it has a composition series in which each composition factor is abelian.

There are several other equivalent definitions of soluble group in the literature.

Examples (a) Examine the composition series for $\boldsymbol{Z}_{24}$ above. Each composition factor is either $\boldsymbol{Z}_2$ or $\boldsymbol{Z}_3$. Since each of these is abelian we conclude that $\boldsymbol{Z}_{24}$ is soluble.

(b) Consider the following subnormal series for S_4 :

$$S_4 \rhd A_4 \rhd V \rhd \{e,(13)(24)\} \rhd \{e\}$$

where V is as above. The composition factors are

$S_4/A_4 \cong \boldsymbol{Z}_2$, $A_4/V \cong \boldsymbol{Z}_3$, $V/\{e,(13)(24)\} \cong \boldsymbol{Z}_2$ and $\{e,(13)(24)\}/\{e\} \cong \boldsymbol{Z}_2$.

Each of these is a simple group, so we have a composition series. Furthermore, each composition factor is abelian and so S_4 is soluble.

(c) Since A_5 is a simple group, its only subnormal series is $A_5 \rhd \{e\}$, which also happens to be a composition series. However the composition factor $A_5/\{e\} \cong A_5$ is not abelian. Hence the group A_5 is not soluble.

By a similar argument, any *non-abelian* finite simple group is not soluble. For example, $PSL(2,7)$ is not soluble.

(d) Consider the following subnormal series for S_5 :

$$S_5 \rhd A_5 \rhd \{e\}$$

This is a composition series, since the composition factors $S_5/A_5 \cong \boldsymbol{Z}_2$ and $A_5/\{e\} \cong A_5$ are simple. However, the second composition factor is not abelian. Unfortunately, this does *not* suffice to show that S_5 is not soluble; although this particular composition series does not have abelian factors maybe some *other* composition series does. Seemingly, we need to check *all* of the composition series.

11.4 Proposition Any finite abelian group is soluble.

Proof: If G is abelian then each group in its composition series is abelian, and hence so are the composition factors. Hence G is soluble.

The difficulty we have encountered in example (d) above can be resolved. It turns out that given any pair of composition series for a finite group, each composition factor in one series is isomorphic to a composition factor in the other series. Hence it is sufficient to examine a single composition series; if one or more of the composition factors fails to be abelian then we can conclude that the group is not soluble. However, before we can prove this we shall need a couple of technical results.

11.5 The Correspondence Theorem Suppose that N is a normal subgroup of G. There is a bijection between the set of subgroups H of G which are such that $N \leq H \leq G$ and the set of subgroups of G/N given by $H \leftrightarrow H/N$.

Furthermore, $H \lhd G$ if and only if $H/N \lhd G/N$.

Proof: Suppose that H is a subgroup such that $N \leq H \leq G$. Notice that $N \lhd H$ and so we may form the quotient group H/N. We show that H/N is a subgroup of G/N:

(i) if $aN, bN \in H/N$ then $aN \cdot bN = abN \in H/N$ since $ab \in H$

(ii) if $aN \in H/N$ then $(aN)^{-1} = a^{-1}N \in H/N$ since $a^{-1} \in H$.

Next we show that the correspondence $H \leftrightarrow H/N$ is injective:

Suppose $H_1/N = H_2/N$. If $h_1 \in H_1$ then $h_1N \in H_1/N$ and

$h_1N = h_2N$ for some $h_2 \in H_2$.

By 8.3 we have $h_2^{-1}h_1 \in N$ and so $h_1 = h_2(h_2^{-1}h_1) \in H_2$.

Hence $H_1 \subseteq H_2$.

By a similar argument, $H_2 \subseteq H_1$ and so $H_1 = H_2$.

Finally we observe that all subgroups of G/N are of the form H/N for some subgroup H of G so that the correspondence $H \leftrightarrow H/N$ is surjective.

To show that the bijection correlates normal subgroups with normal subgroups, first suppose that $H \lhd G$ and consider H/N in G/N. For $g \in G$ and $h \in H$ we have

$$gN \cdot hN \cdot (gN)^{-1} = ghg^{-1}N \in H/N \text{ since } ghg^{-1} \in H.$$

Conversely suppose that $H/N \lhd G/N$. Then for $g \in G$ and $h \in H$ we have $ghg^{-1}N \in H/N$ and hence $ghg^{-1} \in H$ as required.

Example Consider the subgroup $8\boldsymbol{Z}$ of $\boldsymbol{Z}$. We have intermediate subgroups as follows:

$$8\boldsymbol{Z} < 4\boldsymbol{Z} < 2\boldsymbol{Z} < \boldsymbol{Z}.$$

The quotient group $\boldsymbol{Z}/8\boldsymbol{Z} \cong \boldsymbol{Z}_8$ has subgroups

$$\{0,4\} \cong 4\boldsymbol{Z}/8\boldsymbol{Z} \text{ and } \{0,2,4,6\} \cong 2\boldsymbol{Z}/8\boldsymbol{Z}.$$

11.6 Definition We say that a normal subgroup N of G is *maximal* if there is no other normal subgroup H such that $N < H < G$.

Example The subgroup $8\mathbf{Z}$ is not maximal in $\mathbf{Z}$ since for example $8\mathbf{Z} < 4\mathbf{Z} < \mathbf{Z}$.

By contrast, $7\mathbf{Z}$ *is* maximal in $\mathbf{Z}$.

11.7 Corollary Suppose that $N \lhd G$. N is maximal if and only if G/N is a simple group.

Proof: The result follows from the correspondence theorem, 11.5.

11.8 Second Isomorphism Theorem Suppose that N is a normal subgroup of G and that H is a (not necessarily normal) subgroup of G. Then we have:

(i) the set $NH = \{nh : n \in N, h \in H\}$ is a subgroup of G

(ii) $N \cap H$ is a normal subgroup of H

(iii) $H/(N \cap H) \cong NH / N$.

Proof: (i) Suppose $n_1 h_1, n_2 h_2 \in NH$.

Then $n_1 h_1 n_2 h_2 = n_1 h_1 n_2 (h_1^{-1} h_1) h_2 = n_1 (h_1 n_2 h_1^{-1}) h_1 h_2 \in NH$ since $h_1 n_2 h_1^{-1} \in N$.

Also, if $nh \in NH$ then

$$(nh)^{-1} = h^{-1} n^{-1} = h^{-1} n^{-1} (hh^{-1}) = (h^{-1} n^{-1} h) h^{-1} \in NH$$

since $h^{-1} n^{-1} h \in N$.

Hence NH is a subgroup of G.

(ii) Since the intersection of any pair of subgroups is a subgroup, it only remains to prove that $N \cap H$ is a *normal* subgroup of H. Suppose $a \in N \cap H$ and $h \in H$.

We have $a, h \in H$ and so $hah^{-1} \in H$.

But we also have $a \in N$ and so $hah^{-1} \in N$ since N is normal.

Hence $hah^{-1} \in N \cap H$ and so $H \cap N$ is a normal subgroup of H.

(iii) We define a mapping $\phi : H \to NH / N$ by $\phi(h) = hN$.

ϕ is a homomorphism since

$$\phi(h_1 h_2) = h_1 h_2 N = h_1 N \cdot h_2 N = \phi(h_1)\phi(h_2).$$

ϕ is surjective since for $nhN = nN \cdot hN = N \cdot hN = hN$ we have $\phi(h) = hN$.

Finally, the kernel is given by

$$\ker\phi = \{h \in H : \phi(h) = eN\} = \{h \in H : hN = eN\}$$
$$= \{h \in H : h \in N\} = H \cap N.$$

Hence by the first isomorphism theorem 8.13 we have

$$H/(N \cap H) \cong NH / N.$$

Examples (a) Consider the dihedral group D_4. Let N be the normal subgroup $\{I, R^2\}$ and let H be the subgroup $\{I, M_1\}$. Notice that H is not a normal subgroup.

We have $NH = \{I, R^2, M_1, M_3\}$ and $N \cap H = \{I\}$.

Part (i) of the theorem tells us that $\{I, R^2, M_1, M_3\}$ is a subgroup. $\{I\}$ is clearly normal, in accordance with part (ii).

Finally, part (iii) of the theorem tells us that

$$\{I, M_1\} \cong \{I, R^2, M_1, M_3\}/\{I, R^2\}.$$

(b) Consider the subgroups $4\mathbf{Z}$ and $6\mathbf{Z}$ in $\mathbf{Z}$. Since $\mathbf{Z}$ is abelian all subgroups are normal, so we are free to choose which is N. We have $4\mathbf{Z} + 6\mathbf{Z} = 2\mathbf{Z}$ and $4\mathbf{Z} \cap 6\mathbf{Z} = 12\mathbf{Z}$.

By the theorem, we have $6\mathbf{Z}/12\mathbf{Z} \cong 2\mathbf{Z}/4\mathbf{Z}$ and $4\mathbf{Z}/12\mathbf{Z} \cong 2\mathbf{Z}/6\mathbf{Z}$.

We now state and prove the promised result about composition series.

11.9 Jordan-Hölder Theorem Suppose that a non-trivial group G has a pair of composition series:

$$G = G_0 \rhd G_1 \rhd G_2 \rhd \dots \rhd G_m = \{e\}$$

$$G = H_0 \rhd H_1 \rhd H_2 \rhd \dots \rhd H_n = \{e\}.$$

Then $m = n$ and there is a bijection between the set of composition factors G_i/G_{i+1} and the set of composition factors H_j/H_{j+1} such that each G_i/G_{i+1} is isomorphic to the corresponding H_j/H_{j+1}.

Proof: We shall prove the result by induction on the order of the group. For the basis of the induction, note that if $|G|$ is prime then G is simple, and so the only composition series is $G \rhd \{e\}$.

Next comes the inductive step. Suppose that the result has been established for all groups of order that divides and is less than $|G|$ and that G has composition series

$$G \rhd G_1 \rhd G_2 \rhd \dots \rhd \{e\} \qquad \text{(I)}$$

$$G \rhd H_1 \rhd H_2 \rhd \dots \rhd \{e\}. \qquad \text{(II)}$$

Case 1: $G_1 = H_1$

By the inductive hypothesis the result is already established for G_1 and so the composition factors of

$$G_1 \rhd G_2 \rhd \dots \rhd \{e\}$$

$$\text{and } H_1 \rhd H_2 \rhd \dots \rhd \{e\}$$

are isomorphic in pairs, and also $G/G_1 = G/H_1$. Hence the result.

Case 2: $G_1 \neq H_1$

Consider $G_1H_1 = \{gh : g \in G_1, h \in H_1\}$.

It can be shown that $G_1H_1 \lhd G$, and this is left as an exercise.

We also have $G_1 \lhd G_1H_1$ and $H_1 \lhd G_1H_1$.

Because G_1 is a maximal normal subgroup of G we must have $G_1H_1 = G_1$ or $G_1H_1 = G$. Because H_1 is a maximal normal subgroup of G we must have $G_1H_1 = H_1$ or $G_1H_1 = G$.

Hence $G_1H_1 = G$.

Let $D = G_1 \cap H_1$ and suppose D has composition series

$$D \rhd D_1 \rhd D_2 \rhd \ldots \rhd \{e\}.$$

We have a pair of composition series for G_1:

$$G_1 \rhd G_2 \rhd G_3 \rhd \ldots \rhd \{e\} \qquad \text{(III)}$$

$$G_1 \rhd D \rhd D_1 \rhd \ldots \rhd \{e\} \qquad \text{(IV)}$$

By the inductive hypothesis, the composition factors of (III) and (IV) are isomorphic in pairs.

Similarly, we have a pair of composition series for H_1:

$$H_1 \rhd H_2 \rhd H_3 \rhd \ldots \rhd \{e\} \qquad \text{(V)}$$

$$H_1 \rhd D \rhd D_1 \rhd \ldots \rhd \{e\} \qquad \text{(VI)}$$

Again, by the inductive hypothesis, the composition factors of (V) and (VI) are isomorphic in pairs.

By the second isomorphism theorem 11.8 we have

$$G_1H_1/H_1 \cong G_1/(G_1 \cap H_1) \text{ and } G_1H_1/G_1 \cong H_1/(G_1 \cap H_1)$$

and so

$$G/H_1 \cong G_1/D \text{ and } G/G_1 \cong H_1/D.$$

Hence the composition factors of

$$G \rhd G_1 \rhd D \rhd D_1 \rhd \ldots \rhd \{e\}$$

$$G \rhd H_1 \rhd D \rhd D_1 \rhd \ldots \rhd \{e\}$$

are isomorphic in pairs. It follows that the composition factors of (I) and (II) are isomorphic in pairs.

Example Consider these two composition series for $\boldsymbol{Z}_{15}$:

$$\boldsymbol{Z}_{15} \rhd \{0, 5, 10\} \rhd \{0\} \text{ and } \boldsymbol{Z}_{15} \rhd \{0, 3, 6, 9, 12\} \rhd \{0\}.$$

In both cases the composition factors are $\boldsymbol{Z}_3$ and $.\boldsymbol{Z}_5$

11.10 Corollary If G has a compositions series in which one or more of the composition factors is non-abelian, then G is not soluble.

Example S_5 is not soluble.

11.11 Third Isomorphism Theorem Suppose that N and H are normal subgroups of G such that $N \lhd H$. Then we have

$$\frac{G/N}{H/N} \cong G/H.$$

Proof: Define a mapping $\phi : G/N \to G/H$ by $\phi(gN) = gH$.

The proof is broken into four parts.

I. The mapping ϕ is well-defined:

Suppose $g_1N = g_2N$. Then $g_2^{-1}g_1 \in N$ and because $N \subseteq H$ we have $g_2^{-1}g_1 \in H$ and $g_1H = g_2H$.

II. ϕ is a homomorphism:

$$\phi(g_1N \cdot g_2N) = \phi(g_1g_2N) = g_1g_2H = g_1H \cdot g_2H = \phi(g_1N) \cdot \phi(g_2N).$$

III. ϕ is surjective:

For $gH \in G/H$ we may take $gN \in G/N$. Then surely $\phi(gN) = gH$.

IV. The kernel of ϕ is H/N:

$$\ker \phi = \{gN \in G/N : gH = eH\} = \{gN \in G/N : g \in H\} = H/N.$$

The result follows by the first isomorphism theorem 8.13.

Example $2\mathbf{Z}$ and $4\mathbf{Z}$ are normal subgroups of $\mathbf{Z}$ with $4\mathbf{Z} \lhd 2\mathbf{Z}$.

The theorem tells us that $\dfrac{\mathbf{Z}/4\mathbf{Z}}{2\mathbf{Z}/4\mathbf{Z}} \cong \mathbf{Z}/2\mathbf{Z}$.

11.12 Corollary Suppose that N is a normal subgroup of a finite group G.

If N and G/N are both soluble groups then G is soluble.

Proof: Suppose G/N is soluble and has a composition series

$$G/N \rhd G_1/N \rhd G_2/N \rhd \ldots \rhd N/N = \{e\}$$

Note that all subgroups of G/N are quotients by N and that we have

$$G \rhd G_1 \rhd G_2 \rhd \ldots \rhd N \qquad \text{(I)}$$

By the third isomorphism theorem 11.11, $G_i/G_{i+1} \cong \dfrac{G_i/N}{G_{i+1}/N}$, and each of these quotients is simple and abelian.

As N is soluble it has a composition series with each N_i/N_{i+1} abelian:

$$N \rhd N_1 \rhd N_2 \rhd \ldots \rhd \{e\} \qquad \text{(II)}$$

By splicing (I) and (II) together we obtain a composition series for G:

$$G \rhd G_1 \rhd G_2 \rhd \ldots \rhd N \rhd N_1 \rhd N_2 \rhd \ldots \rhd \{e\}.$$

Since each of the quotients in this series is abelian, G is soluble.

11.13 Corollary Suppose that we have a subnormal series

$$G = G_n \rhd G_{n-1} \rhd \ldots \rhd G_1 \rhd G_0 = \{e\}.$$

If G_i/G_{i-1} is soluble for $i = 1, 2, \ldots, n$ then G is soluble.

Proof: The proof is by induction on n and is left as an exercise.

Here is a deeper result about soluble groups:

11.14 Theorem (Feit-Thompson, 1963)

Every finite group of odd order is soluble.

This is an alternative version of Theorem 10.19. It is an interesting exercise to show that each version of the theorem implies the other.

Exercises 11

1. Find composition series for each of the following groups:

   (i) $\boldsymbol{Z}_6$ (ii) $\boldsymbol{Z}_8$ (iii) $\boldsymbol{Z}_{10}$ (iv) $\boldsymbol{Z}_{12}$

   In each case state the composition factors.

2. Show that the dihedral group D_4 is soluble. Extend your argument to show that all dihedral groups D_n are soluble.

3. Show that the quaternion group Q_8 is soluble.

4. Find all of the subgroups H of $\boldsymbol{Z}$ such that $36\boldsymbol{Z} < H < \boldsymbol{Z}$.

   Find all of the proper non-trivial subgroups of $\boldsymbol{Z}_{36} \cong \boldsymbol{Z}/36\boldsymbol{Z}$.

5. Suppose that N is a normal subgroup of G. Show that if G is soluble then G/N is also soluble.

6. Suppose that we have a subnormal series

   $$G = G_n \rhd G_{n-1} \rhd \ldots \rhd G_1 \rhd G_0 = \{e\}.$$

   Prove that if G_i/G_{i-1} is soluble for $i = 1, 2, \ldots, n$ then G is soluble.

   Hint: Use induction on n and 11.12.

7. Suppose that G_1 and H_1 are normal subgroups of G.

   Show that $G_1H_1 = \{gh : g \in G_1, h \in H_1\}$ is also a normal subgroup of G.

8. Show that the groups $\boldsymbol{Z}_6$ and S_3, each of order 6, have the same composition factors.

9. Suppose that H_1 and H_2 are subgroups of an abelian group G.

Show that $H_1 + H_2 = \{a + b : a \in H_1, b \in H_2\}$ is a subgroup of G.

For the subgroups $6\boldsymbol{Z}$ and $9\boldsymbol{Z}$ of $\boldsymbol{Z}$ write each of $6\boldsymbol{Z} + 9\boldsymbol{Z}$ and $6\boldsymbol{Z} \cap 9\boldsymbol{Z}$ in the form $n\boldsymbol{Z}$.

State the two isomorphisms which may be deduced from the second isomorphism theorem 11.8

Chapter 12: Fields and their Extensions

Recall from chapter 3 that a *field* is a set F together with a pair of binary operations $+$ and $\cdot$ on F such that

(i) $(F,+)$ is an abelian group

(ii) $\cdot$ is commutative and associative, has a unity element, written 1, satisfying $1 \cdot a = a = a \cdot 1$ and each non-zero element $a \in F$ has a multiplicative inverse, written a^{-1}, satisfying

$$a \cdot a^{-1} = 1 = a^{-1} \cdot a$$

(iii) $\cdot$ distributes over $+$, so that

$$a \cdot (b + c) = a \cdot b + a \cdot c \text{ for all } a, b, c \in F.$$

Examples The rational numbers $\boldsymbol{Q}$, the real numbers $\boldsymbol{R}$, and the complex numbers $\boldsymbol{C}$ are each a field under addition and multiplication.

The integers modulo p under addition and multiplication modulo p are a field, provided that p is prime. This field is written $\boldsymbol{Z}_p$ or $\boldsymbol{F}_p$.

Suppose that E is a field. A subset $F \subseteq E$ is called a *subfield* if F is itself a field under the same operations as E. For example $\boldsymbol{R}$ is a subfield of $\boldsymbol{C}$, and $\boldsymbol{Q}$ is a subfield of $\boldsymbol{R}$.

If F is a subfield of a field E then we say that E is an *extension field* of F.

For example, $\boldsymbol{C}$ is an extension field of $\boldsymbol{R}$, and $\boldsymbol{R}$ is an extension field of $\boldsymbol{Q}$.

A *polynomial* over a field F is an expression of the form

$$f(x) = a_n x^n + a_{n-1} x^{n-1} + \ldots + a_1 x + a_0$$

where $a_n, a_{n-1}, ..., a_1, a_0 \in F$. These elements of F are called the *coefficients* of the polynomial. We let $f(a)$ denote the result of substituting a field element a for x in the polynomial. If $f(a) = 0$ then we say that a is a *root* of f. The *degree* of a polynomial is the highest power of x with non-zero coefficient. The set of all polynomials over a field F is written $F[x]$.

A polynomial $f(x) \in F[x]$ is called an *irreducible polynomial* over F if it cannot be factorised as a product of polynomials of strictly smaller degree.

Examples Consider the set of polynomials with rational coefficients, $\boldsymbol{Q}[x]$.

The polynomial $x^2 - 5x + 6$ is not irreducible over $\boldsymbol{Q}$ as it may be factorised as $(x-2)(x-3)$.

The polynomial $x^2 - 5x + 5$ is irreducible over $\boldsymbol{Q}$ since it cannot be factorised in $\boldsymbol{Q}[x]$.

The following result is useful in constructing irreducible polynomials:

12.1 Eisenstein's Criterion Let p be a prime and suppose

$$f(x) = a_n x^n + a_{n-1} x^{n-1} + ... + a_1 x + a_0 \in \boldsymbol{Z}[x].$$

If $a_n \not\equiv 0 (\bmod p)$ but $a_i \equiv 0 (\bmod p)$ for $i = 0, 1, ..., n-1$ and $a_0 \not\equiv 0 (\bmod p^2)$ then $f(x)$ is an irreducible polynomial over $\boldsymbol{Q}$.

Example Consider $25x^5 + 9x^4 + 3x^3 + 6x^2 + 12$.

Since $25 \not\equiv 0 (\bmod 3)$ but $9, 3, 6, 12 \equiv 0 (\bmod 3)$ and $12 \not\equiv 0 (\bmod 9)$ this polynomial is irreducible over $\boldsymbol{Q}$.

12.2 Definition Suppose that E is an extension field of F and that $a \in E$. We say that a is *algebraic over* F if a is a root of some polynomial $f(x) \in F[x]$. Otherwise we say that a is *transcendental over* F.

If every element of E is algebraic over F then we say that E is an *algebraic extension* of F. Otherwise we say that E is a *transcendental extension* of F. We shall be concerned only with algebraic extensions.

Examples (a) $\boldsymbol{R}$ is an extension field of $\boldsymbol{Q}$. The real numbers $\frac{3}{4}$, $\sqrt{2}$ and $\sqrt[5]{10}$ are algebraic over $\boldsymbol{Q}$ because they are roots of the polynomials $4x-3$, x^2-2 and x^5-10 in $\boldsymbol{Q}[x]$ respectively. The real numbers π and e are transcendental over $\boldsymbol{Q}$, but it is beyond the scope of this course to show this. It follows that $\boldsymbol{R}$ is a transcendental extension of $\boldsymbol{Q}$.

(b) $\boldsymbol{C}$ is an extension field of $\boldsymbol{R}$. The complex numbers i and $3+2i$ are algebraic over $\boldsymbol{R}$ because they are roots of the polynomials x^2+1 and $x^2-6x+13$ in $\boldsymbol{R}[x]$ respectively. In fact, every complex number is algebraic over $\boldsymbol{R}$: any $a+bi \in \boldsymbol{C}$ is a root of

$$(x-(a+bi))(x-(a-bi)) = x^2-2ax+(a^2+b^2) \in \boldsymbol{R}[x].$$

Hence $\boldsymbol{C}$ is an algebraic extension of $\boldsymbol{R}$.

A polynomial is said to be *monic* if the coefficient of its highest power is 1. For example, x^2+3x-7 is monic whereas $2x^2+x+1$ is not.

12.3 Theorem Suppose that E is an extension field of F, and that $a \in E$ is algebraic over F. Then there is a unique irreducible monic polynomial $p(x) \in F[x]$ of minimum degree such that $p(a)=0$.

We call $p(x)$ the *minimal polynomial* of a. We call the degree of $p(x)$ the *degree* of a.

Sketch of Proof: Since $a \in E$ is algebraic over F it must be a root of *some* polynomial, say $f(x) \in F[x]$. We factorise $f(x)$ into irreducible polynomials one of which, say $p(x)$, has a as a root. Since $p(x)$ is already irreducible, a cannot be a root of a polynomial of smaller degree. Finally, to show uniqueness suppose that a is a root of two monic polynomials, say $p(x)$ and $q(x)$, each of degree n. Then a will also be a root of $p(x)-q(x)$, which is of degree $n-1$ or less. But $p(x)$ was already of minimal degree, so a cannot be a root of a polynomial of smaller degree. Hence we must have $p(x)=q(x)$.

Examples We shall find the minimal polynomial for three examples of real numbers that are algebraic over $\boldsymbol{Q}$.

(a) Let $a = \sqrt{2}$. Then $a^2 = 2$ so we take $p(x) = x^2 - 2$ and so $\sqrt{2}$ is of degree 2.

(b) Let $a = \sqrt{2} + \sqrt{3}$.

Then $a^2 = 5 + 2\sqrt{6}$, hence $(a^2 - 5)^2 = 24$ and so $a^4 - 10a^2 + 1 = 0$, so we take $p(x) = x^4 - 10x^2 + 1$.

Hence $\sqrt{2} + \sqrt{3}$ is of degree 4.

(c) Let $a = \sqrt[3]{\sqrt{5} - 2}$. Then $a^3 = \sqrt{5} - 2$, hence $(a^3 + 2)^2 = 5$ and so $a^6 + 4a^3 - 1 = 0$, so we take $p(x) = x^6 + 4x^3 - 1$.

Hence $a = \sqrt[3]{\sqrt{5} - 2}$ is of degree 6.

Suppose that F is a field and V is a non-empty set, and that we have a binary operation $+$ on V and an operation which combines an element $\lambda \in F$ with an element $v \in V$ to give an element $\lambda v \in V$.

Recall from chapter 3 that V is a *vector space over* F if $(V, +)$ is an abelian group and we have:

$$(\lambda + \mu)v = \lambda v + \mu v$$

$$\lambda(u + v) = \lambda u + \lambda v$$

$$\lambda(\mu v) = (\lambda\mu)v$$

$$1v = v$$

for all $u, v \in V$ and all $\lambda, \mu \in F$.

Examples (a) For the set of ordered pairs $(x, y) \in \boldsymbol{R} \times \boldsymbol{R}$ we may define $+$ by

$$(x_1, y_1) + (x_2, y_2) = (x_1 + x_2, y_1 + y_2)$$

and multiplication by a real number λ by

$$\lambda(x, y) = (\lambda x, \lambda y).$$

Under these operations $\boldsymbol{R} \times \boldsymbol{R}$ is a vector space over $\boldsymbol{R}$.

(b) Let $V = \{a + b\sqrt{2} \, : a, b \in \boldsymbol{Q}\}$ We define vector addition on V by

$$(a_1 + b_1\sqrt{2}) + (a_2 + b_2\sqrt{2}) = (a_1 + a_2) + (b_1 + b_2)\sqrt{2}$$

and multiplication by a scalar $\lambda \in \boldsymbol{Q}$ by

$$\lambda(a + b\sqrt{2}) = \lambda a + \lambda b\sqrt{2}.$$

Under these operations V is a vector space over $\boldsymbol{Q}$.

If E is an extension field of F then we may view E as a vector space over F. Vector addition is the addition operation of E and multiplication by a scalar is the multiplication operation of E. It is easy to check that the conditions above are satisfied. This is a very fruitful way of thinking about extension fields, as we may make use of the ideas that we have developed in linear algebra.

Example $\boldsymbol{C}$ is an extension field of $\boldsymbol{R}$. We can view $\boldsymbol{C}$ as a real vector space: vector addition is addition of complex numbers and multiplication by a scalar is given by

$$\lambda(x + yi) = \lambda x + \lambda yi \text{ for } \lambda \in \boldsymbol{R} \text{ and } x + yi \in \boldsymbol{C}.$$

Suppose that E is an extension field of F, and that $a_1, a_2, \ldots, a_n \in E$.

The "smallest" extension field K of F containing the elements $a_1, a_2, \ldots, a_n$, in the sense that no subfield of K contains all of the a_i, is written $F(a_1, a_2, \ldots, a_n)$.

Where $E = F(a)$ we say that E is a *simple* extension of F. We may say that $F(a)$ is the field obtained by *adjoining* the element a to F.

Examples (a) $\boldsymbol{R}(i) = \boldsymbol{C}$

(b) $\boldsymbol{Q}(\sqrt{2}) = \{a + b\sqrt{2} \, : a, b \in \boldsymbol{Q}\}$

(c) $\boldsymbol{Q}(\sqrt[5]{3})$ is a simple extension of $\boldsymbol{Q}$. However it is more difficult to see what the structure of this field is.

Suppose that V is a vector space over a field F. Recall from chapter 5 that a set of vectors $B = \{v_1, v_2, \ldots, v_n\}$ is a *basis* for V if

(i) B *spans* V: any vector $v \in V$ may be written as

$v = \lambda_1 v_1 + \lambda_2 v_2 + ... + \lambda_n v_n$ for some $\lambda_1, \lambda_2, ... \lambda_n \in F$.

(ii) B is *linearly independent*:

If $\lambda_1 v_1 + \lambda_2 v_2 + ... + \lambda_n v_n = \mathbf{0}$ then $\lambda_1 = \lambda_2 = ... = \lambda_n = 0$.

Examples $\{(1,0),(0,1)\}$ is a basis for $\boldsymbol{R} \times \boldsymbol{R}$.

$\{1, \sqrt{2}\}$ is a basis for $\boldsymbol{Q}(\sqrt{2})$.

12.4 Theorem Suppose that E is an extension field of F, and that $a \in E$ is algebraic over F of degree n. Then the simple extension $F(a)$ as a vector space over F has basis

$$\{1, a, a^2, ..., a^{n-1}\}.$$

Proof: We show that $\{1, a, a^2, ..., a^{n-1}\}$ is linearly independent. The proof that this set spans $F(a)$ is omitted.

Suppose that $\lambda_0 + \lambda_1 a + \lambda_2 a^2 + ... + \lambda_{n-1} a^{n-1} = 0$ with not all of the λ_i's zero. Then a is the root of a polynomial of degree $n-1$ or less. But the minimal polynomial of a is of degree n. Hence we must have $\lambda_0 = \lambda_1 = \lambda_2 = ... = \lambda_{n-1} = 0$.

Examples (a) $\boldsymbol{R}$ is an extension field of $\boldsymbol{Q}$. The real number $\sqrt{5}$ is algebraic over $\boldsymbol{Q}$ since it is a root of $x^2 - 5 \in \boldsymbol{Q}[x]$, and this is the minimal polynomial. Hence $\boldsymbol{Q}(\sqrt{5})$ has basis $\{1, \sqrt{5}\}$.

(b) The real number $\sqrt[5]{3}$ is algebraic over $\boldsymbol{Q}$ since it is a root of $x^5 - 3 \in \boldsymbol{Q}[x]$, and this is the minimal polynomial. Hence $\boldsymbol{Q}(\sqrt[5]{3})$has basis $\left\{1, 3^{\frac{1}{5}}, 3^{\frac{2}{5}}, 3^{\frac{3}{5}}, 3^{\frac{4}{5}}\right\}$ so that

$$\boldsymbol{Q}(\sqrt[5]{3}) = \left\{a + b3^{\frac{1}{5}} + c3^{\frac{2}{5}} + d3^{\frac{3}{5}} + e3^{\frac{4}{5}} : a, b, c, d, e \in \boldsymbol{Q}\right\}.$$

Recall that the *dimension* of a vector space is the number of vectors that form a basis. If E is an extension field of F, then the dimension of E as a vector space over F is written $[E : F]$. If this dimension is finite we say that E is a *finite extension* of F.

Examples (a) C is a finite extension of $\boldsymbol{R}$ with $[\boldsymbol{C}:\boldsymbol{R}]=2$.

(b) $\boldsymbol{Q}(\sqrt[5]{3})$ is a finite extension of $\boldsymbol{Q}$ with $[\boldsymbol{Q}(\sqrt[5]{3}):\boldsymbol{Q}]=5$.

12.5 Theorem If E is a finite extension of a field F, then E is an algebraic extension of F.

Proof:

Suppose that E is a finite extension of a field F and that $[E:F]=n$.

For any $\alpha \in E$ the set of $n+1$ vectors $\{1, \alpha, \alpha^2, ..., \alpha^n\}$ is not linearly independent.

Hence there are $\lambda_0, \lambda_1, ..., \lambda_n \in F$, not all zero, such that

$$\lambda_0 + \lambda_1\alpha + ... + \lambda_n\alpha^n = 0.$$

Now let $f(x) = \lambda_0 + \lambda_1 x + ... + \lambda_n x^n \in F[x]$.

Then $\alpha \in E$ is a root of $f(x)$, and so α is algebraic over F.

Hence E is an algebraic extension of F.

Curiously, the converse of this theorem is *false*: not every algebraic extension is a finite extension. In particular, the field of all complex numbers that are algebraic over $\boldsymbol{Q}$ illustrated on the front cover is an infinite extension of $\boldsymbol{Q}$.

Examples

(a) For the extension $\boldsymbol{Q}(\sqrt{5})$ of $\boldsymbol{Q}$ we have $[\boldsymbol{Q}(\sqrt{5}):\boldsymbol{Q}]=2$.

Hence $\boldsymbol{Q}(\sqrt{5})$ is an algebraic extension of $\boldsymbol{Q}$.

(b) We have already stated that $\boldsymbol{R}$ is a transcendental extension of $\boldsymbol{Q}$.

By the theorem, $\boldsymbol{R}$ is an infinite extension of $\boldsymbol{Q}$

12.6 Theorem Suppose that a field K is a finite extension of a field E, and that E is a finite extension of a field F, so that $K \supseteq E \supseteq F$. Then K is a finite extension of F, and

$$[K:F]=[K:E][E:F].$$

Proof: Suppose that $[K:E]=m$ and that K has basis $\{a_1, a_2, ..., a_m\}$ as a vector space over E, whilst $[E:F]=n$ and that E has basis $\{b_1, b_2, ..., b_n\}$ as a vector space over F.

We claim that the set of mn vectors

$$S=\{a_1b_1, a_1b_2, ..., a_1b_n, a_2b_1, ..., a_2b_n, ..., a_mb_1, ..., a_mb_n\}$$

is a basis for K as a vector space over F.

S spans K: Any element $k \in K$ may be written as

$$k = e_1a_1 + e_2a_2 + ... + e_ma_m \text{ for some } e_1, ..., e_m \in E$$

since $\{a_1, a_2, ..., a_m\}$ spans K. Each of $e_1, ..., e_m$ may be written as

$$e_1 = \lambda_{11}b_1 + \lambda_{12}b_2 + ... + \lambda_{1n}b_n \text{ for some } \lambda_{11}, \lambda_{12}, ..., \lambda_{1n} \in F$$

$$e_2 = \lambda_{21}b_1 + \lambda_{22}b_2 + ... + \lambda_{2n}b_n \text{ for some } \lambda_{21}, \lambda_{22}, ..., \lambda_{2n} \in F$$

$$\vdots$$

$$e_m = \lambda_{m1}b_1 + \lambda_{m2}b_2 + ... + \lambda_{mn}b_n \text{ for some } \lambda_{m1}, \lambda_{m2}, ..., \lambda_{mn} \in F$$

since $\{b_1, b_2, ..., b_n\}$ spans E. Hence

$$\begin{aligned} k &= (\lambda_{11}b_1 + \lambda_{12}b_2 + ... + \lambda_{1n}b_n)a_1 + (\lambda_{21}b_1 + \lambda_{22}b_2 + ... + \lambda_{2n}b_n)a_2 \\ &\quad + ... + (\lambda_{m1}b_1 + \lambda_{m2}b_2 + ... + \lambda_{mn}b_n)a_m \\ &= \lambda_{11}a_1b_1 + \lambda_{12}a_1b_2 + ... + \lambda_{mn}a_mb_n. \end{aligned}$$

S is linearly independent:

Suppose $\lambda_{11}a_1b_1 + \lambda_{12}a_1b_2 + ... + \lambda_{mn}a_mb_n = 0$. Then

$$\begin{aligned} &(\lambda_{11}b_1 + \lambda_{12}b_2 + ... + \lambda_{1n}b_n)a_1 + (\lambda_{21}b_1 + \lambda_{22}b_2 + ... + \lambda_{2n}b_n)a_2 \\ &\quad + ... + (\lambda_{m1}b_1 + \lambda_{m2}b_2 + ... + \lambda_{mn}b_n)a_m \\ &= 0 \end{aligned}$$

Because $\{a_1, a_2, ..., a_m\}$ is a linearly independent set we have

$$\lambda_{11}b_1 + \lambda_{12}b_2 + \ldots + \lambda_{1n}b_n = 0$$
$$\lambda_{21}b_1 + \lambda_{22}b_2 + \ldots + \lambda_{2n}b_n = 0$$
$$\vdots$$
$$\lambda_{m1}b_1 + \lambda_{m2}b_2 + \ldots + \lambda_{mn}b_n = 0.$$

Because $\{b_1, b_2, ..., b_n\}$ is a linearly independent set we have

$$\lambda_{11} = \lambda_{12} = \ldots = \lambda_{1n} = 0$$
$$\lambda_{21} = \lambda_{22} = \ldots = \lambda_{2n} = 0$$
$$\vdots$$
$$\lambda_{m1} = \lambda_{m2} = \ldots = \lambda_{mn} = 0.$$

Examples (a) Consider the fields

$$\boldsymbol{Q} \subset \boldsymbol{Q}(\sqrt{2}\,) \subset \boldsymbol{Q}(\sqrt{2}, \sqrt{3}\,).$$

The minimal polynomial of $\sqrt{2}$ over $\boldsymbol{Q}$ is $x^2 - 2$ and so $[\boldsymbol{Q}(\sqrt{2}\,) : \boldsymbol{Q}] = 2$, with basis $\{1, \sqrt{2}\,\}$.

The minimal polynomial of $\sqrt{3}$ over $\boldsymbol{Q}(\sqrt{2}\,)$ is $x^2 - 3$ and so $[\boldsymbol{Q}(\sqrt{2}, \sqrt{3}\,) : \boldsymbol{Q}(\sqrt{2}\,)] = 2$, with basis $\{1, \sqrt{3}\,\}$.

Hence by the theorem we have $[\boldsymbol{Q}(\sqrt{2}, \sqrt{3}\,) : \boldsymbol{Q}] = 2 \times 2 = 4$, with basis $\{1, \sqrt{2}, \sqrt{3}, \sqrt{6}\,\}$.

(b) Consider the fields

$$\boldsymbol{Q} \subset \boldsymbol{Q}(\sqrt{2}\,) \subset \boldsymbol{Q}(\sqrt{2}, \sqrt[3]{5}\,).$$

The minimal polynomial of $\sqrt{2}$ over $\boldsymbol{Q}$ is $x^2 - 2$ and so

$$[\boldsymbol{Q}(\sqrt{2}\,) : \boldsymbol{Q}] = 2.$$

The minimal polynomial of $\sqrt[3]{5}$ over $\boldsymbol{Q}(\sqrt{2}\,)$ is $x^3 - 5$ and so

$$[\boldsymbol{Q}(\sqrt{2}, \sqrt[3]{5}\,) : \boldsymbol{Q}(\sqrt{2}\,)] = 3.$$

Hence by the theorem we have $[\boldsymbol{Q}(\sqrt{2}, \sqrt[3]{5}\,) : \boldsymbol{Q}] = 2 \times 3 = 6$.

12.7 Definition In the field $\boldsymbol{C}$ of complex numbers the *nth roots of unity* are

$$1\angle\frac{2\pi k}{n} = \cos\frac{2\pi k}{n} + i\sin\frac{2\pi k}{n}$$

for $k = 0, 1, 2, ..., n-1$.

An *n*th root ω is said to be *primitive* if each of the *n*th roots of unity can be expressed as ω^k for some positive integer *k*.

Examples (a) The square roots of unity are -1 and 1. The primitive square root is -1.

(b) The fourth roots of unity are $i, -1, -i, 1$. Of these, i and $-i$ are primitive fourth roots of unity.

(c) The 6th roots of unity are

$$1 \qquad 1\angle\frac{\pi}{3} \qquad 1\angle\frac{2\pi}{3} \qquad -1 \qquad 1\angle\frac{-2\pi}{3} \qquad 1\angle\frac{-\pi}{3}$$

Of these, only $1\angle\frac{\pi}{3}$ and $1\angle\frac{-\pi}{3}$ are primitive 6th roots of unity.

We may adjoin a primitive *n*th root of unity to a field:

Examples (a) Let ω be a primitive 6th root of unity. Then ω is a root of $x^6 - 1$ in $\boldsymbol{Q}[x]$.

This factorises as

$$x^6 - 1 = (x^3 - 1)(x^3 + 1) = (x-1)(x^2 + x + 1)(x+1)(x^2 - x + 1).$$

Hence the minimal polynomial of ω is quadratic and so

$$[\boldsymbol{Q}(\omega) : \boldsymbol{Q}] = 2.$$

By adjoining ω to $\boldsymbol{Q}$ we get $\boldsymbol{Q}(\omega) = \{a + b\omega : a, b \in \boldsymbol{Q}\}$.

(b) Let ω be a primitive 5th root of unity. Then ω is a root of $x^5 - 1$ in $\boldsymbol{Q}[x]$.

This factorises as $x^5 - 1 = (x-1)(x^4 + x^3 + x^2 + x + 1)$.

Hence the minimal polynomial of ω is quartic and so $[\boldsymbol{Q}(\omega) : \boldsymbol{Q}] = 4$.

By adjoining ω to $\boldsymbol{Q}$ we get

$$Q(\omega) = \{a + b\omega + c\omega^2 + d\omega^3 : a, b, c, d \in Q\}.$$

12.8 Definition Suppose that F is a field and $f(x) \in F[x]$. We say that $f(x)$ *splits* over a field E if $f(x)$ may be factorised as

$$(x - a_1)(x - a_2)...(x - a_n)$$

for $a_1, a_2, ..., a_n \in E$. We say that E is the *splitting field* for $f(x)$ if

(i) $f(x)$ splits over E

(ii) $f(x)$ does not split over any subfield of E.

Examples (a) Consider $x^2 - 3x + 2 \in Q[x]$. The polynomial splits over Q as $(x-1)(x-2)$. Since Q has no subfields, it must be the splitting field.

(b) Consider $x^2 - 3 \in Q[x]$. The polynomial splits as $(x - \sqrt{3})(x + \sqrt{3})$ over R.

However, it also splits over a subfield of R, namely $Q(\sqrt{3})$. Since no subfield of $Q(\sqrt{3})$ contains $\sqrt{3}$, it is the splitting field.

(c) Consider $x^3 - 5 \in Q[x]$. Our first guess might be that the splitting field is $Q(\sqrt[3]{5})$.

However $Q(\sqrt[3]{5})$ contains only the real cube root of 5 but does not contain the two complex cube roots $\omega\sqrt[3]{5}$ and $\omega^2\sqrt[3]{5}$, where ω is a primitive cube root of 1.

Hence the splitting field is $Q(\omega, \sqrt[3]{5})$. We find the dimension of the splitting field:

$$Q \subset Q(\omega) \subset Q(\omega, \sqrt[3]{5}).$$

The minimal polynomial of ω over Q is $x^2 + x + 1$ and so by 12.4

$$[Q(\omega) : Q] = 2$$

The minimal polynomial of $\sqrt[3]{5}$ over $Q(\omega)$ is $x^3 - 5$ and so

$$[Q(\omega, \sqrt[3]{5}) : Q(\omega)] = 3.$$

Hence by theorem 12.6 we have $[Q(\omega, \sqrt[3]{5}) : Q] = 2 \times 3 = 6$.

Exercises 12

1. Apply Eisenstein's Criterion to show that each of the following is an irreducible polynomial over $\mathbf{Q}$:

   (i) $3x^2 + 8x - 2$

   (ii) $2x^4 + 6x^3 - 3x^2 + 9x + 12$

   (iii) $4x^3 - 14x^2 + 7x + 21$

   Explain why Eisenstein's Criterion *fails* for each of the following polynomials:

   (iv) $3x^2 + 8x - 4$

   (v) $9x^4 + 6x^3 - 3x^2 + 9x + 12$

   Does this show that these polynomials fail to be irreducible over $\mathbf{Q}$?

2. Show that the field of rational numbers $\mathbf{Q}$ does not have any proper subfields.

   Hint: Suppose that F is a subfield of $\mathbf{Q}$.

   Show that each integer is in F, and hence deduce that $F = \mathbf{Q}$.

3. Verify that $V = \mathbf{Q}(\sqrt{2})$ is a vector space over $\mathbf{Q}$ under the operations given in the chapter.

4. (i) Find all of the 8th roots of unity. Which of these are primitive 8th roots?

   (ii) Find all of the 7th roots of unity. Which of these are primitive 7th roots?

5. (i) Let ω be a primitive 8th root of unity. What is the degree of ω over $\boldsymbol{Q}$?

What is the dimension $[\boldsymbol{Q}(\omega) : \boldsymbol{Q}]$?

Give a basis for $\boldsymbol{Q}(\omega)$ as a vector space over $\boldsymbol{Q}$.

(ii) Repeat this exercise for a primitive 7th root of unity.

6. For each of the following polynomials in $\boldsymbol{Q}[x]$ state the splitting field:

(i) $x^2 - 7$

(ii) $x^2 - 9$

(iii) $x^3 - 4$

(iv) $x^2 - x - 1$

(v) $x^2 + 1$

(vi) $x^4 - 2$

In each case find the dimension of the splitting field E as a vector space over $\boldsymbol{Q}$.

7. Show that $\boldsymbol{Q}(\sqrt{2} + \sqrt{3}) = \boldsymbol{Q}(\sqrt{2}, \sqrt{3})$.

Is $\boldsymbol{Q}(\sqrt{2}, \sqrt{3})$ a simple extension of $\boldsymbol{Q}$?

8. What is the dimension of $\boldsymbol{Q}(\sqrt{3}, \sqrt[3]{4})$ as a vector space over $\boldsymbol{Q}$?

Write down a basis.

Chapter 13: The Galois Group

Suppose that F_1 and F_2 are fields. A mapping $\sigma : F_1 \to F_2$ is a *homomorphism* of fields if

$$\sigma(a+b) = \sigma(a) + \sigma(b)$$

and $$\sigma(ab) = \sigma(a)\sigma(b) \text{ for all } a, b \in F_1.$$

13.1 Proposition If σ is a homomorphism of fields then we have

(i) $\sigma(0) = 0$

(ii) $\sigma(1) = 0$ or 1

(iii) $\sigma(-a) = -\sigma(a)$

(iv) $\sigma(a^{-1}) = \sigma(a)^{-1}$ provided that $a \neq 0, \sigma(a) \neq 0$.

The proof of these is left as an easy exercise.

A homomorphism that is bijective is called an *isomorphism* of fields. An isomorphism which has the same field F as both its domain and codomain is called an *automorphism* of F.

Example The mapping $\sigma : \boldsymbol{C} \to \boldsymbol{C}$ given by $\sigma(z) = \overline{z}$, the complex conjugate, is an automorphism. It is easily verified that

$$\sigma(z_1 + z_2) = \overline{z_1 + z_2} = \overline{z_1} + \overline{z_2} = \sigma(z_1) + \sigma(z_2)$$

$$\text{and } \sigma(z_1 z_2) = \overline{z_1 z_2} = \overline{z_1}\ \overline{z_2} = \sigma(z_1)\sigma(z_2).$$

For a field F, the set of automorphisms of F forms a group under composition, denoted $Aut\ F$. To see this, first observe that if $\sigma, \pi \in Aut\ F$ then

$$\sigma \circ \pi(a+b) = \sigma(\pi(a+b)) = \sigma(\pi(a) + \pi(b)) = \sigma \circ \pi(a) + \sigma \circ \pi(b)$$

$$\sigma \circ \pi(ab) = \sigma(\pi(ab)) = \sigma(\pi(a)\pi(b)) = \sigma \circ \pi(a)\sigma \circ \pi(b)$$

so that composition of automorphisms yields an automorphism. Composition of mappings is associative, the identity mapping

$1 : F \to F$ given by $1(a) = a$ is the identity for $Aut\ F$, and finally if $\sigma \in Aut\ F$ then the inverse mapping $\sigma^{-1} \in Aut\ F$ gives an inverse for each element.

13.2 Definition Suppose that E is an extension field of F and that $\sigma \in Aut\ E$. We say that σ *leaves F fixed* if $\sigma(a) = a$ for all $a \in F$. The set of such automorphisms forms a subgroup of $Aut\ E$ called the *Galois group of E over F*, denoted $Gal(E/F)$.

The notation E/F is not intended to indicate a quotient structure.

Examples (a) We compute $Gal(\boldsymbol{C}/\boldsymbol{R})$, the Galois group of $\boldsymbol{C}$ over $\boldsymbol{R}$. If $\sigma \in Aut\ \boldsymbol{C}$ leaves $\boldsymbol{R}$ fixed then

$$\sigma(a+bi) = \sigma(a) + \sigma(b)\sigma(i) = a + b\sigma(i).$$

Hence such an automorphism is determined by specifying $\sigma(i)$. Now since $\sigma(i)\sigma(i) = \sigma(i^2) = \sigma(-1) = -1$ we must have either $\sigma(i) = i$ or $\sigma(i) = -i$. In the first case we get the identity automorphism and in the second case the automorphism sending each complex number to its conjugate:

$$\sigma_0(a+bi) = a+bi \text{ and } \sigma_1(a+bi) = a-bi.$$

Hence $Gal(\boldsymbol{C}/\boldsymbol{R}) \cong \boldsymbol{Z}_2$.

(b) We compute $Gal(\boldsymbol{Q}(\sqrt{2}, \sqrt{3})/\boldsymbol{Q})$. For a typical element of $\boldsymbol{Q}(\sqrt{2}, \sqrt{3})$ we have

$$\sigma(a + b\sqrt{2} + c\sqrt{3} + d\sqrt{6}) = a + b\sigma(\sqrt{2}) + c\sigma(\sqrt{3}) + d\sigma(\sqrt{2})\sigma(\sqrt{3}).$$

Hence such an automorphism may be determined by specifying $\sigma(\sqrt{2})$ and $\sigma(\sqrt{3})$.

We have $\sigma(\sqrt{2})\sigma(\sqrt{2}) = \sigma(2) = 2$ and so we must have $\sigma(\sqrt{2}) = \pm\sqrt{2}$, and similarly $\sigma(\sqrt{3}) = \pm\sqrt{3}$.

Hence the Galois group comprises four automorphisms:

$$a+b\sqrt{2}+c\sqrt{3}+d\sqrt{6} \mapsto a+b\sqrt{2}+c\sqrt{3}+d\sqrt{6}$$
$$a+b\sqrt{2}+c\sqrt{3}+d\sqrt{6} \mapsto a-b\sqrt{2}+c\sqrt{3}-d\sqrt{6}$$
$$a+b\sqrt{2}+c\sqrt{3}+d\sqrt{6} \mapsto a+b\sqrt{2}-c\sqrt{3}-d\sqrt{6}$$
$$a+b\sqrt{2}+c\sqrt{3}+d\sqrt{6} \mapsto a-b\sqrt{2}-c\sqrt{3}+d\sqrt{6}.$$

Each non identity element is of order 2, so we have

$$Gal(\boldsymbol{Q}(\sqrt{2}, \sqrt{3})/\boldsymbol{Q}) \cong \boldsymbol{Z}_2 \times \boldsymbol{Z}_2.$$

(c) We compute $Gal(\boldsymbol{Q}(\sqrt[3]{5})/\boldsymbol{Q})$. For a typical element of $\boldsymbol{Q}(\sqrt[3]{5})$ we have

$$\sigma(a+b5^{\frac{1}{3}}+c5^{\frac{2}{3}}) = a+b\sigma(5^{\frac{1}{3}})+c\sigma(5^{\frac{1}{3}})^2.$$

Hence such an automorphism may be determined by specifying $\sigma(\sqrt[3]{5})$. Now since $\sigma(\sqrt[3]{5})^3 = \sigma(5) = 5$, we must have that $\sigma(\sqrt[3]{5})$ is one of the three cube roots of five.

But only the real cube root of 5 is in $\boldsymbol{Q}(\sqrt[3]{5})$, and hence $Gal(\boldsymbol{Q}(\sqrt[3]{5})/\boldsymbol{Q})$ is the trivial group.

13.3 Definition

Suppose $f(x) \in F[x]$ and let E be the splitting field of f.

The *Galois group* of the polynomial f is $Gal(E/F)$.

Examples

(a) $x^2-7x+12 = (x-3)(x-4) \in \boldsymbol{Q}[x]$ has splitting field $\boldsymbol{Q}$.

Hence the Galois group of this polynomial is $Gal(\boldsymbol{Q}/\boldsymbol{Q}) = \{e\}$.

(b) $x^2-3 \in \boldsymbol{Q}[x]$ factorises as $(x-\sqrt{3})(x+\sqrt{3})$ and so has splitting field $\boldsymbol{Q}(\sqrt{3})$. Hence the Galois group of this polynomial is $Gal(\boldsymbol{Q}(\sqrt{3})/\boldsymbol{Q}) \cong \boldsymbol{Z}_2$.

13.4 Theorem Suppose that $f(x) \in F[x]$ has splitting field E and that $\sigma \in Gal(E/F)$.

If λ is a root of f then $\sigma(\lambda)$ is also a root of f.

Proof: Suppose that $f(x) = a_n x^n + ... + a_1 x + a_0$. If λ is a root of f then

$$a_n \lambda^n + ... + a_1 \lambda + a_0 = 0.$$

Since σ is an automorphism of E that leaves F fixed, we have

$$\sigma(a_n \lambda^n + ... + a_1 \lambda + a_0) = a_n \sigma(\lambda)^n + ... + a_1 \sigma(\lambda) + a_0 = \sigma(0) = 0,$$

hence the result

This theorem shows that each automorphism in the Galois group performs a permutation of the roots of the polynomial. We now show that if an automorphism gives the identity permutation of the roots then it is the identity automorphism.

13.5 Lemma Let $F(a)$ be a simple algebraic extension of F, and suppose $\sigma \in Gal(F(a)/F)$.

If $\sigma(a) = a$ then σ is the identity automorphism.

Proof: Suppose that the minimal polynomial of a is $f(x)$ of degree m.

By 12.4 $F(a)$ has basis $\{1, a, a^2, ..., a^{m-1}\}$ as a vector space over F.

If $\sigma(a) = a$ then for a typical element of $F(a)$ we have

$$\begin{aligned}\sigma(b_0 + b_1 a + ... + b_{m-1} a^{m-1}) &= b_0 + b_1 \sigma(a) + ... + b_{m-1} \sigma(a)^{m-1} \\ &= b_0 + b_1 a + ... + b_{m-1} a^{m-1}\end{aligned}$$

Hence σ is the identity automorphism.

13.6 Corollary Suppose $f(x) \in F[x]$ has splitting field E.

If $\sigma \in Gal(E/F)$ satisfies $\sigma(\lambda) = \lambda$ for each root of $f(x)$ then σ is the identity automorphism.

Proof: Observe that $E = F(\lambda_1, ..., \lambda_n)$ where the λ_i's are the roots of $f(x)$. We can obtain E via a finite sequence of simple extensions:

$$F \subseteq F(\lambda_1) \subseteq F(\lambda_1, \lambda_2) \subseteq ... \subseteq F(\lambda_1, \lambda_2, ..., \lambda_n).$$

The result follows by n applications of the previous lemma.

13.7 Corollary Suppose $f(x) \in F[x]$ has roots $\lambda_1, \lambda_2, ..., \lambda_n$ in the splitting field E.

If $\sigma_1, \sigma_2 \in Gal(E/F)$ satisfy $\sigma_1(\lambda_i) = \sigma_2(\lambda_i)$ for each root then $\sigma_1 = \sigma_2$.

Proof: If $\sigma_1(\lambda_i) = \sigma_2(\lambda_i)$ then $\sigma_2^{-1} \circ \sigma_1(\lambda_i) = \lambda_i$ for each root.

By the previous result $\sigma_2^{-1} \circ \sigma_1$ is the identity automorphism.

Hence $\sigma_1 = \sigma_2$, as required.

We see from these results that the Galois group of a polynomial of degree n with distinct roots is isomorphic to a subgroup of the symmetric group S_n.

13.8 Theorem Suppose that $f(x)$ is a polynomial in $F[x]$ with distinct roots and splitting field E. Then the order of the Galois group of E over F is equal to the dimension of E as a vector space over F, that is

$$|Gal(E/F)| = [E : F].$$

The proof is omitted.

Examples (a) We have already seen that $Gal(\boldsymbol{C}/\boldsymbol{R}) \cong \boldsymbol{Z}_2$, since there are two automorphisms of $\boldsymbol{C}$ that fix $\boldsymbol{R}$, namely the identity and complex conjugation. We also have $[\boldsymbol{C} : \boldsymbol{R}] = 2$.

(b) $Gal(\boldsymbol{Q}(\sqrt{2}, \sqrt{3})/\boldsymbol{Q}) = \boldsymbol{Z}_2 \times \boldsymbol{Z}_2$, which is of order 4, and

$$[\boldsymbol{Q}(\sqrt{2}, \sqrt{3}) : \boldsymbol{Q}] = 4.$$

13.9 Definition Let E be an algebraic extension of a field F. If every irreducible polynomial in $F[x]$ which has a root in E has *all* of its roots in E then E is called a *normal extension* of F.

Examples (a) $\boldsymbol{Q}(\sqrt[3]{2})$ is not a normal extension of $\boldsymbol{Q}$. Consider the irreducible polynomial $x^3 - 2 \in \boldsymbol{Q}[x]$. The real cube root of 2 lies in $\boldsymbol{Q}(\sqrt[3]{2})$ but the pair of complex cube roots of 2 do not.

(b) $\mathbf{Q}(\sqrt[3]{2}, \omega)$ where ω is a primitive cube root of 1 is a normal extension of $\mathbf{Q}$.

13.10 Lemma Let F be a field and let G be a group of automorphisms of F. We define

$$F_G = \{a \in F : \sigma(a) = a \text{ for all } \sigma \in G\}.$$

Then F_G is a subfield of F.

Proof: Suppose that $\alpha, \beta \in F_G$ so that

$$\sigma(\alpha) = \alpha \text{ and } \sigma(\beta) = \beta \text{ for all } \sigma \in G.$$

Then $\sigma(\alpha + \beta) = \sigma(\alpha) + \sigma(\beta) = \alpha + \beta$ so that $\alpha + \beta \in F_G$.

Similarly $\sigma(\alpha\beta) = \sigma(\alpha)\sigma(\beta) = \alpha\beta$ so that $\alpha\beta \in F_G$.

Also $\sigma(-\alpha) = -\sigma(\alpha) = -\alpha$ so that $-\alpha \in F_G$ and $\sigma(\alpha^{-1}) = \sigma(\alpha)^{-1} = \alpha^{-1}$ so that $\alpha^{-1} \in F_G$ for $a \neq 0$.

13.11 Fundamental Theorem of Galois Theory Suppose that E is a finite normal extension of a field F with Galois group $Gal(E/F)$. Then we have:

(i) There is a bijection between the set of fields K such that $F \subseteq K \subseteq E$ and the set of subgroups of $Gal(E/F)$, given by $K \mapsto Gal(E/K)$;

(ii) If K such that $F \subseteq K \subseteq E$ is a normal extension of F then $Gal(E/K)$ is a normal subgroup of $Gal(E/F)$. In this case $Gal(K/F) \cong Gal(E/F) \,/\, Gal(E/K)$.

Proof: (i) Since $F \subseteq K$, any automorphism of E that fixes K will also fix F, so clearly $Gal(E/K)$ is a subgroup of $Gal(E/F)$.

Surjectivity: Suppose that G is a subgroup of $Gal(E/F)$. Let $K = E_G$, the subfield of E fixed by each automorphism in G, as in 13.10. Since G fixes F we have $F \subseteq K \subseteq E$.

Then $Gal(E/K) = G$, as required.

Injectivity: Suppose that K_1 and K_2 are fields such that $Gal(E/K_1) = Gal(E/K_2)$. Call this group G.

Then $E_G = K_1$ and $E_G = K_2$, so that $K_1 = K_2$ as required.

(ii) Suppose that K is a normal extension of F, and that K is the splitting field of a polynomial $f(x) \in F[x]$. Let a be a root of $f(x)$. For $\sigma \in Gal(E/F)$ by 13.4 $\sigma(a)$ is also a root of $f(x)$. Because K is a normal extension, $\sigma(a) \in K$.

Now suppose $\tau \in Gal(E/K)$.

Then for any $x \in K$ we have $\tau(x) = x$, and in particular

$$\tau \circ \sigma(a) = \sigma(a).$$

Hence $\sigma^{-1} \circ \tau \circ \sigma(a) = a$, so that $\sigma^{-1} \circ \tau \circ \sigma \in Gal(E/K)$.

Hence $Gal(E/K) \lhd Gal(E/F)$.

Finally to show that $Gal(K/F) \cong Gal(E/F) \,/\, Gal(E/K)$ we define a group homomorphism

$$\phi : Gal(E/F) \to Gal(K/F)$$

by restricting each automorphism of E to an automorphism of K. The kernel is

$$\ker\phi = \{\sigma \in Gal(E/F) : \sigma \text{ fixes } K\} = Gal(E/K).$$

The result follows by an application of the first isomorphism theorem, 8.13.

Example $\boldsymbol{Q}(\sqrt{2}, \sqrt{3})$ is a finite normal extension of $\boldsymbol{Q}$.

We have already seen that

$$Gal(\boldsymbol{Q}(\sqrt{2}, \sqrt{3})/\boldsymbol{Q}) \cong \boldsymbol{Z}_2 \times \boldsymbol{Z}_2 \cong \{e, a, b, c\},$$

the Klein four group.

Here are the Hasse diagrams of the fields K such that $\boldsymbol{Q} \subseteq K \subseteq \boldsymbol{Q}(\sqrt{2}, \sqrt{3})$ and the subgroups of the Klein four group, each ordered by inclusion:

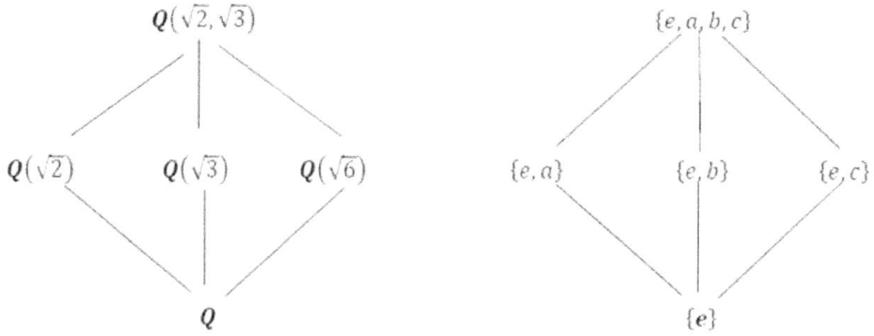

13.12 Lemma Let E be the splitting field of $x^n - a \in F[x]$. Then $Gal(E/F)$ is a soluble group.

Proof: Case 1: F contains all of the nth roots of 1:

The roots of $x^n - a$ are $\sqrt[n]{a}$, $\sqrt[n]{a}\,\omega$, $\sqrt[n]{a}\,\omega^2, \ldots, \sqrt[n]{a}\,\omega^{n-1}$, where ω is a primitive nth root of 1.

Hence the splitting field is $E = F(\sqrt[n]{a})$. Suppose that $\sigma, \tau \in Gal(E/F)$.

Since by Theorem 13.4 automorphisms in the Galois group carry roots to roots, we must have

$$\sigma(\sqrt[n]{a}) = \sqrt[n]{a}\,\omega^i \text{ and } \tau(\sqrt[n]{a}) = \sqrt[n]{a}\,\omega^j \text{ for some } i \text{ and } j.$$

Hence we have

$$\begin{aligned}\sigma \circ \tau(\sqrt[n]{a}) &= \sigma(\sqrt[n]{a}\,\omega^j) = \sigma(\sqrt[n]{a})\omega^j = \sqrt[n]{a}\,\omega^i\omega^j \\ &= \sqrt[n]{a}\,\omega^j\omega^i = \tau(\sqrt[n]{a})\omega^i = \tau(\sqrt[n]{a}\,\omega^i) = \tau \circ \sigma(\sqrt[n]{a}).\end{aligned}$$

It follows that $Gal(E/F)$ is abelian, and hence by 11.4 is a soluble group.

Case 2 : F does not contain all of the nth roots of 1:

Since the splitting field E must contain all of the roots of $x^n - a$ we require $\sqrt[n]{a}$, $\sqrt[n]{a}\,\omega \in E$ and hence $\omega \in E$ where ω is a primitive nth root of 1. Hence the splitting field is $E = F(\omega, \sqrt[n]{a})$.

Let $K = F(\omega)$ and consider $Gal(K/F)$.

If $\sigma, \tau \in Gal(K/F)$ then we have $\sigma(\omega) = \omega^i$ and $\tau(\omega) = \omega^j$ for some i and j.

Hence we have

$$\sigma \circ \tau(\omega) = \sigma(\omega^j) = (\omega^j)^i = \omega^{ij} = \omega^{ji} = (\omega^j)^i = \tau(\omega^j) = \tau \circ \sigma(\omega).$$

Hence $Gal(K/F)$ is abelian, and so once again is a soluble group. Now since K contains all of the nth roots of 1, by the argument used for Case 1 we have that $Gal(E/K)$ is abelian, and hence soluble.

By 13.11(ii) we have that $Gal(K/F) \cong Gal(E/F) \,/\, Gal(E/K)$, and hence by 11.12 $Gal(E/F)$ is soluble.

13.13 Definition An extension E of a field F is an *extension by radicals* if there are $a_1, a_2, ..., a_r \in E$ such that $E = F(a_1, a_2, ..., a_r)$ and there are natural numbers $n_1, n_2, ..., n_r$ such that

$$a_1^{n_1} \in F \text{ and } a_i^{n_i} \in F(a_1, a_2, ..., a_{i-1}) \text{ for } i = 2, ..., r.$$

A polynomial $f(x) \in F[x]$ is *solvable by radicals* if its splitting field is an extension of F by radicals.

Examples (a) $x^2 - 3 \in \boldsymbol{Q}[x]$ is solvable by radicals since its splitting field $\boldsymbol{Q}(\sqrt{3})$ is an extension of $\boldsymbol{Q}$ by radicals.

(b) $\boldsymbol{Q}\left(\sqrt{3}, \sqrt[3]{1+\sqrt{3}}\right)$ is an extension of $\boldsymbol{Q}$ by radicals since $(\sqrt{3})^2 = 3 \in \boldsymbol{Q}$ and $\left(\sqrt[3]{1+\sqrt{3}}\right)^3 = 1 + \sqrt{3} \in \boldsymbol{Q}(\sqrt{3})$.

13.14 Theorem If $f(x) \in F[x]$ is solvable by radicals and has splitting field E then the Galois group $Gal(E/F)$ is a soluble group.

Proof: Since $f(x)$ is solvable by radicals its splitting field E is an extension of F by radicals.

So there are $a_1, a_2, \ldots, a_r \in E$ such that $E = F(a_1, a_2, \ldots, a_r)$ and there are natural numbers $n_1, n_2, \ldots, n_r$ such that

$$a_1^{n_1} \in F \text{ and } a_i^{n_i} \in F(a_1, a_2, \ldots, a_{i-1}) \text{ for } i = 2, \ldots, r.$$

Let $K_0 = F$, and let K_1 be the splitting field of $x^{n_1} - a_1^{n_1}$ over K_0.

Similarly let K_i be the splitting field of $x^{n_i} - a_i^{n_i}$ over K_{i-1} for $i = 2, \ldots, r$. We have $K_r = E$.

Each K_i is a normal extension of K_{i-1}

By Lemma 13.12, $Gal(K_i/K_{i-1})$ is soluble for $i = 1, 2, \ldots, r$.

Now consider the subnormal series

$$Gal(E/K_0) \triangleright Gal(E/K_1) \triangleright Gal(E/K_2) \triangleright \ldots \triangleright Gal(E/K_r) = \{e\}$$

The quotients are soluble, and so by 11.13 $Gal(E/F)$ is soluble.

The converse of this theorem is also true: if the splitting field of $f(x) \in F[x]$ has a soluble Galois group then $f(x)$ is solvable by radicals.

Example - The Insolvability by Radicals of a Quintic Equation

Consider the polynomial $f(x) = x^5 - 6x^3 - 27x - 3 \in \mathbb{Q}[x]$.

By Eisenstein's criterion with $p = 3$ we see that $f(x)$ is irreducible.

Differentiating we obtain $f'(x) = 5x^4 - 18x^2 - 27$. Setting $f'(x) = 0$ to locate the stationary points we find

$$x^2 = \frac{18 \pm \sqrt{864}}{10} = \frac{9 \pm 6\sqrt{6}}{5} \text{ and hence } x = \sqrt{\frac{9 \pm 6\sqrt{6}}{5}} \approx \pm 2$$

(ignoring complex solutions) Hence $f(x)$ has just two stationary points.

Here is a table of values of $f(x)$ for integer values of x between -3 and 4:

x	-3	-2	-1	0	1	2	3	4
$f(x)$	-3	67	29	-3	-35	-73	-3	529

From this we can make a sketch of the graph of $f(x)$

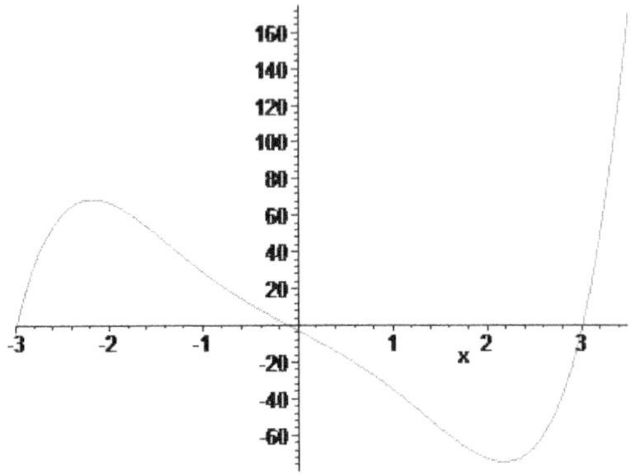

Thus $f(x)$ has three distinct real roots and a pair of complex roots that are complex conjugates of each other. Let E be the splitting field of $f(x)$. We know that $Gal(E/\boldsymbol{Q})$ is a subgroup of S_5 and that each $\sigma \in Gal(E/\boldsymbol{Q})$ performs a permutation of the roots of $f(x)$. We shall show that $Gal(E/\boldsymbol{Q}) \cong S_5$.

We have the automorphism $\sigma(x+yi) = x - yi$ which leaves fixed the real roots and swaps the complex roots. Now let a be one of the roots of $f(x)$ and consider $\boldsymbol{Q}(a)$.

Because $f(x)$ is irreducible, a is algebraic of degree 5 over $\boldsymbol{Q}$ and so $\{1, a, a^2, a^3, a^4\}$ is a basis for $\boldsymbol{Q}(a)$ and by 12.4, $[\boldsymbol{Q}(a) : \boldsymbol{Q}] = 5$.

From 12.6 we have $[E : \boldsymbol{Q}] = [E : \boldsymbol{Q}(a)][\boldsymbol{Q}(a) : \boldsymbol{Q}] = 5[E : \boldsymbol{Q}(a)]$ and hence $[E : \boldsymbol{Q}]$ is divisible by 5. But $|Gal(E/\boldsymbol{Q})| = [E : \boldsymbol{Q}]$ by 13.8 and so $|Gal(E/\boldsymbol{Q})|$ is divisible by 5.

By the First Sylow Theorem 9.10, $Gal(E/\boldsymbol{Q})$ has a subgroup of order 5. This must be a cyclic subgroup of S_5 generated by a cycle of length 5. By 10.11, a transposition together with a cycle of length 5 generates S_5. Hence $Gal(E/\boldsymbol{Q}) \cong S_5$.

But S_5 is not a soluble group and hence $x^5 - 6x^3 - 27x - 3$ is not solvable by radicals.

Exercises 13

1. Suppose that $\sigma : F_1 \longrightarrow F_2$ is a homomorphism of fields. Show that:

$$\sigma(0) = 0$$

$$\sigma(-a) = -\sigma(a) \text{ for all } a \in F_1$$

$$\sigma(a^{-1}) = \sigma(a)^{-1} \text{ provided } a \neq 0, \sigma(a) \neq 0.$$

2. (i) Show that $Gal(\boldsymbol{Q}(\sqrt{3})/\boldsymbol{Q}) \cong \boldsymbol{Z}_2$

   (ii) Show that $Gal(\boldsymbol{Q}(\sqrt{3}, \sqrt{5})/\boldsymbol{Q}) \cong \boldsymbol{Z}_2 \times \boldsymbol{Z}_2$.

   (iii) Show that $Gal(\boldsymbol{Q}(\sqrt{2}, \sqrt{3}, \sqrt{5})/\boldsymbol{Q}) \cong \boldsymbol{Z}_2 \times \boldsymbol{Z}_2 \times \boldsymbol{Z}_2$.

3. Define $\phi : \boldsymbol{Q}(\sqrt{2}) \longrightarrow \boldsymbol{Q}(\sqrt{3})$ by $\phi(a + b\sqrt{2}) = a + b\sqrt{3}$. Is ϕ an isomorphism of fields?

4. For each of the splitting fields E in Question 6 of Exercises 12, determine the Galois group $Gal(E/\boldsymbol{Q})$.

5. Draw Hasse diagrams for:

   (i) fields F such that $\boldsymbol{Q}(\sqrt{2}, \sqrt{3}, \sqrt{5}) \supseteq F \supseteq \boldsymbol{Q}$

   (ii) the subgroups of $\boldsymbol{Z}_2 \times \boldsymbol{Z}_2 \times \boldsymbol{Z}_2$

   both ordered by inclusion.

Chapter 14: The Ring Axioms and Examples

A *ring* is an algebraic structure in which we can add and multiply elements. We begin with the integers $\mathbf{Z}$. Addition and multiplication of integers satisfy well known properties:

Commutative:	$a+b=b+a$
Associative:	$(a+b)+c=a+(b+c)$
Zero:	$a+0=a=0+a$
Inverses:	$a+(-a)=0=(-a)+a$
Commutative:	$a\cdot b=b\cdot a$
Associative:	$(a\cdot b)\cdot c=a\cdot(b\cdot c)$
Unity:	$a\cdot 1=a=1\cdot a$
Distributive:	$a\cdot(b+c)=a\cdot b+a\cdot c$
	$(a+b)\cdot c=a\cdot c+b\cdot c$

Zero-divisor law: If $a\cdot b=0$ then $a=0$ or $b=0$ (or both).

for all $a,b,c\in\mathbf{Z}$. Taking the integers as a motivating example, we define a broad category of algebraic structures which satisfy similar laws to those of the integers. We make the following definition:

14.1 Definition A *ring* is a set R with binary operations $+$ and $\cdot$ on R such that:

(i) $(R,+)$ is an abelian group.

The identity, written 0, is called the *zero*. The inverse of each $a\in R$ is written $-a$.

(ii) $\cdot$ is associative: $(a\cdot b)\cdot c=a\cdot(b\cdot c)$ for all $a,b,c\in R$

(iii) $\cdot$ distributes over $+$, so that

$$a\cdot(b+c)=a\cdot b+a\cdot c$$

$$(a+b)\cdot c=a\cdot c+b\cdot c \text{ for all } a,b,c\in R$$

These statements are called the *ring axioms.*

Although a ring is strictly speaking a triple $(R, +, \cdot)$, we shall simply refer to "the ring R" when it is clear what the two binary operations are.

A ring R in which $a \cdot b = b \cdot a$ for all $a, b \in R$ is called a *commutative ring.*

An element $1 \in R$ such that $a \cdot 1 = a = 1 \cdot a$ for all $a \in R$ is called a *unity.*

A ring with such an element is called a *ring with unity.*

Examples (a) The most familiar example of a ring is the set of integers $\boldsymbol{Z}$ under addition and multiplication. The rational numbers $\boldsymbol{Q}$, the real numbers $\boldsymbol{R}$, and the complex numbers $\boldsymbol{C}$ under addition and multiplication are also commutative rings with unity.

(b) The set $\{0, 1, 2, ..., n-1\}$ is a ring under addition and multiplication modulo *n*, denoted $\boldsymbol{Z}_n$.

These rings are examples of *quotient rings*, which are considered in detail in chapter 16.

(c) The set of 2×2 matrices with integer entries, denoted $Mat(2, \boldsymbol{Z})$, is a ring under matrix addition and multiplication. For $Mat(2, \boldsymbol{Z})$ the zero and unity are

$$\begin{pmatrix} 0 & 0 \\ 0 & 0 \end{pmatrix} \text{ and } \begin{pmatrix} 1 & 0 \\ 0 & 1 \end{pmatrix}.$$

$Mat(2, \boldsymbol{Z})$ is *not* a commutative ring - there are matrices $A, B \in Mat(2, \boldsymbol{Z})$ such that $AB \neq BA$.

More generally, for any ring R the set of $n \times n$ matrices with entries in R is a ring under matrix addition and multiplication, denoted $Mat(n, R)$.

(d) The set of polynomials with integer coefficients, denoted $\boldsymbol{Z}[x]$, is a ring under polynomial addition and multiplication.

More generally, for any ring R the set of polynomials with coefficients in R is a ring under polynomial addition and multiplication, denoted $R[x]$.

(e) Consider the set of functions $\boldsymbol{R} \to \boldsymbol{R}$. For a pair of functions $f, g : \boldsymbol{R} \to \boldsymbol{R}$ we define addition and multiplication as follows:

$$\begin{aligned}(f+g)(x) &= f(x)+g(x) \\ (f \cdot g)(x) &= f(x) \cdot g(x)\end{aligned}$$

The operations $+$ and $\cdot$ on the right hand sides are those of $\boldsymbol{R}$.

For example, if $f(x) = x^2$ and $g(x) = \sin x$ then

$(f+g)(x) = x^2 + \sin x$ and $(f \cdot g)(x) = x^2 \sin x$.

The set of functions is a ring under these operations. Zero and unity functions $z, e : \boldsymbol{R} \to \boldsymbol{R}$ are defined by

$$z(x) = 0 \text{ for all } x \in \boldsymbol{R}$$

$$e(x) = 1 \text{ for all } x \in \boldsymbol{R}.$$

(f) The set of even integers $2\boldsymbol{Z}$ is a ring under addition and multiplication. This ring has no unity.

More generally, for each natural number n the set $n\boldsymbol{Z}$ of integer multiples of n is a ring.

For example the set $3\boldsymbol{Z} = \{\ldots, -3, 0, 3, 6, 9, \ldots\}$ is a ring.

(g) The set $\boldsymbol{Z}[\sqrt{3}\,] = \{m + n\sqrt{3} \ : m, n \in \boldsymbol{Z}\}$ is a ring under the following binary operations:

$$(m + n\sqrt{3}\,) + (m' + n'\sqrt{3}\,) = (m + m') + (n + n')\sqrt{3}$$

$$(m + n\sqrt{3}\,) \cdot (m' + n'\sqrt{3}\,) = (mm' + 3nn') + (mn' + nm')\sqrt{3}\,.$$

Similar rings may be defined with other square-free integers in place of 3.

(h) The ring $\boldsymbol{Z}[i] = \{m+ni : m,n \in \boldsymbol{Z}\}$ where $i^2 = -1$, is known as the *Gaussian integers.*

This is the analogue of the ring of integers in the field of complex numbers.

We now present some basic facts about rings. It is useful to establish these properties for rings in general because we then know that these hold for any particular example of a ring that we wish to work with.

14.2 Proposition A ring *R* has exactly one zero and at most one unity.

For each $a \in R$ there is exactly one element $-a \in R$ which is the additive inverse of a.

Proof: Suppose that there are a pair of elements 0_1 and 0_2 in *R*, which are both zeroes.

Then $0_1 + 0_2 = 0_1$ because 0_2 is a zero, and $0_1 + 0_2 = 0_2$ because 0_1 is a zero.

Hence $0_1 = 0_2$ and so there is only a single zero.

The result for the unity is similar and is left as an exercise.

Now suppose that for some $a \in R$ there are a pair of elements $-a$ and $\sim a$, each of which is an additive inverse of a.

Then $-a = -a + 0 = -a + (a + (\sim a)) = (-a + a) + \sim a = 0 + \sim a = \sim a$

Hence $-a = \sim a$ and so a has a unique additive inverse.

14.3 Proposition For a ring R and elements $a, b, c \in R$ we have

(i) if $a+b=a+c$ then $b=c$

(ii) if $b+a=c+a$ then $b=c$.

These are called the *additive cancellation laws.*

Proof: We prove (i). By commutativity of addition (ii) follows.

Suppose that $a+b=a+c$. The element a has an additive inverse, and so

$$-a+(a+b)=-a+(a+c)$$

By associativity $(-a+a)+b=(-a+a)+c$ and so $0+b=0+c$ and hence $b=c$.

Not every ring has multiplicative cancellation laws. For example in the ring $\boldsymbol{Z}_4$ we have

$$2\times 1=2\times 3.$$

but we cannot cancel the 2's, since $1 \neq 3$.

To carry out multiplicative cancellation a commutative ring needs to satisfy an extra axiom, the *zero divisor law.* This type of ring, known as an *integral domain*, is discussed in chapter 17.

Next we state and prove from the ring axioms of 14.1 some important properties of the zero and additive inverses of a ring:

14.4 Theorem For any ring R and any elements $a, b \in R$ we have:

(i) $0\cdot a=0=a\cdot 0$

(ii) $-(-a)=a$

(iii) $(-a)\cdot b=-(a\cdot b)=a\cdot(-b)$

(iv) $(-a)\cdot(-b)=a\cdot b$

Proof: (i) Since $0+0=0$ we have $(0+0)\cdot a=0\cdot a$.

By the distributive law we have $0\cdot a+0\cdot a=0\cdot a$.

Hence $0\cdot a+0\cdot a=0\cdot a+0$, and by additive cancellation from 14.3 $0\cdot a=0$ as required.

The right-hand equality is proved similarly.

(ii) a is the additive inverse of $-a$ and $-(-a)$ is the additive inverse of $-a$.

By uniqueness of additive inverses from 14.2 we have $-(-a)=a$.

(iii) $-(a\cdot b)+a\cdot b=0$, but also

$$(-a)\cdot b+a\cdot b=(-a+a)\cdot b \text{ by the distributive laws}$$
$$=0\cdot b=0 \text{ by (i)}$$

Again using uniqueness of additive inverses from 14.2 we have $(-a)\cdot b=-(a\cdot b)$. The right-hand equality is proved similarly.

(iv)

$$(-a)\cdot(-b)=-(a\cdot(-b)) \text{ by (iii)}$$
$$=-(-(a\cdot b)) \text{ by (iii) again}$$
$$=a\cdot b \text{ by (ii).}$$

To simplify notation we write $a\cdot b$ as ab and $a+(-b)$ as $a-b$.

We have distributive laws for subtraction:

$$(a-b)c = (a+(-b))c = ac+(-b)c = ac+(-bc)=ac-bc$$

$$a(b-c) = a(b+(-c)) = ab+a(-c) = ab+(-ac)=ab-ac.$$

14.5 Definition For a pair of rings R_1 and R_2 we may define a ring structure on the Cartesian product

$$R_1 \times R_2 = \{(x, y) : x \in R_1 \text{ and } y \in R_2\}.$$

The binary operations are defined as follows:

$$\begin{aligned}(x_1, y_1) + (x_2, y_2) &= (x_1 + x_2, y_1 + y_2) \\ (x_1, y_1) \cdot (x_2, y_2) &= (x_1 x_2, y_1 y_2)\end{aligned}$$

for any $x_1, x_2 \in R_1$ and $y_1, y_2 \in R_2$.

The new ring defined in this way is called the *direct sum* of R_1 and R_2 and is denoted by $R_1 \oplus R_2$. It is left as an exercise to verify that $R_1 \oplus R_2$ satisfies the ring axioms.

Examples (a) If R_1 and R_2 are both the ring of integers, then $\boldsymbol{Z} \oplus \boldsymbol{Z}$ is the ring of ordered pairs of integers, in which for example

$$(3,4) + (2,5) = (5,9)$$

$$(3,4) \cdot (2,5) = (6,20).$$

(b) $\boldsymbol{Z}_2 \oplus \boldsymbol{Z}_3 = \{(0,0), (0,1), (0,2), (1,0), (1,1), (1,2)\}$

14.6 Definition Let R be a ring with unity. For each natural number n let

$$n \cdot 1 = \underbrace{1 + 1 + \ldots + 1}_{n}$$

The smallest n such that $n \cdot 1 = 0$ is called the *characteristic* of R, denoted char R.

If $n \cdot 1 \neq 0$ for all n then we say that R has characteristic 0.

Examples char $\boldsymbol{Z}_3 = 3$, and more generally char $\boldsymbol{Z}_n = n$.

char $\boldsymbol{Z}_2 \oplus \boldsymbol{Z}_3 = 6$, and more generally char $\boldsymbol{Z}_m \oplus \boldsymbol{Z}_n = \text{lcm}(m, n)$.

char $\boldsymbol{Z} = 0$.

Exercises 14

1. Convince yourself that each of the examples (a) through (h) on pages 193-195 satisfy the ring axioms of 14.1. You may assume that these axioms hold for the rings $\boldsymbol{Z}$ and $\boldsymbol{R}$.

2. Prove that any ring has at most one unity.

   Hint: Suppose that there are two.

3. Let R be a ring.

   Prove the right hand equalities of 14.4 (i) and (iii):

   $$a \cdot 0 = 0 \qquad\qquad a \cdot (-b) = -(a \cdot b)$$

   for all $a, b \in R$.

4. Prove that in a ring R we have:

   $$\text{if } a + c = b + c \text{ then } a = b.$$

   for $a, b, c \in R$. Hence show that the equation $a + x = b$ has a unique solution given by $x = -a + b$.

5. Suppose that a ring R satisfies $a \cdot a = a$ for all $a \in R$.

   Prove that:

   (i) $a + a = 0$ for all $a \in R$

   (ii) R is a commutative ring.

   Hint: Expand $(a+a)^2$ and $(a+b)^2$. Such a ring is known as a *Boolean ring*.

6. Prove that in a ring R we have

   $$(a+b)+(c+d) = (a+c)+(b+d)$$

   for $a, b, c, d \in R$. State which axiom you are using at each step.

7. For rings R_1 and R_2 show that the direct sum $R_1 \oplus R_2$ satisfies the ring axioms.

8. List the elements of the ring $\boldsymbol{Z}_3 \oplus \boldsymbol{Z}_4$.

   What is the characteristic of this ring?

Chapter 15: Subrings, Ideals and Ring Homomorphisms

We have already met the idea of a *subset* included within another set, and a *subgroup* included within another group. It should therefore come as no surprise that along with rings comes the concept of a *subring*. We begin with the definition:

15.1 Definition Let $(R,+,\cdot)$ be a ring and suppose that S is a subset of R.

S is a *subring* of the ring R if $(S,+,\cdot)$ is itself a ring.

Notice that a subring has the *same* binary operations as the ring that it lies inside.

Examples (a) $\boldsymbol{Z}$ is a subring of $\boldsymbol{Q}$, and $\boldsymbol{Q}$ is a subring of $\boldsymbol{R}$.

(b) $Mat(2,\boldsymbol{Z})$ is a subring of $Mat(2,\boldsymbol{Q})$, and $Mat(2,\boldsymbol{Q})$ is a subring of $Mat(2,\boldsymbol{R})$.

(c) $\boldsymbol{Z}[x]$ is a subring of $\boldsymbol{Q}[x]$, and $\boldsymbol{Q}[x]$ is a subring of $\boldsymbol{R}[x]$.

(d) For each natural number n the ring $n\boldsymbol{Z}$ of integer multiples of n is a subring of $\boldsymbol{Z}$.

(e) Each ring R is a subring of $R[x]$, by identifying each element $a \in R$ with a degree 0 polynomial in $R[x]$.

(f) The set of matrices $\left\{ \begin{pmatrix} a & b & 0 \\ c & d & 0 \\ 0 & 0 & e \end{pmatrix} : a,b,c,d,e \in \boldsymbol{Z} \right\}$ is a subring of $Mat(3,\boldsymbol{Z})$.

(g) The set of odd integers, $S = \{\ldots,-3,-1,1,3,5,\ldots\}$ is *not* a subring of $\boldsymbol{Z}$ because for example $1+3=4 \notin S$.

Notice that for any ring R, the trivial ring $\{0\}$ and R itself are subrings of R.

To check whether or not a subset of a ring is a subring we apply the following test:

15.2 Test for a subring Let $(R,+,\cdot)$ be a ring and let S be a non-empty subset of R. Then S is a subring of R if and only if:

(i) if $a,b \in S$ then $a+b \in S$, $ab \in S$

(ii) if $a \in S$ then $-a \in S$

Notice that the test implies that any subring must contain the zero element:

Suppose $a \in S$. By (ii) we have $-a \in S$, and by (i) $0 = a + (-a) \in S$.

Examples (a) $\boldsymbol{Z}$ is a subset of $\boldsymbol{Q}$.

For $m,n \in \boldsymbol{Z}$ we have $m+n \in \boldsymbol{Z}$, $mn \in \boldsymbol{Z}$ and $-m \in \boldsymbol{Z}$.

Hence $\boldsymbol{Z}$ is a subring of $\boldsymbol{Q}$, by 15.2.

(b) Let $S = \left\{ \begin{pmatrix} a & 0 \\ 0 & b \end{pmatrix} : a,b \in \boldsymbol{Z} \right\}$

We have $\begin{pmatrix} a & 0 \\ 0 & b \end{pmatrix} + \begin{pmatrix} a' & 0 \\ 0 & b' \end{pmatrix} = \begin{pmatrix} a+a' & 0 \\ 0 & b+b' \end{pmatrix} \in S$ and

$$\begin{pmatrix} a & 0 \\ 0 & b \end{pmatrix} \begin{pmatrix} a' & 0 \\ 0 & b' \end{pmatrix} = \begin{pmatrix} aa' & 0 \\ 0 & bb' \end{pmatrix} \in S$$

Finally, $-\begin{pmatrix} a & 0 \\ 0 & b \end{pmatrix} = \begin{pmatrix} -a & 0 \\ 0 & -b \end{pmatrix} \in S$

Hence S is a subring of $Mat(2,\boldsymbol{Z})$ by 15.2.

We now introduce a special type of subring known as an *ideal*, which is important in the chapters that follow. Ideals play a role in ring theory analogous to the role of normal subgroups in group theory.

15.3 Definition A non-empty subset I of a ring R is called an *ideal* of R if it satisfies:

(i) if $a, b \in I$ then $a + b \in I$

(ii) if $a \in I$ then $-a \in I$

(iii) if $a \in I$ and $r \in R$ then $ar \in I$, $ra \in I$.

Notice that in any ring R the trivial ring $\{0\}$ and R itself are ideals. We call an ideal that is not equal to R a *proper ideal.* Every ideal is a subring, but not every subring is an ideal.

Examples (a) For any natural number n the subset $n\boldsymbol{Z}$ is an ideal in $\boldsymbol{Z}$.

Suppose $a, b \in n\boldsymbol{Z}$ so that $a = \lambda n, b = \mu n$ for some $\lambda, \mu \in \boldsymbol{Z}$. Then

(i) $a + b = \lambda n + \mu n = (\lambda + \mu)n \in n\boldsymbol{Z}$

(ii) $-a = -(\lambda n) = (-\lambda)n \in n\boldsymbol{Z}$

(iii) $ra = r(\lambda n) = (r\lambda)n \in n\boldsymbol{Z}$ for all $r \in \boldsymbol{Z}$

(b) Let R be a commutative ring with unity. The ideal *generated* by $a \in R$ is

$$\langle a \rangle = \{ra : r \in R\}$$

Suppose $x, y \in \langle a \rangle$ so that $x = r_1 a, y = r_2 a$ for some $r_1, r_2 \in R$. Then

(i) $x + y = r_1 a + r_2 a = (r_1 + r_2)a \in \langle a \rangle$

(ii) $-x = -(r_1 a) = (-r_1)a \in \langle a \rangle$

(iii) $rx = r(r_1 a) = (rr_1)a \in \langle a \rangle$ for all $r \in R$

Such an ideal is called a *principal ideal.*

(c) Let $E \subset \boldsymbol{Z}[x]$ be the set of all polynomials with even constant. E is an ideal in $\boldsymbol{Z}[x]$.

(d) For each natural number m, $Mat(n, m\mathbf{Z})$ is an ideal in $Mat(n, \mathbf{Z})$.

(e) $\mathbf{Z}$ is *not* an ideal in $\mathbf{Q}$, despite being a subring, because condition (iii) fails:

$$1 \in \mathbf{Z},\ \tfrac{1}{2} \in \mathbf{Q},\ \text{but } 1 \times \tfrac{1}{2} = \tfrac{1}{2} \notin \mathbf{Z}.$$

15.4 Proposition Suppose that I and J are ideals in a ring R. The following are also ideals in R:

(i) $I \cap J$

(ii) $I + J = \{a + b : a \in I \text{ and } b \in J\}$

Proof: (i) Suppose that $x, y \in I \cap J$.

Then $x, y \in I$ and hence $x + y \in I, -x \in I$ and $rx, xr \in I$ since I is an ideal.

Similarly $x, y \in J$ and hence $x + y \in J, -x \in J$ and $rx, xr \in J$ since J is an ideal.

It follows that $x + y,\ -x,\ rx,\ xr \in I \cap J$, so that $I \cap J$ is an ideal.

The proof of (ii) is left as an exercise.

A pair of natural numbers a and b are said to be *relatively prime* or *coprime* if $\gcd(a, b) = 1$.

Examples (a) 8 and 25 are relatively prime, since $\gcd(8, 25) = 1$.

(b) 8 and 26 are *not* relatively prime, since $\gcd(8, 26) = 2$.

The following fact arises from Euclid's algorithm:

15.5 Proposition Suppose that a and b are natural numbers.

If $\gcd(a, b) = t$ then there are integers λ and μ such that $t = \lambda a + \mu b$.

For integers a and b we write $a \mid b$ if a divides b exactly without remainder.

Here is an important property of prime numbers:

15.6 Proposition Suppose that p is a prime number, and that a and b are natural numbers.

If $p \mid ab$ then $p \mid a$ or $p \mid b$.

Proof: Suppose that p divides ab but does *not* divide a. Then $\gcd(a,p) = 1$.

By the Euclidean algorithm there are integers λ and μ such that

$$1 = \lambda a + \mu p.$$

Then $b = \lambda ab + \mu pb$ and since $p \mid ab$ we have $p \mid b$.

By a similar argument, if p does *not* divide b then we have $p \mid a$. Hence the result.

By induction, a corollary of 15.6 is that if $p \mid a_1 a_2 \ldots a_n$ then $p \mid a_i$ for some i.

15.7 Proposition For natural numbers a and b let $s = \operatorname{lcm}(a,b)$ and $t = \gcd(a,b)$. Then

(i) $a\boldsymbol{Z} \cap b\boldsymbol{Z} = s\boldsymbol{Z}$

(ii) $a\boldsymbol{Z} + b\boldsymbol{Z} = t\boldsymbol{Z}$

Proof: (i) Since $a \mid s$ and $b \mid s$ we have $s\boldsymbol{Z} \subseteq a\boldsymbol{Z}$ and $s\boldsymbol{Z} \subseteq b\boldsymbol{Z}$. Hence $s\boldsymbol{Z} \subseteq a\boldsymbol{Z} \cap b\boldsymbol{Z}$.

Now suppose $x \in a\boldsymbol{Z} \cap b\boldsymbol{Z}$.

Then x is a common multiple of a and b, so that $x \in s\boldsymbol{Z}$.

(ii) Since $t \mid a$ and $t \mid b$ we have $a\boldsymbol{Z} \subseteq t\boldsymbol{Z}$ and $b\boldsymbol{Z} \subseteq t\boldsymbol{Z}$.

Hence $a\boldsymbol{Z} + b\boldsymbol{Z} \subseteq t\boldsymbol{Z}$.

By 15.5 there are integers λ and μ such that $t = \lambda a + \mu b$.

Hence $t\boldsymbol{Z} \subseteq a\boldsymbol{Z} + b\boldsymbol{Z}$.

Example $4\boldsymbol{Z} \cap 6\boldsymbol{Z} = 12\boldsymbol{Z}$ $\quad$ $4\boldsymbol{Z} + 6\boldsymbol{Z} = 2\boldsymbol{Z}$

15.8 Proposition Suppose I is an ideal in a ring R and $a, b \in R$. We may define a relation on R by

$$a \sim b \text{ iff } a - b \in I.$$

Then $\sim$ is an equivalence relation.

Proof: *Reflexive:* $a - a = 0 \in I$ and so $a \sim a$ for all $a \in R$.

Symmetric: If $a \sim b$ then $a - b \in I$.

Then $b - a = -(a - b) \in I$ so that $b \sim a$.

Transitive: If $a \sim b$ and $b \sim c$ then $a - b \in I$, $b - c \in I$.

Then $a - c = (a - b) + (b - c) \in I$ and so $a \sim c$.

The equivalence classes are known as *cosets*. The set of cosets is denoted R/I. In the next chapter we show how R/I becomes a ring by defining a suitable pair of binary operations.

As in the case of groups, a *homomorpism* is a mapping which preserves structure. But whereas with a group there is only a single binary operation to preserve, for rings we need to preserve two binary operations. This leads us to the following definition:

15.9 Definition Suppose that R and S are rings.

A mapping $f: R \to S$ is called a *ring homomorphism* if

(i) $f(a + b) = f(a) + f(b)$

(ii) $f(a \cdot b) = f(a) \cdot f(b)$

for all $a, b \in R$.

Monomorphisms, epimorphisms and isomorphisms are defined exactly as for groups:

If f is injective then it is called a *monomorphism.*

If f is surjective then it is called an *epimorphism.*

If f is bijective (both injective and surjective) then it is called an *isomorphism.*

If there is an isomorphism $f: R \to S$ then we say that the rings R and S are *isomorphic* and we write $R \cong S$.

Where a pair of rings are isomorphic, they may be thought of as essentially *the same ring*.

Examples (a) The mapping $f: \boldsymbol{C} \to Mat(2, \boldsymbol{R})$ given by

$$f(x+yi) = \begin{pmatrix} x & y \\ -y & x \end{pmatrix}$$

is a ring monomorphism. To show this, let

$$z_1 = x_1 + y_1 i \text{ and } z_2 = x_2 + y_2 i$$

Then we have

$$f(z_1 + z_2) = f((x_1 + x_2) + (y_1 + y_2)i) = \begin{pmatrix} x_1 + x_2 & y_1 + y_2 \\ -y_1 - y_2 & x_1 + x_2 \end{pmatrix}$$

$$= \begin{pmatrix} x_1 & y_1 \\ -y_1 & x_1 \end{pmatrix} + \begin{pmatrix} x_2 & y_2 \\ -y_2 & x_2 \end{pmatrix} = f(z_1) + f(z_2)$$

$$f(z_1 z_2) = f((x_1 + y_1 i)(x_2 + y_2 i))$$

$$= f((x_1 x_2 - y_1 y_2) + (x_1 y_2 + x_2 y_1)i)$$

$$= \begin{pmatrix} x_1 x_2 - y_1 y_2 & x_1 y_2 + x_2 y_1 \\ -x_1 y_2 - x_2 y_1 & x_1 x_2 - y_1 y_2 \end{pmatrix} = \begin{pmatrix} x_1 & y_1 \\ -y_1 & x_1 \end{pmatrix} \begin{pmatrix} x_2 & y_2 \\ -y_2 & x_2 \end{pmatrix}$$

$$= f(z_1) f(z_2).$$

Hence f is a ring homomorphism.

Also if $f(z_1) = f(z_2)$ so that $\begin{pmatrix} x_1 & y_1 \\ -y_1 & x_1 \end{pmatrix} = \begin{pmatrix} x_2 & y_2 \\ -y_2 & x_2 \end{pmatrix}$ then $x_1 = x_2$ and $y_1 = y_2$ so that $z_1 = z_2$.

Hence f is injective and so is a ring monomorphism. Notice that f is *not* a ring epimorphism, for example there is no $z \in \boldsymbol{C}$ such that

$$f(z) = \begin{pmatrix} 1 & 0 \\ 0 & 0 \end{pmatrix}.$$

Notice that this last example allows us to define the complex numbers without making use of the square root of minus one - we think of complex numbers as real matrices of the form

$$\begin{pmatrix} x & y \\ -y & x \end{pmatrix}.$$

(b) We may define a pair of binary operations on $\boldsymbol{R}^2$ by:

$$\begin{pmatrix} x_1 \\ y_1 \end{pmatrix} + \begin{pmatrix} x_2 \\ y_2 \end{pmatrix} = \begin{pmatrix} x_1 + x_2 \\ y_1 + y_2 \end{pmatrix}$$

$$\begin{pmatrix} x_1 \\ y_1 \end{pmatrix} \cdot \begin{pmatrix} x_2 \\ y_2 \end{pmatrix} = \begin{pmatrix} x_1 x_2 - y_1 y_2 \\ x_1 y_2 + x_2 y_1 \end{pmatrix}$$

Then it may be shown that $(\boldsymbol{R}^2, +, \cdot)$ is a ring.

The mapping $f : \boldsymbol{R}^2 \to \boldsymbol{C}$ given by $f\begin{pmatrix} x \\ y \end{pmatrix} = x + yi$ is a ring isomorphism.

For $\begin{pmatrix} x_1 \\ y_1 \end{pmatrix}, \begin{pmatrix} x_2 \\ y_2 \end{pmatrix} \in \boldsymbol{R}^2$ we have:

$$f\left(\begin{pmatrix} x_1 \\ y_1 \end{pmatrix} + \begin{pmatrix} x_2 \\ y_2 \end{pmatrix}\right) = f\begin{pmatrix} x_1 + x_2 \\ y_1 + y_2 \end{pmatrix} = (x_1 + x_2) + (y_1 + y_2)i$$

$$= (x_1 + y_1 i) + (x_2 + y_2 i) = f\begin{pmatrix} x_1 \\ y_1 \end{pmatrix} + f\begin{pmatrix} x_2 \\ y_2 \end{pmatrix}$$

and

$$f\left(\begin{pmatrix} x_1 \\ y_1 \end{pmatrix} \cdot \begin{pmatrix} x_2 \\ y_2 \end{pmatrix}\right) = f\begin{pmatrix} x_1x_2 - y_1y_2 \\ x_1y_2 + x_2y_1 \end{pmatrix}$$

$$= (x_1x_2 - y_1y_2) + (x_1y_2 + x_2y_1)i$$

$$= (x_1 + y_1 i)(x_2 + y_2 i) = f\begin{pmatrix} x_1 \\ y_1 \end{pmatrix} \cdot f\begin{pmatrix} x_2 \\ y_2 \end{pmatrix}.$$

Hence *f* is a ring homomorphism.

Also if $f\begin{pmatrix} x_1 \\ y_1 \end{pmatrix} = f\begin{pmatrix} x_2 \\ y_2 \end{pmatrix}$ then $x_1 + y_1 i = x_2 + y_2 i$ so that $x_1 = x_2$ and $y_1 = y_2$.

Hence $\begin{pmatrix} x_1 \\ y_1 \end{pmatrix} = \begin{pmatrix} x_2 \\ y_2 \end{pmatrix}$ and so *f* is injective.

Finally, for any $x + yi \in \boldsymbol{C}$ we have $f\begin{pmatrix} x \\ y \end{pmatrix} = x + yi$, so *f* is surjective.

Hence *f* is a ring isomorphism. In this way, we may think of the complex numbers as the vector space $\boldsymbol{R}^2$ equipped with a suitable multiplication.

This is the second way of defining the complex numbers without making use of the square root of minus one. We shall meet yet a third when we study quotient rings in chapter 16.

(c) Let

$$S = \left\{ \begin{pmatrix} a & b & 0 \\ c & d & 0 \\ 0 & 0 & e \end{pmatrix} : a, b, c, d, e \in \boldsymbol{Z} \right\} \subseteq Mat(3, \boldsymbol{Z})$$

as in 15.1 example (f). Then there is a ring isomorphism $f: S \to Mat(2, \boldsymbol{Z}) \oplus \boldsymbol{Z}$ given by

$$f\begin{pmatrix} a & b & 0 \\ c & d & 0 \\ 0 & 0 & e \end{pmatrix} = \left(\begin{pmatrix} a, b \\ c, d \end{pmatrix}, e\right).$$

It is left as an exercise to verify that *f* is a ring isomorphism.

(d) The innocent looking mapping $f: \mathbf{Z} \to \mathbf{Z}$ given by $f(a) = 2a$ is not a ring homomorphism. For example

$$\begin{aligned} f(2 \times 2) &= f(4) = 8 \\ \text{whereas} \quad f(2) \times f(2) &= 4 \times 4 = 16. \end{aligned}$$

15.10 Evaluation Homomorphisms

Suppose that S is a subring of R, let $\lambda \in R$, and consider the ring of polynomials $S[x]$.

The *evaluation homomorphism* $\phi_\lambda : S[x] \longrightarrow R$ is defined by replacing the indeterminate x by λ and evaluating in the ring R.

Hence if $p(x) = a_n x^n + ... + a_2 x^2 + a_1 x + a_0$ then

$$\phi_\lambda(p(x)) = a_n \lambda^n + ... + a_2 \lambda^2 + a_1 \lambda + a_0.$$

Examples (a) For $\phi_2 : \mathbf{Z}[x] \longrightarrow \mathbf{Z}$ we have

$$\phi_2(x^2 + 5x + 6) = 2^2 + 5 \times 2 + 6 = 20.$$

(b) For $\phi_i : \mathbf{Z}[x] \longrightarrow \mathbf{Z}[i]$ we have

$$\phi_i(x^2 + 5x + 6) = i^2 + 5i + 6 = 5 + 5i.$$

15.11 Proposition Let $\phi_\lambda : R[x] \longrightarrow R$ be an evaluation homomorphism.

$\phi_\lambda(p(x)) = 0$ iff $p(x) = (x - \lambda)q(x)$ for some $q(x) \in R[x]$.

Proof: "$\Leftarrow$" should be clear.

"$\Rightarrow$" Divide $p(x)$ by $x - \lambda$ to give $p(x) = (x - \lambda)q(x) + r$ for some remainder $r \in R$. If $\phi_\lambda(p(x)) = 0$ then $r = 0$. Hence the result.

We denote the zero elements of rings R and S by 0_R and 0_S respectively. We have the following useful result:

15.12 Proposition If $f: R \to S$ is a ring homomorphism then

(i) $f(0_R) = 0_S$

(ii) $f(-a) = -f(a)$ for all $a \in R$

Hence a homomorphism of rings always carries the zero of one ring to the zero of the other, and also preserves additive inverses.

Proof: (i) $0_R + 0_R = 0_R$ and so $f(0_R + 0_R) = f(0_R)$.

Because f is a ring homomorphism $f(0_R) + f(0_R) = f(0_R)$.

Hence $f(0_R) + f(0_R) = f(0_R) + 0_S$, and so by cancellation $f(0_R) = 0_S$.

(ii) For any $a \in R$ we have $a + (-a) = 0_R$

and so $f(a + (-a)) = f(0_R) = 0_S$ by (i) above.

Because f is a ring homomorphism $f(a) + f(-a) = 0_S$.

Similarly, $f(-a) + f(a) = 0_S$

We have shown that $f(a)$ and $f(-a)$ are additive inverses in S.

Hence by the uniqueness of additive inverses from 14.2 we have $f(-a) = -f(a)$ as required.

Notice that it follows that

$$f(a-b) = f(a+(-b)) = f(a) + f(-b) = f(a) - f(b)$$

15.13 Definition Suppose $f: R \to S$ is a homomorphism of rings.

The *image* of f is $\text{im}f = \{f(a) : a \in R\}$.

The *kernel* of f is $\ker f = \{a \in R : f(a) = 0\}$.

These are similar to the definitions of image and kernel of a group homomorphism from 2.14.

15.14 Proposition For a ring homomorphism $f: R \to S$ we have:

(i) $\mathrm{im} f$ is a subring of S

(ii) f is an epimorphism iff $\mathrm{im} f = S$

(iii) $\ker f$ is a subring of R

(iv) f is a monomorphism iff $\ker f = \{0\}$.

Proof: (i) If $b_1, b_2 \in \mathrm{im} f$ then there are $a_1, a_2 \in R$ such that $f(a_1) = b_1$ and $f(a_2) = b_2$.

Then $f(a_1 + a_2) = f(a_1) + f(a_2) = b_1 + b_2$ so that $b_1 + b_2 \in \mathrm{im} f$.

Similarly $f(a_1 a_2) = f(a_1) f(a_2) = b_1 b_2$ so that $b_1 b_2 \in \mathrm{im} f$.

Also by 15.12(ii) we have $f(-a_1) = -f(a_1) = -b_1$ so that $-b_1 \in \mathrm{im} f$.

Hence by 15.2 we conclude that $\mathrm{im} f$ is a subring of S.

(ii) f is an epimorphism iff $\mathrm{im} f = S$ follows immediately from the definitions.

(iii) If $a_1, a_2 \in \ker f$ then $f(a_1) = 0, f(a_2) = 0$.

Then $f(a_1 + a_2) = f(a_1) + f(a_2) = 0 + 0 = 0$ so that $a_1 + a_2 \in \ker f$

and $f(a_1 a_2) = f(a_1) f(a_2) = 0 \cdot 0 = 0$ so that $a_1 a_2 \in \ker f$.

Also by 15.12(ii) we have $f(-a_1) = -f(a_1) = 0$ so that $-a_1 \in \ker f$

Hence by 15.2 we conclude that $\ker f$ is a subring of R.

(iv) f is a monomorphism iff $\ker f = \{0\}$ is left as an exercise.

Examples (a) $f: \boldsymbol{Z} \to \boldsymbol{Z}_n$ given by $f(a) = [a]$, where $[a]$ is the congruence class modulo n.

$\mathrm{im} f = \boldsymbol{Z}_n$, $\ker f = n\boldsymbol{Z}$ and so f is a ring epimorphism but not a monomorphism.

(b) $f: \boldsymbol{C} \to Mat(2, \boldsymbol{R}), \quad f(x + yi) = \begin{pmatrix} x & y \\ -y & x \end{pmatrix}$

$$\operatorname{im} f = \left\{ \begin{pmatrix} a & b \\ c & d \end{pmatrix} \in Mat(2, \boldsymbol{R}) : a = d, b = -c \right\}$$

Since the image is a proper subring of $Mat(2, \boldsymbol{R})$ the mapping is not an epimorphism.

$\ker f = \{0\}$ and so f is a ring monomorphism.

Exercises 15

1. Convince yourself that each of the examples (a) through (f) of 15.1 is a subring by using 15.2

   Is $\boldsymbol{Z}_3$ a subring of $\boldsymbol{Z}_4$?

2. (i) Let S be the following subset of $Mat(2,\boldsymbol{R})$:

$$S=\left\{\begin{pmatrix} x & y \\ -y & x \end{pmatrix} : x, y \in \boldsymbol{R}\right\}$$

   Using 15.2 show that S is a subring of $Mat(2,\boldsymbol{R})$.

   (ii) Let Q be the following subset of $Mat(2,\boldsymbol{C})$:

$$Q=\left\{\begin{pmatrix} w & z \\ -\overline{z} & \overline{w} \end{pmatrix} : w, z \in \boldsymbol{C}\right\}$$

   Using 15.2 show that Q is a subring of $Mat(2,\boldsymbol{C})$.

   Q is isomorphic to the ring of quaternions, explored in chapter 24.

3. State, with reasons, whether or not each of the following subsets of $Mat(2,\boldsymbol{Z})$ is a subring:

   (i) $\left\{\begin{pmatrix} a & b \\ 0 & c \end{pmatrix} : a, b, c \in \boldsymbol{Z}\right\}$

   (ii) $\{A \in Mat(2,\boldsymbol{Z}) : \det A = 1\}$

   (iii) For a fixed $B \in Mat(2,\boldsymbol{Z})$ the subset

   $\{A \in Mat(2,\boldsymbol{Z}) : AB = BA\}$.

4. State, with reasons, whether or not each of the following subsets of $\boldsymbol{R}[x]$ is an ideal:

(i) $\boldsymbol{R}$

(ii) $\{a_n x^n + \ldots + a_2 x^2 + a_1 x + a_0 : a_0 + a_1 + \ldots + a_n = 0\}$

(iii) $\boldsymbol{Q}[x]$

(iv) $\{a_n x^n + \ldots + a_2 x^2 + a_1 x + a_0 : a_0 = 0\}$

(v) $\langle x-2 \rangle = \{(x-2)p(x) : p(x) \in \boldsymbol{R}[x]\}$

(vi) $\boldsymbol{R}[x]$.

5. Find all of the ideals in the ring $\boldsymbol{Z}_{20}$.

6. For an ideal I in a ring R with unity show that if $1 \in I$ then

$$I = R.$$

7. Suppose that I and J are ideals in a ring R.

Show that $I + J$ is also an ideal in R.

8. State, with reasons, whether or not each of the following is a ring homomorphism:

(a) $f: \boldsymbol{Z} \to \boldsymbol{Z}$ $\quad f(a) = -a$

(b) $f: \boldsymbol{Z} \oplus \boldsymbol{Z} \to \boldsymbol{Z} \oplus \boldsymbol{Z}$ $\quad f(a,b) = (a,0)$

(c) $f: \boldsymbol{Z} \oplus \boldsymbol{Z} \to \boldsymbol{Z} \oplus \boldsymbol{Z}$ $\quad f(a,b) = (a,1)$

(d) $f: \boldsymbol{Z} \oplus \boldsymbol{Z} \to Mat(2,\boldsymbol{Z})$ $\quad f(a,b) = \begin{pmatrix} a & 0 \\ 0 & b \end{pmatrix}$

(e) $f: \boldsymbol{Z}[i] \to \boldsymbol{Z}[i]$ $\quad f(a+bi) = a - bi$

(f) $f: \boldsymbol{Z}[i] \to \boldsymbol{Z}[i]$ $\quad f(a+bi) = b + ai$

9. For a ring homomorphism $f: R \to S$ show that

f is a monomorphism iff $\ker f = \{0\}$

10. Let S be the ring of 15.1 example (f).

Show that the mapping $f: S \to Mat(2, \boldsymbol{Z}) \oplus \boldsymbol{Z}$ given by

$$f\begin{pmatrix} a & b & 0 \\ c & d & 0 \\ 0 & 0 & e \end{pmatrix} = \left(\begin{pmatrix} a, b \\ c, d \end{pmatrix}, e \right)$$

is a ring isomorphism.

Chapter 16: Quotient Rings

In this chapter we generalise *modular arithmetic*, which was introduced in chapter 1. The construction is applicable to any ring, so that we can say for example what it means to perform arithmetic modulo a polynomial.

Suppose I is an ideal in a ring R and that $a \in R$.

The *coset* of a is $a+I=\{a+x : x \in I\}$.

Examples (a) Consider the ideal $I=\{0,3,6\}$ in the ring $\boldsymbol{Z}_9$.

Cosets include $0+I=\{0,3,6\}, 1+I=\{1,4,7\}, 2+I=\{2,5,8\}$ and $5+I=\{5,8,2\}$.

Notice that the last two of these are the same. Here is the criterion for two cosets being equal:

16.1 Proposition $a+I=b+I$ if and only if $a-b \in I$.

Proof: "⇒" Clearly $a \in a+I$. If $a+I=b+I$ then $a \in b+I$

So $a=b+x$ for some $x \in I$. Hence $a-b=x \in I$.

"⇐" Suppose $a-b \in I$.

If $y \in a+I$ the $y=a+x$ for some $x \in I$.

Hence $y=b+a-b+x \in b+I$.

Similarly if $y \in b+I$ the $y=b+x$ for some $x \in I$.

Hence $y=a-(a-b)+x \in a+I$.

Next, we show that distinct cosets are disjoint:

16.2 Proposition If $a+I \neq b+I$ then $a+I \cap b+I=\varnothing$.

Proof: Suppose $y \in a+I \cap b+I$.

Then $y=a+x$ and $y=b+x'$ for some $x, x' \in I$.

Then $a+x=b+x'$ and so $a-b=x'-x \in I$.

Hence by 16.1 we have $a+I=b+I$.

16.3 Proposition Let I be an ideal in a ring R. The set of cosets R/I is a ring under the following binary operations:

$$(a+I)+(b+I)=a+b+I$$

$$(a+I)\cdot(b+I)=ab+I$$

Proof: We need to show that the binary operations are well-defined.

Suppose that $a+I=a'+I$ so that $a-a' \in I$ and $b+I=b'+I$ so that $b-b' \in I$.

For addition we have $(a+b)-(a'+b')=(a-a')+(b-b') \in I$

so that $a+b+I=a'+b'+I$.

For multiplication we have

$$\begin{aligned} ab-a'b' &= ab-ab'+ab'-a'b' \\ &= a(b-b')+(a-a')b' \in I \end{aligned}$$

so that $ab+I=a'b'+I$.

The verification of the ring axioms is tedious but easy. For example,

$$(a+I)+(b+I)=a+b+I=b+a+I=(b+I)+(a+I).$$

The zero element is $0+I$. The additive inverse of $a+I$ is $-a+I$, and so on.

A ring $(R/I,+,\cdot)$ constructed in this way is called a *quotient ring*.

16.4 Corollary The *quotient map* $q: R \to R/I$ given by

$$q(a)=a+I$$

for each $a \in R$ is a ring epimorphism with kernel I.

Proof: q is a ring homomorphism:

$$q(a+b)=a+b+I=(a+I)+(b+I)=q(a)+q(b)$$

$$q(a\cdot b)\ =ab+I\ =(a+I)\cdot(b+I)=q(a)\cdot q(b)$$

q is surjective because for each $a+I\in R/I$ we have $q(a)=a+I.$

Hence q is an epimorphism.

$$\ker q=\{a\in R:q(a)=0+I\}=\{a\in R:a+I=0+I\}$$
$$=\{a\in R:a-0\in I\}=I.$$

Examples (a) $3\boldsymbol{Z}$ is an ideal in $\boldsymbol{Z}$. The quotient ring $\boldsymbol{Z}/3\boldsymbol{Z}$ is the ring of integers modulo 3, written $\boldsymbol{Z}_3$.

(b) $\langle x+1\rangle=\{(x+1)p(x):p(x)\in\boldsymbol{Z}[x]\}$ is an ideal in $\boldsymbol{Z}[x]$.

Consider the quotient ring $\boldsymbol{Z}(x)/\langle x+1\rangle$. For a pair of polynomials $p(x),\ q(x)\in\boldsymbol{Z}[x]$

$$p(x)+\langle x+1\rangle=q(x)+\langle x+1\rangle \text{ iff } p(x)-q(x)\in\langle x+1\rangle$$

So a pair of polynomials are in the same equivalence class if and only if $p(x)-q(x)$ is a polynomial multiple of $x+1$.

Thus $\boldsymbol{Z}[x]/\langle x+1\rangle$ is the ring of polynomials with integer coefficients working "modulo" the polynomial $x+1$.

(c) $Mat(2,3\boldsymbol{Z})$ is an ideal in $Mat(2,\boldsymbol{Z})$. The quotient ring $Mat(2,\boldsymbol{Z})/Mat(2,3\boldsymbol{Z})$ is isomorphic to $Mat(2,\boldsymbol{Z}_3)$, the ring of 2×2 matrices with integer modulo 3 entries.

The following important theorem allows us to prove isomorphism between quotient rings, arising from the construction above, and other familiar rings.

16.5 First Isomorphism Theorem Suppose that $f: R \to S$ is a ring homomorphism and is surjective. Then $\ker f$ is an ideal in R, and the mapping $\bar{f}: R/\ker f \to S$ given by

$$\bar{f}(a + \ker f) = f(a)$$

is an isomorphism, so that $R/\ker f \cong S$.

Proof: The proof is presented in four sections:

(I) $\ker f$ is an ideal in R:

We have already shown in 15.14 that $\ker f$ is a subring of R. To show that it is an ideal, suppose $a \in \ker f$ and $r \in R$ so that $f(a) = 0$.

Then $f(ar) = f(a)f(r) = 0f(r) = 0$ and $f(ra) = f(r)f(a) = f(r)0 = 0$ and hence $ar,\ ra \in \ker f$.

(II) The mapping $\bar{f}$ is well-defined:

We show that $\bar{f}$ is independent of choice of representatives of cosets.

Suppose that $a + \ker f = a' + \ker f$ so that $a - a' \in \ker f$.

By the definition of the kernel, $f(a - a') = 0$.

Because f is a homomorphism we have $f(a) - f(a') = 0$ and so $f(a) = f(a')$.

Hence $\bar{f}(a + \ker f) = \bar{f}(a' + \ker f)$ as required.

(III) The mapping $\bar{f}$ is a homomorphism of rings:

$$\begin{aligned}\bar{f}((a + \ker f) + (b + \ker f)) &= \bar{f}(a + b + \ker f)\\ &= f(a + b) = f(a) + f(b)\\ &= \bar{f}(a + \ker f) + \bar{f}(b + \ker f)\end{aligned}$$

$$\begin{aligned}\bar{f}((a + \ker f) \cdot (b + \ker f)) &= \bar{f}(a \cdot b + \ker f)\\ &= f(a \cdot b) = f(a) \cdot f(b)\\ &= \bar{f}(a + \ker f) \cdot \bar{f}(b + \ker f)\end{aligned}$$

(IV) The mapping $\bar{f}$ is an isomorphism of rings:

Because f is surjective, for $b \in S$ there is $a \in R$ such that $f(a) = b$.

Then $\bar{f}(a + \ker f) = f(a) = b$. Hence $\bar{f}$ is also surjective.

Suppose $\bar{f}(a + \ker f) = \bar{f}(b + \ker f)$.

Then $f(a) = f(b)$, so that $f(a - b) = 0$.

Hence $a - b \in \ker f$, so that $a + \ker f = b + \ker f$, so that $\bar{f}$ is injective, and so is an isomorphism.

Examples (a) The evaluation homomorphism $\phi_2 : \boldsymbol{Z}[x] \to \boldsymbol{Z}$ is surjective.

We have $\ker \phi_2 = \langle x - 2 \rangle$ and hence $\boldsymbol{Z}[x]/\langle x - 2 \rangle \cong \boldsymbol{Z}$.

(b) The evaluation homomorphism $\phi_i : \boldsymbol{R}[x] \to \boldsymbol{C}$ is surjective.

We have $\ker \phi_i = \langle x^2 + 1 \rangle$ and hence $\boldsymbol{R}[x]/\langle x^2 + 1 \rangle \cong \boldsymbol{C}$.

This gives us our third definition of the complex numbers, this time as a quotient ring of polynomials.

(c) We may define a mapping f: $Mat(2, \boldsymbol{Z}) \to Mat(2, \boldsymbol{Z}_3)$ by

$$f\begin{pmatrix} a & b \\ c & d \end{pmatrix} = \begin{pmatrix} [a] & [b] \\ [c] & [d] \end{pmatrix}.$$

where $[a]$ denotes the congruence class of a modulo 3, and so on.

Then f is surjective and $\ker f = Mat(2, 3\boldsymbol{Z})$. Hence by the isomorphism theorem we have

$$Mat(2, \boldsymbol{Z})/Mat(2, 3\boldsymbol{Z}) \cong Mat(2, \boldsymbol{Z}_3).$$

The next example shows why we work with *ideals* rather than just subrings:

16.6 Attempted Construction of the ring $\boldsymbol{Q}/\boldsymbol{Z}$

Recall that $\boldsymbol{Z}$ is a subring of $\boldsymbol{Q}$. We may define the cosets of $a \in \boldsymbol{Q}$ as

$$a+\boldsymbol{Z}=\{a+n : n \in \boldsymbol{Z}\}.$$

By 16.1 we have $a+\boldsymbol{Z}=b+\boldsymbol{Z}$ iff $a-b \in \boldsymbol{Z}$ for rational numbers $a, b \in \boldsymbol{Q}$.

For example, $\frac{1}{2}+\boldsymbol{Z}=\frac{3}{2}+\boldsymbol{Z}$ and $\frac{1}{3}+\boldsymbol{Z}=\frac{4}{3}+\boldsymbol{Z}$.

We attempt to define binary operations on the set of equivalence classes of rational numbers as before by

$$(a+\boldsymbol{Z})+(b+\boldsymbol{Z})=a+b+\boldsymbol{Z}$$

$$(a+\boldsymbol{Z})\cdot(b+\boldsymbol{Z})=ab+\boldsymbol{Z}$$

So for example $(\frac{1}{2}+\boldsymbol{Z})\cdot(\frac{1}{3}+\boldsymbol{Z})=\frac{1}{6}+\boldsymbol{Z}$.

By choosing alternative representatives for the cosets we also have

$$(\tfrac{3}{2}+\boldsymbol{Z})\cdot(\tfrac{4}{3}+\boldsymbol{Z})=2+\boldsymbol{Z}.$$

Since $\frac{1}{6}+\boldsymbol{Z} \neq 2+\boldsymbol{Z}$ the multiplication operation on $\boldsymbol{Q}/\boldsymbol{Z}$ is not well-defined and our construction fails.

We conclude the chapter with two further isomorphism theorems.

16.7 Second Isomorphism Theorem Let I and J be ideals in a ring R.

We have $I/(I \cap J) \cong (I+J)/J$

Proof: Define $\phi : I \to (I+J)/J$ by $\phi(a)=a+J$.

Clearly ϕ is surjective: for $a+b+J \in (I+J)/J$ we have $a+b+J=a+J=\phi(a)$.

$$\ker\phi=\{a \in I : \phi(a)=0+J\}=\{a \in I : a+J=0+J\}$$
$$=\{a \in I : a \in J\}=I \cap J.$$

The result follows by the first isomorphism theorem, 16.5.

Example

$4\mathbf{Z}$ and $6\mathbf{Z}$ are ideals in $\mathbf{Z}$, with $4\mathbf{Z} \cap 6\mathbf{Z} = 12\mathbf{Z}$ and $4\mathbf{Z} + 6\mathbf{Z} = 2\mathbf{Z}$, by 15.7

Hence by the theorem $4\mathbf{Z}/12\mathbf{Z} \cong 2\mathbf{Z}/6\mathbf{Z}$ and $6\mathbf{Z}/12\mathbf{Z} \cong 2\mathbf{Z}/4\mathbf{Z}$.

16.8 Third Isomorphism Theorem

Let I and J be ideals in a ring R with $I \subseteq J$.

We have

$$\frac{R/I}{J/I} \cong R/J$$

Proof: Define $\phi : R/I \to R/J$ by $\phi(a+I) = a+J$.

ϕ is well-defined: if $a+I = a' + I$ then $a - a' \in I \subseteq J$.

Then $a+J = a' + J$ so that $\phi(a+I) = \phi(a' + I)$

Clearly ϕ is surjective: for $a+J \in R/J$ we have $a+J = \phi(a+I)$.

$$\begin{aligned} \ker\phi &= \{a+I : \phi(a+I) = 0+J\} = \{a+I : a+J = 0+J\} \\ &= \{a+I : a \in J\} = J/I. \end{aligned}$$

The result follows by the first isomorphism theorem, 16.5.

Examples (a) $2\mathbf{Z}$ and $4\mathbf{Z}$ are ideals in $\mathbf{Z}$, with $4\mathbf{Z} \subset 2\mathbf{Z}$.

Hence by the theorem $\dfrac{\mathbf{Z}/4\mathbf{Z}}{2\mathbf{Z}/4\mathbf{Z}} \cong \mathbf{Z}/2\mathbf{Z}$.

(b) $\langle x^2 - 1\rangle$ and $\langle x - 1\rangle$ are ideals in $\mathbf{Z}[x]$ with $\langle x^2 - 1\rangle \subset \langle x - 1\rangle$.

By the theorem,

$$\frac{\mathbf{Z}[x]/\langle x^2 - 1\rangle}{\langle x - 1\rangle/\langle x^2 - 1\rangle} \cong \mathbf{Z}[x]/\langle x - 1\rangle.$$

Exercises 16

1. Verify that in a quotient ring R/I the following hold:

   (i) Associativity of multiplication:

   $$((a+I)\cdot(b+I))\cdot(c+I)=(a+I)\cdot((b+I)\cdot(c+I))$$

   (ii) Distributive law:

   $$((a+I)+(b+I))\cdot(c+I)=(a+I)\cdot(c+I)+(b+I)\cdot(c+I)$$

2. Under what conditions will a quotient ring R/I

   (i) have a unity?

   (ii) be a commutative ring?

3. For the ring $\boldsymbol{Z}[x]$ describe the quotient ring $\boldsymbol{Z}[x]/I$ for each of the following ideals:

(i) $I=\{0\}$

(ii) $I=\{a_nx^n+\ldots+a_2x^2+a_1x+2a_0 : a_0,a_1,\ldots,a_n\in\boldsymbol{Z}\}$
the ideal of polynomials with even constant.

(iii) $I=2\boldsymbol{Z}[x]=\{a_nx^n+\ldots+a_2x^2+a_1x+a_0 : a_i\in 2\boldsymbol{Z} \text{ for all } i\}$
the ideal of polynomials with all coefficients even.

(iv) $I=\langle x\rangle$

(v) $I=\langle x^2+1\rangle$

(vi) $I=\boldsymbol{Z}[x]$.

Chapter 17: Integral Domains and Fields

In this chapter we consider a special type of ring known as an *integral domain,* and return to the study of fields. We show how each of these may arise as a quotient ring.

17.1 Definition A commutative ring with unity is an *integral domain* if for $a \cdot b = 0$ we have $a = 0$ or $b = 0$ (or both).

Examples (a) $\boldsymbol{Z}, \boldsymbol{Q}, \boldsymbol{R}$ and $\boldsymbol{C}$ are integral domains.

(b) $\boldsymbol{Z}[x], \boldsymbol{Q}[x]$ and $\boldsymbol{R}[x]$ are integral domains.

(c) $\boldsymbol{Z}_6[x]$ is not an integral domain since, for example

$$(2x+2)(3x+3) = 6x^2 + 12x + 6 = 0.$$

(d) For a pair of commutative rings R_1 and R_2 the direct sum $R_1 \oplus R_2$ is not an integral domain because

$$(x,0)\cdot(0,y) = (0,0) \text{ for all } x \in R_1, y \in R_2.$$

17.2 Definition A pair of non-zero elements a and b in a ring R are called *divisors of zero* if $a \cdot b = 0$.

Examples (a) In the ring $\boldsymbol{Z}_{16}$ we have $2 \times 8 = 0$ and $4 \times 4 = 0$ and so 2, 4 and 8 are divisors of zero.

(b) In the ring $\boldsymbol{Z}_4[x]$ we have $(2x+2)(2x+2) = 4x^2 + 8x + 4 \equiv 0$ and so the polynomial $2x + 2$ is a divisor of zero.

So we have another way of characterising an integral domain: an integral domain is a commutative ring with unity that has no divisors of zero.

17.3 Proposition In $\boldsymbol{Z}_n$ a non-zero element x is a divisor of zero iff $\gcd(x,n) \neq 1$.

Proof: Suppose $\gcd(x,n)=d\neq 1$.

Then $x\neq 0$ and $\frac{n}{d}\neq 0$ in $\boldsymbol{Z}_n$ but $x\times\frac{n}{d}=\frac{x}{d}\times n=0$ so that x is a divisor of zero.

Now suppose $\gcd(x,n)=1$, so that there are $\lambda,\mu\in\boldsymbol{Z}$ such that

$$\lambda x+\mu n=1.$$

For $y\in\boldsymbol{Z}_n$, if we have $xy=0$ then $n\mid xy$ in $\boldsymbol{Z}$.

Since $\lambda xy+\mu ny=y$, it follows that $n\mid y$ so that $y=0$ in $\boldsymbol{Z}_n$.

Hence x is not a divisor of zero.

Since if n is prime $\gcd(n,x)=1$ for all integers $1\leq x<n$, we have:

17.4 Corollary $\boldsymbol{Z}_n$ is an integral domain iff n is prime.

The lack of divisors of zero is necessary and sufficient to carry out *multiplicative cancellation*, as the following result shows:

17.5 Proposition Let R be a commutative ring with unity. Then R is an integral domain if and only if it satisfies the following:

$$\text{if } a\cdot b=a\cdot c \text{ then } b=c$$

for $b,c\in R$ and non-zero $a\in R$.

Proof: "$\Rightarrow$" Suppose that R is an integral domain and that $a\cdot b=a\cdot c$ with $a\neq 0$. Then $a\cdot b-a\cdot c=0$.

Factorising gives $a\cdot(b-c)=0$ and so $b-c=0$ since $a\neq 0$. Hence we have $b=c$.

"$\Leftarrow$" Suppose that if $a\cdot b=a\cdot c$ then $b=c$ for $b,c\in R$ and non-zero $a\in R$.

Suppose that non-zero elements $a,b\in R$ are such that $a\cdot b=0$.

We have $a\cdot b=a\cdot 0$ and so if $a\neq 0$ then by cancellation $b=0$.

Similarly $a\cdot b=0\cdot b$ and so if $b\neq 0$ then by cancellation $a=0$.

Hence R is an integral domain.

17.6 Proposition The characteristic of an integral domain is either 0 or a prime.

Proof: Clearly $\boldsymbol{Z}$ is an integral domain with characteristic 0. Now suppose that D is an integral domain of non-zero characteristic with $\text{char } D = mn$, with $m, n > 1$.

Then $\underbrace{(1+1+...+1)}_{m} \cdot \underbrace{(1+1+...+1)}_{n} = \underbrace{(1+1+...+1)}_{mn} = 0.$

Hence either $\underbrace{(1+1+...+1)}_{m} = 0$ or $\underbrace{(1+1+...+1)}_{n} = 0$, which contradicts $\text{char } D = mn$.

Hence $\text{char } D$ must be prime.

We recall the definition of a field:

17.7 Definition A commutative ring R with a unity is a *field* if every *non-zero* $a \in R$ has a *multiplicative inverse* $a^{-1} \in R$ satisfying

$$a \cdot a^{-1} = 1 = a^{-1} \cdot a.$$

It is left as an exercise to show that the multiplicative inverse of each non-zero element is unique.

Examples (a) The number systems $\boldsymbol{R}$, $\boldsymbol{Q}$ and $\boldsymbol{C}$ are the familiar examples of fields. $\boldsymbol{Z}$ is *not* a field.

(b) $\boldsymbol{Q}[x]$ and $\boldsymbol{R}[x]$ are not fields, even though both are integral domains and both $\boldsymbol{Q}$ and $\boldsymbol{R}$ are fields.

(c) $\boldsymbol{Q}(\sqrt{2}\,) = \{r + s\sqrt{2} : r, s \in \boldsymbol{Q}\}$ is a field. For a non-zero element $r + s\sqrt{2} \in \boldsymbol{Q}(\sqrt{2}\,)$ the multiplicative inverse is given by

$$\frac{r - s\sqrt{2}}{r^2 - 2s^2}.$$

17.8 Proposition Any field is an integral domain.

Proof: Suppose that F is a field, and $a \cdot b = 0$ in F.

Then if a is non-zero it has a multiplicative inverse and

$$b = (a^{-1} \cdot a) \cdot b = a^{-1} \cdot (a \cdot b) = a^{-1} \cdot 0 = 0.$$

Similarly, if b is non-zero then $a = 0$. Hence F is an integral domain.

Note that the converse is false: there are integral domains that are not fields. For example the ring of integers $\boldsymbol{Z}$ is an integral domain but is *not* a field. However, we do have the following:

17.9 Proposition Any *finite* integral domain is a field.

Proof: Suppose that D is a finite integral domain, and that $a \in D$ is non-zero.

Consider the elements $a, a^2, a^3, a^4, \ldots$

Since D is finite these cannot all be distinct. Suppose that $a^i = a^j$, where $i > j$. By multiplicative cancellation we have $a^{i-j} = 1$ where $i - j \geq 1$. This may be rewritten as

$$a \cdot a^{i-j-1} = 1$$

So a has a multiplicative inverse a^{i-j-1}. Hence D is a field.

17.10 Corollary $\boldsymbol{Z}_n$ is a field iff n is prime.

Proof: If n is prime then $\boldsymbol{Z}_n$ is an integral domain by 17.4, and since $\boldsymbol{Z}_n$ is finite, by 17.9 is a field.

If n is not prime then by 17.4 $\boldsymbol{Z}_n$ is not an integral domain, and so by 17.8 is not a field.

It turns out that a field has no proper non-trivial ideals:

17.11 Proposition A field F has no ideals, except for the trivial ideal $\{0\}$ and F itself.

Proof: Suppose that $I \neq \{0\}$ is an ideal in a field F. Let $x \in F$ and choose a non-zero element $a \in I$. Because I is an ideal we have $xa \in I$. F is a field, and so $a \in F$ must have a multiplicative inverse $a^{-1} \in F$. Again, because I is an ideal we have $(xa)a^{-1} \in I$. Then

$$(xa)a^{-1} = x(aa^{-1}) = x1 = x \in I.$$

Hence if $x \in F$ then $x \in I$ so that $F \subseteq I$. It follows that $I = F$.

The converse is also true: If a commutative ring F with unity has no ideals except $\{0\}$ and F itself, then F is a field. It is left as an exercise to show this.

Next we define two special types of ideal known as *prime* ideals and *maximal* ideals. The quotient rings which arise from these are *integral domains* and *fields* respectively.

17.12 Definition A proper ideal P in a commutative ring R is called a *prime ideal* if for any $a, b \in R$ we have:

$$\text{if } ab \in P \text{ then } a \in P \text{ or } b \in P.$$

Examples (a) The ideal $2\boldsymbol{Z}$ in the ring $\boldsymbol{Z}$ is a prime ideal: for $m, n \in \boldsymbol{Z}$ if mn is even then at least one of m and n must be even. $p\boldsymbol{Z}$ is a prime ideal in $\boldsymbol{Z}$ whenever p is a prime number, as we show below.

(b) The ideal $\langle x \rangle$ is prime in the ring $\boldsymbol{Z}[x]$. Notice that

$$p(x) \in \langle x \rangle \text{ iff } p(x) \text{ has constant term } 0.$$

If $p(x)q(x) \in \langle x \rangle$ then $p(x)q(x)$ has constant term 0. Hence at least one of $p(x)$ and $q(x)$ must have constant term 0 and so lie in $\langle x \rangle$.

By contrast, the ideal $\langle x^2 \rangle$ is *not* prime in $\boldsymbol{Z}[x]$:

Clearly $x^2 = x \cdot x \in \langle x^2 \rangle$ but $x \notin \langle x^2 \rangle$.

17.13 Proposition In the ring of integers $\boldsymbol{Z}$ the ideal $n\boldsymbol{Z}$ is a prime ideal iff n is prime.

Proof: Let n be prime and $ab \in n\boldsymbol{Z}$.

Then $n \mid ab$ so that by 15.6 we have $n \mid a$ or $n \mid b$. If $n \mid a$ then $a \in n\boldsymbol{Z}$ and if $n \mid b$ then $b \in n\boldsymbol{Z}$

Hence $n\boldsymbol{Z}$ is a prime ideal.

If n is not prime then $n = ab$ for $a, b < n$. Now $ab \in n\boldsymbol{Z}$ but $a, b \notin n\boldsymbol{Z}$.

Hence $n\boldsymbol{Z}$ is not a prime ideal.

17.14 Definition A proper ideal M in a commutative ring R is called a *maximal ideal* if there is no other ideal I in R such that

$$M \subset I \subset R$$

Thus M is maximal if there is no other proper ideal that strictly contains M.

Examples (a) The ideal $4\boldsymbol{Z}$ is *not* maximal in the ring of integers $\boldsymbol{Z}$ because

$$4\boldsymbol{Z} \subset 2\boldsymbol{Z} \subset \boldsymbol{Z}.$$

An ideal $p\boldsymbol{Z}$ is maximal in $\boldsymbol{Z}$ iff p is prime.

(b) The ideals in $\boldsymbol{Z}_{12}$ are $\{0\}$, $\{0,6\}$, $\{0,4,8\}$, $\{0,3,6,9\}$, $\{0,2,4,6,8,10\}$ and $\boldsymbol{Z}_{12}$ itself.

We may order these ideals by inclusion, as shown in the Hasse diagram below.

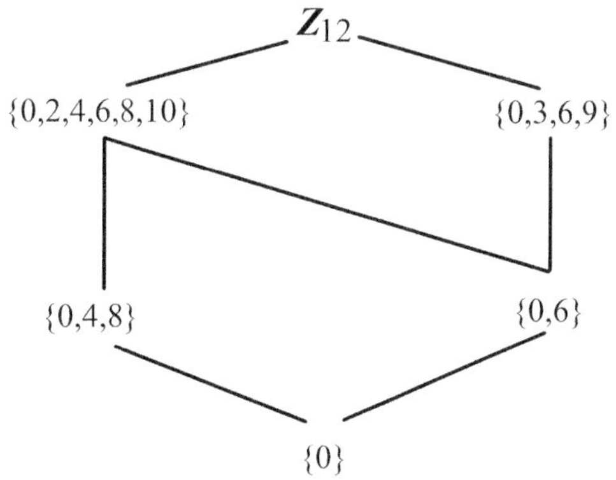

The edges in the diagram show inclusion of one ideal inside another. The maximal ideals are $\{0,2,4,6,8,10\}$ and $\{0,3,6,9\}$.

(c) The ideal $\langle x^2+2x+1\rangle$ is *not* maximal in $\boldsymbol{Z}[x]$ because

$$\langle x^2+2x+1\rangle \subset \langle x+1\rangle \subset \boldsymbol{Z}[x].$$

A polynomial $p(x) \in F[x]$ is an *irreducible polynomial* over F if the following holds:

$$\text{if } p(x) = q(x)r(x) \text{ then } \deg q(x) = 0 \text{ or } \deg r(x) = 0.$$

Thus an irreducible polynomial is one that cannot be factorised as a product of polynomials of strictly smaller degree. For example, $x^2 - 7x + 12 \in \boldsymbol{Q}[x]$ is not an irreducible polynomial over $\boldsymbol{Q}$ as it can be factorised as $(x-3)(x-4)$, whereas $x^2 - 7x + 11$ is an irreducible polynomial over $\boldsymbol{Q}$.

17.15 Proposition Let F be a field and let $p(x) \in F[x]$.

The ideal $\langle p(x) \rangle$ is maximal if and only if $p(x)$ is an irreducible polynomial over F.

We defer the proof of this until chapter 19.

The following lemma will be used twice:

17.16 Lemma Let M be a maximal ideal in a commutative ring R with unity, and suppose $a \notin M$. Let $I = \{m + ra : m \in M, r \in R\}$. Then

(i) I is an ideal in R

(ii) $I = R$.

Proof: (i) is left as an exercise.

(ii) For $m \in M$ we have $m + 0a \in I$ and so $M \subseteq I$. To show that $M \subset I$ observe that $a = 0 + 1 \cdot a \in I$ but $a \notin M$. So we have $M \subset I \subseteq R$, and because M is maximal $I = R$.

17.17 Proposition Let R be a commutative ring with unity.

Each maximal ideal of R is also a prime ideal.

Proof: Let M be a maximal ideal in a commutative ring R with unity.

Suppose that $ab \in M$ but $a \notin M$. We construct the ideal $I = \{m + ra : m \in M, \ r \in R\}$ as above.

By the previous lemma, $I = R$.

In particular, because R has a unity we have $1 \in I$. So there are elements $m \in M$ and $r \in R$ such that $1 = m + ra$. Then

$$b = 1 \cdot b = (m + ra)b = mb + rab.$$

Now because $m \in M$ and $ab \in M$ it follows from the properties of an ideal that $b \in M$.

In a similar way, we can show that if $ab \in M$ but $b \notin M$ then $a \in M$.

Hence if $ab \in M$ then $a \in M$ or $b \in M$, and so M is a prime ideal.

Notice that the converse is false: in a commutative ring with unity a prime ideal may not be maximal.

Example The ideal $\langle x \rangle$ is prime in the ring $\boldsymbol{Z}[x]$. However $\langle x \rangle$ is not maximal because for example

$$\langle x \rangle \subset E \subset \boldsymbol{Z}[x]$$

where E is the ideal of all polynomials in $\boldsymbol{Z}[x]$ with even constant.

We now state and prove two important theorems about quotient rings arising from prime and maximal ideals:

17.18 Theorem Let R be a commutative ring with unity and let P be an ideal in R.

Then P is a prime ideal iff R/P is an integral domain.

Proof: "$\Rightarrow$" Suppose that P is a prime ideal in R and that we have $a + P, b + P \in R/P$ such that $(a + P) \cdot (b + P) = 0 + P$. Then we have $ab + P = 0 + P$ so that

$$ab = ab - 0 \in P.$$

Because P is a prime ideal we have $a \in P$ or $b \in P$. If $a \in P$ then $a + P = 0 + P$ and if $b \in P$ then $b + P = 0 + P$. Hence R/P is an integral domain.

"$\Leftarrow$" Suppose that R/P is an integral domain and that $ab \in P$ for some elements $a, b \in R$.

Then

$$(a+P)\cdot(b+P) = ab+P = 0+P \text{ in } R/P.$$

Since R/P is an integral domain we must have $a+P=0+P$, in which case $a \in P$, or $b+P=0+P$, in which case $b \in P$. Hence P is a prime ideal in R.

17.19 Theorem Let R be a commutative ring with unity and let M be an ideal in R.

Then M is a maximal ideal iff R/M is a field.

Proof: "$\Rightarrow$" Since R is commutative and has a unity, the quotient ring R/M is also commutative and has a unity. To prove that R/M is a *field*, we need to show the existence of a multiplicative inverse for each non-zero element.

Let $a+M \in R/M$ be a non-zero element, so that $a \notin M$. We construct the ideal $I = \{m+ra \ : m \in M,\ r \in R\}$ as before. Once again by lemma 17.16 we have $I = R$.

Hence $1 \in I$ and so $1 = m+ra$ for some elements $m \in M$ and $r \in R$. Therefore $1-ra = m \in M$ so that

$$1+M = ra+M = (r+M)\cdot(a+M),$$

so that $r+M$ is the multiplicative inverse of $a+M$. Hence R/M is a field.

"$\Leftarrow$" Suppose that R/M is a field, and suppose there is an ideal I of R such that $M \subset I \subset R$.

Let $q : R \to R/M$ be the quotient mapping of 16.4 given by $q(a) = a+M$.

Consider $q(I) = \{q(a) : a \in I\} = \{a+M : a \in I\}$.

If $a+M \in q(I)$ and $r+M \in R/M$ then

$$(a+M)\cdot(r+M) = ar+M \in q(I)$$

since $ar \in I$. Hence $q(I)$ is an ideal in R/M.

But since R/M is a field, it has no non-trivial proper ideals.

So $q(I) = R/M$, in which case $I = R$, or $q(I)$ is trivial, in which case $I = M$.

Hence no such ideal I exists, and M is a maximal ideal.

Notice that these two theorems are consistent with the facts that any maximal ideal is a prime ideal and any field is an integral domain.

Examples (a) The ideal $p\boldsymbol{Z}$ in $\boldsymbol{Z}$ is maximal iff p is a prime number, and so $\boldsymbol{Z}/p\boldsymbol{Z} \cong \boldsymbol{Z}_p$ is a field if and only if p is a prime number.

(b) The ideal $I = \{0, 3, 6, 9\}$ in $\boldsymbol{Z}_{12}$ is maximal and so the quotient ring $\boldsymbol{Z}_{12}/I$ is a field.

(c) $\langle x^2 + 1\rangle$ is maximal in $\boldsymbol{R}[x]$ since $x^2 + 1$ is irreducible.

Hence $\boldsymbol{R}[x]/\langle x^2 + 1\rangle$ is a field. This is of course the field of complex numbers, $\boldsymbol{C}$.

Exercises 17

1. Find all of the divisors of zero in $\boldsymbol{Z}_8$ and in $\boldsymbol{Z}_{11}$.

2. Prove that in a field F, each non-zero element $a \in F$ has a *unique* multiplicative inverse $a^{-1} \in F$.

3. Show that the ideal $\langle x^2 - 1 \rangle$ in $\boldsymbol{Z}[x]$ is *not* maximal.

   What does this tell us about the quotient ring $\boldsymbol{Z}[x]/\langle x^2 - 1 \rangle$?

4. Describe the quotient ring $\boldsymbol{Z}[x]/\langle x - 3 \rangle$.

   What does this tell us about the ideal $\langle x - 3 \rangle$ in $\boldsymbol{Z}[x]$?

5. Find all of the maximal ideals of $\boldsymbol{Z}_{30}$.

6. For any element $a \in R$ and any ideal $I \subseteq R$, prove that the set

   $$J = \{b + ra : b \in I,\ r \in R\}$$

   is an ideal in R.

7. Let R be a commutative ring with unity. Prove that if R has no proper non-trivial ideals then R is a field.

   Hint: Choose a non zero element $a \in R$ *and consider* $\langle a \rangle$.

Chapter 18: Finite Fields

Let F be a field and consider the ring of polynomials $F[x]$. For a polynomial $f(x) \in F[x]$ we can form the ideal $\langle f(x) \rangle$ in $F[x]$ and construct the quotient ring $F[x]/\langle f(x) \rangle$. We are interested in the case where this quotient ring turns out to be a field.

As stated in 17.15, the ideal $\langle f(x) \rangle$ is maximal in $F[x]$ if and only if $f(x)$ is irreducible over F.

Hence by 17.19, the ring $F[x]/\langle f(x) \rangle$ is a field if and only if $f(x)$ is irreducible over F.

Example The quadratic polynomial $x^2 + 1$ is clearly irreducible in $\boldsymbol{R}[x]$. Hence $\boldsymbol{R}[x]/\langle x^2 + 1 \rangle$ must be a field. In fact, we have already seen that this field is $\boldsymbol{C}$, the complex numbers.

Now suppose that we begin with a finite field $\boldsymbol{Z}_p$, where p is a prime, and construct the ring of polynomials $\boldsymbol{Z}_p[x]$. Next, we choose a monic irreducible polynomial $f(x)$ of degree n in $\boldsymbol{Z}_p[x]$. We form the ideal $\langle f(x) \rangle$ and construct the quotient ring $\boldsymbol{Z}_p[x]/\langle f(x) \rangle$. Because $f(x)$ is irreducible, $F = \boldsymbol{Z}_p[x]/\langle f(x) \rangle$ is a field.

If $f(x) = x^n + a_{n-1}x^{n-1} + \ldots + a_1x + a_0$ then since we are working modulo $f(x)$ we have

$$x^n = -a_{n-1}x^{n-1} - \ldots - a_1x - a_0.$$

Hence we can think of the elements of F as polynomials of degree not exceeding $n - 1$. Since there are p^n such polynomials, the field F has p^n elements.

Example Consider the field $\boldsymbol{Z}_2$. It is very easy to see that the polynomial $x^2 + x + 1 \in \boldsymbol{Z}_2[x]$ is irreducible - we just check that neither 0 nor 1 is a root:

$$0^2 + 0 + 1 = 1 \neq 0$$

$$1^2 + 1 + 1 = 1 \neq 0$$

We construct the quotient ring $\boldsymbol{Z}_2[x]/\langle x^2+x+1\rangle$. We know from proposition 17.19 that this is a field. We wish to determine the structure of this field by giving Cayley tables for addition and multiplication.

Since we are working with polynomials modulo x^2+x+1, we have $x^2+x+1=0$ or equivalently $x^2=x+1$ (remember we are working modulo 2). Hence whenever x^2 appears we can replace it by $x+1$. Our field has four elements: $0, 1, x$ and $x+1$.

Here is the Cayley table for addition:

$+$	0	1	x	$x+1$
0	0	1	x	$x+1$
1	1	0	$x+1$	x
x	x	$x+1$	0	1
$x+1$	$x+1$	x	1	0

Here is the Cayley table for multiplication:

$\cdot$	0	1	x	$x+1$
0	0	0	0	0
1	0	1	x	$x+1$
x	0	x	$x+1$	1
$x+1$	0	$x+1$	1	x

For example $x\cdot(x+1)=x^2+x=x+1+x=1$.

We have constructed a new field with four elements. This is written as $\boldsymbol{F}_4$.

These finite fields were first studied by Galois in the early nineteenth century. Here is a remarkable fact about finite fields:

18.1 Proposition In any finite field F, the non-zero elements form a cyclic group under multiplication.

The proof of this result is omitted.

Examples (a) The multiplicative group of the field $\boldsymbol{F}_7$ is cyclic and generated by 3:

$$3^0 = 1 \quad 3^1 = 3 \quad 3^2 = 2 \quad 3^3 = 6 \quad 3^4 = 4 \quad 3^5 = 5$$

(b) For the field $\boldsymbol{F}_4$ the three non-zero elements are $1, x$ and $x+1$. By taking x as a generator we get

$$x^0 = 1 \qquad x^1 = x \qquad x^2 = x+1$$

18.2 Definition Suppose that F is a field of characteristic p. We define a mapping $\sigma_p : F \to F$ by

$$\sigma_p(a) = a^p \text{ for each } a \in F.$$

This mapping is called the *Frobenius automorphism.*

18.3 Lemma For $\alpha, \beta \in \boldsymbol{Z}_p$ we have $(\alpha+\beta)^p = \alpha^p + \beta^p$.

Proof: The binomial coefficients $\binom{p}{i}$ are all congruent to zero modulo p for $1 \le i \le p-1$.

18.4 Proposition The mapping σ_p is an automorphism.

Proof: For $\alpha, \beta \in F$ we have

$$\sigma_p(\alpha+\beta) = (\alpha+\beta)^p = \alpha^p + \beta^p = \sigma_p(\alpha) + \sigma_p(\beta) \text{ by 18.3}$$

$$\sigma_p(\alpha\beta) = (\alpha\beta)^p = \alpha^p\beta^p = \sigma_p(\alpha)\sigma_p(\beta)$$

Hence ϕ is a homomorphism. Now $\ker \sigma_p = \{0\}$, and so by 15.14(iv) σ_p is injective.

Since F is finite, by counting we see that σ_p is surjective, and hence is an automorphism.

Example For the field $\boldsymbol{F}_4$ the automorphism σ_2 gives

$$\sigma_2(0) = 0^2 = 0 \qquad \sigma_2(1) = 1^2 = 1$$

$$\sigma_2(x) = x^2 = x + 1 \qquad \sigma_2(x+1) = (x+1)^2 = x^2 + 1 = x$$

18.5 Proposition Let F be a finite field with p^n elements.

The Frobenius automorphism $\sigma_p : F \to F$ is of order n.

That is, n is the smallest natural number such that

$$\sigma_p^n = \underbrace{\sigma_p \circ \sigma_p \circ \ldots \circ \sigma_p}_{n}$$

is the identity mapping.

Proof: The multiplicative group of non-zero elements of F is of order $p^n - 1$.

Hence $a^{p^n - 1} = 1$ for all non-zero $a \in F$.

It follows that $\sigma_p^n(a) = a^{p^n} = a$ for all $a \in F$.

18.6 Fermat's Theorem If p is a prime that does not divide a then

$$a^{p-1} \equiv 1(\operatorname{mod} p).$$

Proof: If p is a prime that does not divide a then $a \not\equiv 0(\operatorname{mod} p)$, so that a is congruent to an element of the multiplicative group of the field $\boldsymbol{Z}_p$.

The multiplicative group of $\boldsymbol{Z}_p$ is cyclic by 18.1 and of order $p - 1$, and hence $a^{p-1} \equiv 1(\operatorname{mod} p)$.

Example Let $p = 3$. We have $7^2 = 49 \equiv 1(\operatorname{mod} 3)$ and $8^2 = 64 \equiv 1(\operatorname{mod} 3)$, and so on.

A finite field of p^n elements has $\boldsymbol{Z}_p$ as a subfield. The Frobenius automorphism fixes each element of this subfield, as the next result show:

18.7 Proposition Let F be a finite field of p^n elements.

For each $a \in \mathbf{Z}_p$ we have $\sigma_p(a) = a$.

Proof: For $a = 0$ we have $\sigma_p(0) = 0$, and for $a \neq 0$ by 18.6 we have $a^{p-1} = 1$ and so $\sigma_p(a) = a^p = a$.

18.8 Proposition Let F be a finite field with $q = p^n$ elements. The elements of F are all of the roots of the polynomial

$$x^q - x \in \mathbf{Z}_p[x].$$

Proof: First, we observe that 0 is clearly a root of the polynomial.

By 18.1 the multiplicative group of F is cyclic of order $q - 1$.

Hence for each non-zero element of F we have $a^{q-1} = 1$, and so $a^q = a$. Hence $a^q - a = 0$ so that a is a root of $x^q - x$.

Since a polynomial of degree q cannot have more than q roots, the result follows.

18.9 Corollary Any two finite fields with the same number of elements are isomorphic.

Example These finite fields are isomorphic:

$E = \mathbf{Z}_3[x]/\langle x^2 + 1 \rangle$ $\qquad$ $F = \mathbf{Z}_3[x]/\langle x^2 + x + 2 \rangle$

Here is one isomorphism:

E	0	1	$x+1$	$2x$	$2x+1$	2	$2x+2$	x	$x+2$
F	0	1	x	$2x+1$	$2x+2$	2	$2x$	$x+2$	$x+1$

Exercises 18

1. Determine which of the following polynomials are irreducible over $\boldsymbol{Z}_2$

   (i) x^2+1 (ii) x^3+x^2+x+1

   (iii) x^3+x^2+1 (iv) x^4+x^2+1

2. Using the irreducible polynomial x^2+1 of degree 2 over $\boldsymbol{Z}_3$, construct the field of nine elements, $\boldsymbol{F}_9$.

   Draw Cayley tables for addition and multiplication in this field.

3. Using the irreducible polynomial x^3+x+1 of degree 3 over $\boldsymbol{Z}_2$, construct the field of eight elements, $\boldsymbol{F}_8$.

   Draw Cayley tables for addition and multiplication in this field.

4. By finding a generator, show that the multiplicative group of non-zero elements in $\boldsymbol{F}_9$ is cyclic.

5. By finding a generator, show that the multiplicative group of non-zero elements in $\boldsymbol{F}_8$ is cyclic.

6. For the Frobenius automorphism, find $\sigma_3(a)$ for each $a \in \boldsymbol{F}_9$.

   Similarly, find $\sigma_2(a)$ for each $a \in \boldsymbol{F}_8$.

Chapter 19: Factorisation in an Integral Domain

A natural number p (other than 1) is said to be *prime* if whenever we have $p = ab$ it follows that either $a = 1$ or $b = 1$.

Thus p is prime if it is divisible only by 1 and by itself.

Examples (a) The thirty smallest prime numbers are 2, 3, 5, 7, 11, 13, 17, 19, 23, 29, 31, 37, 41, 43, 47, 53, 59, 61, 67, 71, 73, 79, 83, 89, 97, 101, 103, 107, 109 and 113.

(b) 12 is not prime because, for example, $12 = 3 \times 4$.

For ease of reference, we restate a key result from chapter 15:

19.1 Proposition Suppose that p is prime.

If $p \mid ab$ then $p \mid a$ or $p \mid b$.

By induction, we may prove this corollary to 19.1:

For a prime p, if $p \mid a_1 a_2 ... a_n$ then $p \mid a_i$ for some i.

19.2 The Fundamental Theorem of Arithmetic

Each natural number (except 1) may be factorised uniquely as a product of prime numbers.

Proof: First we prove the existence of a factorisation into primes: if n is prime then we are done. Otherwise, $n = ab$ where $a, b \neq 1$. Hence $a < n$ and $b < n$.

If a and b are prime, then we are done. If not then we repeat the process above with each of a and b, and so on. Since the numbers get smaller at each step, this process must terminate.

To prove that the factorisation into primes is unique, suppose that a natural number n has a pair of factorisations:

$$n = p_1 p_2 ... p_r \text{ and } n = q_1 q_2 ... q_s.$$

We may suppose that $r \leq s$.

Since p_1 divides n we have p_1 divides $q_1q_2...q_s$. Now by 19.1 we have that p_1 divides q_i for some i. We may relabel the q_i's so that p_1 divides q_1. But because q_1 is prime we have $p_1 = q_1$.

Hence $p_1p_2...p_r = p_1q_2...q_s$, and by cancellation

$$p_2p_3...p_r = q_2q_3...q_s.$$

By repeating the argument above we have $p_2 = q_2$.

Hence $p_2p_3...p_r = p_2q_3...q_s$, and by cancellation $p_3...p_r = q_3...q_s$.

We continue in this way until all of the p_i's are exhausted. We have shown $p_i = q_i$ for $i = 1, 2, ..., r$. Finally by cancellation we have $1 = q_{r+1}...q_s$. Hence there can by no further primes q_i and so $r = s$.

Examples (a) 30 factorises into primes as $2\times3\times5$. Of course, it also factorises as $5\times3\times2$ and $3\times2\times5$ and so on. But each of these factorisations is obtained from another by changing the order of the primes, and so the factorisations are essentially the same.

(b) Notice that the same prime can occur several times in a factorisation:

$$504 = 2\times2\times2\times3\times3\times7.$$

19.3 Definition Suppose that D is an integral domain and $a, b \in D$.

We say that *a divides* b, denoted $a \mid b$, if there is $x \in D$ such that $b = xa$.

Examples (a) For the integers $\mathbf{Z}$ we have $2 \mid 8$ since $8 = 4\times2$.

(b) In the ring $\mathbf{Z}[x]$ we have $x+1 \mid x^2+3x+2$ since $x^2+3x+2 = (x+2)(x+1)$.

(c) In the ring $\mathbf{Z}[\sqrt{-2}\,]$ we have $1+3\sqrt{-2} \mid 10+11\sqrt{-2}$ since

$$10+11\sqrt{-2} = (4-\sqrt{-2}\,)(1+3\sqrt{-2}\,).$$

19.4 Definition Suppose that D is an integral domain.

We call $a \in D$ a *unit* if $a \mid 1$

A pair of elements $a, b \in D$ are called *associates* if both

$$a \mid b \text{ and } b \mid a.$$

Examples (a) For the integers $\boldsymbol{Z}$ the units are -1 and 1.

A pair of integers m and n are associates if and only if $m = \pm n$.

(b) For the Gaussian integers $\boldsymbol{Z}[i]$ the elements $i, -1, -i$ and 1 are units. We show below that these are the only units.

The associates of 2 are $2i, -2, -2i$ and 2 itself. The associates of $1-3i$ are $3+i, -3-i, -1+3i$ and $1-3i$ itself.

(c) In $\boldsymbol{Z}[\sqrt{-3}\,] = \{a + b\sqrt{-3} \;: a, b \in \boldsymbol{Z}\}$ the units are -1 and 1.

(d) In $F[x]$, for F a field, the units are the degree 0 polynomials. In $\boldsymbol{Z}[x]$ the only units are -1 and 1, and the other degree 0 polynomials are not units.

(e) For the real numbers $\boldsymbol{R}$, and indeed for any field, all non-zero elements are units and any pair of non-zero elements are associates.

19.5 Lemma Suppose that D is an integral domain. Elements $a, b \in D$ are associates if and only if $a = ub$ for some unit $u \in D$.

Proof: "$\Rightarrow$" Suppose that a and b are associates.

Then $a \mid b$ so that $b = xa$ for some $x \in D$ and $b \mid a$ so that $a = yb$ for some $y \in D$.

Hence $a = y(xa) = (yx)a$ so that $yx = 1$, and so x and y are units.

"$\Leftarrow$" Suppose $a = ub$, where u is a unit. Then $b \mid a$.

Because u is a unit we have $uv = 1$ for some unit v.

Then $va = v(ub) = (vu)b = b$, and so $b \mid a$.

Hence a and b are associates.

19.6 Definition Suppose that D is an integral domain. A *multiplicative norm* on D is a mapping $N : D \to \boldsymbol{Z}$ satisfying:

$$N(\alpha\beta) = N(\alpha)N(\beta) \text{ for all } \alpha, \beta \in D.$$

Examples (a) $N : \boldsymbol{Z}[i] \to \boldsymbol{Z}$ given by $N(a+bi) = a^2 + b^2$.

(b) $N : \boldsymbol{Z}[\sqrt{-3}\,] \to \boldsymbol{Z}$ given by $N(a + b\sqrt{-3}\,) = a^2 + 3b^2$.

19.7 Proposition Suppose that N is a multiplicative norm on an integral domain D.

(i) $N(1) = 1$ (ii) If u is a unit then $N(u) = \pm 1$.

Proof: (i) $N(1) = N(1 \times 1) = N(1)N(1)$. The result follows by cancellation.

(ii) Suppose that u is a unit, so that $uv = 1$ for some $v \in D$

Then $N(u)N(v) = N(uv) = N(1) = 1$ and so $N(u) = \pm 1$.

Example Consider $\boldsymbol{Z}[i]$ with the multiplicative norm above.

If $N(a+bi) = a^2 + b^2 = 1$ then we have

$$a = \pm 1, b = 0 \text{ or } a = 0, b = \pm 1.$$

Hence 1, i, -1 and -i are the only units in $\boldsymbol{Z}[i]$.

19.8 Definition Suppose that $a \in D$ is non-zero and is not a unit. Then a is called an *irreducible* if in each factorisation $a = bc$ either b is a unit or c is a unit.

Examples (a) In $\boldsymbol{Z}$ the irreducibles are the prime numbers and their negatives:

$$\pm 2, \pm 3, \pm 5, \pm 7, \pm 11, \ldots$$

(b) In the Gaussian integers $\boldsymbol{Z}[i]$ notice that 2 is not an irreducible:

$2 = (1+i)(1-i)$ but neither $1+i$ or $1-i$ is a unit.

However, 3 is irreducible in $\boldsymbol{Z}[i]$. For if $3=(a+bi)(c+di)$, then by applying the multiplicative norm above we get

$$9=(a^2+b^2)(c^2+d^2) \text{ in } \boldsymbol{Z}.$$

But $a^2+b^2 \neq 3$ for any choice of integers, and so either $a^2+b^2=1$ or $c^2+d^2=1$, in each case leading to one factor being a unit.

(c) For the real numbers **R**, and indeed for any field, there are no irreducibles, since every non-zero element is a unit.

(d) Consider the ring $\boldsymbol{Z}[\sqrt{-3}\,]=\{a+b\sqrt{-3}\ :a,b\in\boldsymbol{Z}\}$. The units are -1 and 1.

The element $1+\sqrt{-3}$ is irreducible:

Suppose that $1+\sqrt{-3}=(a+b\sqrt{-3}\,)(c+d\sqrt{-3}\,)$ for some $a,b,c,d\in\boldsymbol{Z}$.

We define a multiplicative norm $N:\boldsymbol{Z}[\sqrt{-3}\,]\to\boldsymbol{Z}$ by

$$N(a+b\sqrt{-3}\,)=a^2+3b^2.$$

Then $N(1+\sqrt{-3}\,)=N(a+b\sqrt{-3}\,)N(c+d\sqrt{-3}\,)$ so that

$$4=(a^2+3b^2)(c^2+3d^2) \text{ in } \boldsymbol{Z}.$$

But $a^2+3b^2,\ c^2+3d^2\neq 2$ for any $a,b,c,d\in\boldsymbol{Z}$, so we must have either $a^2+3b^2=1$, in which case $a+b\sqrt{-3}=\pm 1$ is a unit, or else $c^2+3d^2=1$, in which case $c+d\sqrt{-3}=\pm 1$ is a unit.

Hence the element $1+\sqrt{-3}$ is an irreducible. Similarly, $1-\sqrt{-3}$ is also an irreducible in $\boldsymbol{Z}[\sqrt{-3}\,]$.

(e) For a field F, an irreducible polynomial over F is also an irreducible in the sense of 19.8. In the case where a polynomial has coefficients in a ring that is not a field we need to be more careful. For example, $2x+2\in\boldsymbol{Z}[x]$ is an irreducible polynomial over $\boldsymbol{Z}$. However, it is not an irreducible in the sense of 19.8 as it factorises as $2(x+1)$, and neither 2 or $x+1$ is a unit in $\boldsymbol{Z}[x]$.

Following from 19.2, we have:

19.9 Proposition Each non-zero integer that is not a unit can be factorised as a product of irreducibles. This factorisation is unique in the following sense:

If an integer n has a pair of factorisations $n = p_1 p_2 \ldots p_r$ and $n = q_1 q_2 \ldots q_s$ then $r = s$ and there is a one-to-one correspondence between the p_i's and the q_j's such that each p_i is an associate of the corresponding q_j.

Example The integer -30 has four factorisations into irreducibles:

$$-2 \times 3 \times 5 \qquad 2 \times -3 \times 5 \qquad 2 \times 3 \times -5 \qquad -2 \times -3 \times -5$$

However these are essentially the same. Each can be obtained from another by replacing each irreducible by an associate.

Unique factorisation of integers motivates us to make the following definition:

19.10 Definition Suppose that D is an integral domain in which every non-zero non-unit element can be factorised a product of irreducibles. Suppose further that if such an element has two such factorisations

$$p_1 p_2 \ldots p_r \text{ and } q_1 q_2 \ldots q_s$$

then $r = s$ and there is a one-to-one correspondence between the p_i's and the q_j's such that each p_i is an associate of the corresponding q_j.

Then D is called a *unique factorisation domain*, or UFD.

Examples (a) The integers $\boldsymbol{Z}$ are a UFD by 19.9.

(b) $\boldsymbol{Z}[\sqrt{-3}\,]$ is *not* a UFD, because for example there are two distinct factorisations into irreducibles for the element $4 + 0\sqrt{-3}$:

$$4 + 0\sqrt{-3} = (2 + 0\sqrt{-3}\,)(2 + 0\sqrt{-3}\,)$$

$$4 + 0\sqrt{-3} = (1 + \sqrt{-3}\,)(1 - \sqrt{-3}\,).$$

(c) $\boldsymbol{Q}[x]$ and $\boldsymbol{R}[x]$ are UFD's. We shall prove this later.

19.11 Definition An element p of an integral domain D is called a *prime* if it satisfies:

$$\text{if } p \mid ab \text{ then } p \mid a \text{ or } p \mid b \text{ for all } a, b \in D.$$

This terminology is rather confusing: *irreducibles* are what we think of in $\mathbf{Z}$ as prime numbers, whereas *primes* are elements satisfying the property of prime numbers expressed by 19.1

Example In $\mathbf{Z}$ the primes are the prime numbers and their negatives.

However, irreducibles and primes coincide in this case only because in $\mathbf{Z}$ we have unique factorisation.

19.12 Proposition In an integral domain every prime is an irreducible.

Proof: Suppose that in an integral domain D an element $p \in D$ is a prime and that $p = ab$ for some $a, b \in D$. Then clearly $p \mid ab$, and so it follows that either $p \mid a$ or $p \mid b$ because p is a prime.

If $p \mid a$ then $a = xp$ for some $x \in D$, so that $a = xab$.

Then $a(1 - xb) = 0$. Since $a \neq 0$ we have $xb = 1$.

Hence $b \mid 1$ so b is a unit.

Similarly, if $p \mid b$ it follows that a is a unit.

We have shown that if $p = ab$ then either a or b is a unit.

Hence p is an irreducible.

However, the converse is *not* true: in an integral domain an irreducible may not be a prime, as the following example shows:

Example We have already shown that $1 + \sqrt{-3}$ is an irreducible in $\mathbf{Z}[\sqrt{-3}\,]$.

We claim $1 + \sqrt{-3}$ is *not* prime in $\mathbf{Z}[\sqrt{-3}\,]$:

$(1 + \sqrt{-3}\,)(1 - \sqrt{-3}\,) = 4$, and so $(1 + \sqrt{-3}\,) \mid 4$. Since $4 = 2 \times 2$, if $1 + \sqrt{-3}$ is prime then $(1 + \sqrt{-3}\,) \mid 2$. If this were the case, then

$$(1+\sqrt{-3})(a+b\sqrt{-3})=2$$

for some $a, b \in \boldsymbol{Z}$. Then

$$(a-3b)+(a+b)\sqrt{-3} = 2+0\sqrt{-3}$$

leading to the system of equations

$$\left.\begin{aligned} a-3b &= 2 \\ a+b &= 0 \end{aligned}\right\}$$

But the solutions $a=\frac{1}{2}$, $b=-\frac{1}{2}$ are not in $\boldsymbol{Z}$.

Hence $1+\sqrt{-3}$ is not prime in $\boldsymbol{Z}[\sqrt{-3}]$.

19.13 Lemma $\langle a\rangle \subseteq \langle b\rangle$ if and only if $b \mid a$

Proof: "$\Rightarrow$" If $\langle a\rangle \subseteq \langle b\rangle$ then $a \in \langle b\rangle$ and so $a = rb$ for some $r \in D$ so that $b \mid a$.

"$\Leftarrow$" If $b \mid a$ then $a = rb$ for some $r \in D$.

If $x \in \langle a\rangle$ then $s = r'a$ for some $r' \in D$ and so

$$s = r'(rb) = (rr')b \in \langle b\rangle.$$

Hence $\langle a\rangle \subseteq \langle b\rangle$.

19.14 Corollary $\langle a\rangle = \langle b\rangle$ if and only if a and b are associates.

19.15 Lemma If u is a unit in an integral domain R then $\langle u\rangle = R$.

Proof: Recall that $\langle 1\rangle = R$. If u is a unit then $u \mid 1$.

By 19.13 we have $\langle 1\rangle \subseteq \langle u\rangle$ and hence $\langle u\rangle = R$.

19.16 Definition An ideal I in a commutative ring R is called a *principal ideal* if $I = \langle a\rangle$ for some $a \in R$.

Examples (a) For each natural number n the ideal $n\boldsymbol{Z} = \langle n\rangle$ is a principal ideal in $\boldsymbol{Z}$.

(b) In $\boldsymbol{Z}[x]$ the ideal $\langle x-2\rangle$ is principal.

(c) In $\boldsymbol{Z}[x]$ the ideal E of all polynomials with an even constant is *not* principal.

To see this, suppose that $E = \langle p(x) \rangle$ for some $p(x) \in \boldsymbol{Z}[x]$.

Then $\langle 2 \rangle \subseteq \langle p(x) \rangle$ and $\langle x \rangle \subseteq \langle p(x) \rangle$ and so $p(x) \mid 2$ and $p(x) \mid x$.

Hence $p(x) = \pm 1$. But $\langle 1 \rangle = \langle -1 \rangle = \boldsymbol{Z}[x]$. Hence E is not principal.

(d) In any integral domain D the trivial ideal $\{0\} = \langle 0 \rangle$ and $D = \langle 1 \rangle$ are principal.

19.17 Definition An integral domain is called a *principal ideal domain*, or PID, if every ideal is a principal ideal.

Examples (a) $\boldsymbol{Z}$ is a PID. We prove this below.

(b) $F[x]$ is a PID for each field F.

(c) $\boldsymbol{Z}[x]$ is *not* a PID, by example (c) above.

19.18 Proposition $\boldsymbol{Z}$ is a principal ideal domain.

Proof: Let I be an ideal in $\boldsymbol{Z}$, and let a be the smallest positive integer in I.

Then $an \in I$ for each $n \in \boldsymbol{Z}$ so that $a\boldsymbol{Z} \subseteq I$.

Now suppose $x \in I$. Then $x = qa + r$, where $0 \leq r < a$.

Since $x \in I$ and $qa \in I$ we have $r \in I$. Because a is the smallest positive integer in I we have $r = 0$. So a divides x, and $x \in a\boldsymbol{Z}$.

Hence $I \subseteq a\boldsymbol{Z}$ and so $I = a\boldsymbol{Z} = \langle a \rangle$.

19.19 Proposition If F is a field then $F[x]$ is a principal ideal domain.

Proof: Suppose that I is an ideal in $F[x]$. Since $\{0\} = \langle 0 \rangle$ is principal, we may suppose that I is non-trivial. Let $p(x)$ be a polynomial in I of least degree. Then $\langle p(x) \rangle \subseteq I$.

Now suppose $f(x) \in I$. By division of polynomials we have $f(x) = q(x)p(x) + r(x)$ for some $q(x), r(x) \in F[x]$, with $\deg r < \deg p$.

Because $f(x), p(x) \in I$ we also have $r(x) \in I$. But $p(x)$ is a polynomial of least degree in I, so $r(x) = 0$. Hence $p(x) \mid f(x)$ and so $f(x) \in \langle p(x) \rangle$.

Hence $I \subseteq \langle p(x) \rangle$ and so $I = \langle p(x) \rangle$, so that I is principal.

19.20 Proposition Suppose that $I_1 \subseteq I_2 \subseteq I_3 \subseteq \ldots$ are ideals in an integral domain D.

Then $I = \bigcup_{n \geq 1} I_n = I_1 \cup I_2 \cup I_3 \cup \ldots$ is also an ideal in D.

Proof: Suppose that $a, b \in I$. Then $a \in I_i$ and $b \in I_j$ for some i and j.

Let $n = \max\{i, j\}$ so that $a, b \in I_n$. Then $a + b \in I_n$ and so $a + b \in I$ and $-a \in I_n$ so that $-a \in I$.

Finally, $ra \in I_n$ and so $ra \in I$ for all $r \in D$. Hence I is an ideal.

19.21 Lemma Let D be a PID. If $I_1 \subseteq I_2 \subseteq I_3 \subseteq \ldots$are ideals in D, then there is a natural number N such that $I_n = I_N$ for all $n \geq N$.

Proof: $I = \bigcup_{n \geq 1} I_n$ is an ideal in D, so because D is a PID we must have $I = \langle a \rangle$ for some $a \in D$.

Since $a \in I$ we have $a \in I_N$ for some N.

Hence for $n \geq N$ we have $\langle a \rangle \subseteq I_N \subseteq I_n \subseteq I = \langle a \rangle$, and so $I_n = I_N$ for all $n \geq N$.

19.22 Proposition In a PID the ideal $\langle a \rangle$ is maximal iff a is irreducible.

Proof: "$\Rightarrow$" Suppose that $\langle a \rangle$ is maximal and $a = bc$.

By 19.20 we have $\langle a \rangle \subseteq \langle b \rangle \subseteq D$ and $\langle a \rangle \subseteq \langle c \rangle \subseteq D$.

Then either $\langle a \rangle = \langle b \rangle$ so that a and b are associates and c is a unit, or $\langle b \rangle = D$ so that b is a unit. Hence a is an irreducible.

"$\Leftarrow$" Suppose that a is an irreducible, and suppose $\langle a \rangle \subseteq \langle b \rangle \subseteq D$. Then $b \mid a$ so that $a = bc$ for some $c \in D$, and either b or c is a unit.

If b is a unit then $\langle b \rangle = D$. If c is a unit then a and b are associates, so $\langle a \rangle = \langle b \rangle$.

Hence $\langle a \rangle$ is a maximal ideal.

This also proves 17.15, that for a field F, the maximal ideals of $F[x]$ are the principal ideals generated by polynomials that are irreducible over F.

19.23 Proposition In a PID every irreducible is prime.

Proof: Suppose that D is a PID and that $a \in D$ is an irreducible.

Suppose that $a \mid bc$ for some $b, c \in D$ so that $bc = xa$ for some $x \in D$

Then $bc \in \langle a \rangle$. Because a is irreducible, by 19.22 the ideal $\langle a \rangle$ is maximal.

By 17.17 the ideal $\langle a \rangle$ is prime, and so either $b \in \langle a \rangle$ so that $a \mid b$ or $c \in \langle a \rangle$ so that $a \mid c$.

Hence $a \in D$ is a prime.

19.24 Theorem Every PID is a UFD.

Proof: Suppose that D is a PID and that $a \in D$ is non-zero and not a unit.

First we prove the existence of a factorisation into irreducibles: if a is irreducible then we are done. Otherwise, $a = bc$ where neither b nor c is a unit.

Hence $\langle a \rangle \subseteq \langle b \rangle$ and $\langle a \rangle \subseteq \langle c \rangle$. Now if $\langle a \rangle = \langle b \rangle$ then $b = ua$ for some unit u.

Then $a = bc = uac$, so that $1 = uc$. But c is not a unit, so we must have $\langle a \rangle \subset \langle b \rangle$, and by similar reasoning $\langle a \rangle \subset \langle c \rangle$.

If b and c are irreducible, then we are done. If not then we repeat the process above for each of b and c. Since we are working in a PID, by 19.21 this process must terminate.

Next, we prove the uniqueness of factorisation: Suppose that $a \in D$ has a pair of factorisations into irreducibles:

$$a = p_1 p_2 ... p_r \text{ and } a = q_1 q_2 ... q_s.$$

We may suppose that $r \leq s$. By 19.23 the p_i's and q_j's are prime.

Since p_1 divides a we have p_1 divides $q_1 q_2 ... q_s$. Hence p_1 divides q_i for some i. We may relabel the q_i's so that p_1 divides q_1. But because q_1 is irreducible we have $q_1 = u_1 p_1$ for some unit u_1.

Hence $p_1 p_2 ... p_r = u_1 p_1 q_2 ... q_s$, and by cancellation

$$p_2 p_3 ... p_r = u_1 q_2 q_3 ... q_s.$$

By repeating the argument above we have $q_2 = u_2 p_2$ for some unit u_2.

Hence $p_2 p_3 ... p_r = u_1 u_2 p_2 q_3 ... q_s$, and by cancellation

$$p_3 ... p_r = u_1 u_2 q_3 ... q_s.$$

We continue in this way until all of the p_i's are exhausted. We have shown each p_i is an associate of the corresponding q_i for $i = 1, 2, ..., r$. Finally by cancellation we have $1 = u_1 u_2 ... u_r q_{r+1} ... q_s$. Hence there can by no further irreducibles q_i and so $r = s$.

Examples (a) We have shown in 19.18 that $\boldsymbol{Z}$ is a PID. By the theorem, $\boldsymbol{Z}$ is a UFD.

(b) By 19.19, $F[x]$ is a PID for each field F. By the theorem, $F[x]$ is a UFD.

Note that the converse of 19.24 is *false*: $\boldsymbol{Z}[x]$ is a UFD, although it is not a PID, as we observed on page 249.

19.25 Definition A *Euclidean valuation* is a mapping $\delta : D \to \boldsymbol{Z}$ satisfying:

(i) $\delta(a) \geq 0$ for all $a \in D$

(ii) For each $a, b \in D$ with $b \neq 0$ there are $q, r \in D$ such that $a = qb + r$ and either $r = 0$ or $\delta(r) < \delta(b)$

(iii) $\delta(a) \leq \delta(ab)$ for all non-zero $a, b \in D$

An integral domain with a Euclidean valuation is called a *Euclidean domain.*

Examples (a) For any field F the mapping $\delta : F[x] \to \boldsymbol{Z}$ given by $\delta(p(x)) = \deg p(x)$ is a Euclidean valuation.

Clearly (i) is satisfied: $\deg p(x) \geq 0$.

To see that (ii) is satisfied, for $a(x), b(x) \in F[x]$ with $b(x) \neq 0$, by division of polynomials there are $q(x), r(x) \in F[x]$ such that $a(x) = q(x)b(x) + r(x)$, and $r(x) = 0$ or $\deg r(x) < \deg b(x)$.

For (iii), $\deg a(x) \leq \deg a(x) + \deg b(x) = \deg a(x)b(x)$.

(b) For the Gaussian integers the multiplicative norm $\delta : \boldsymbol{Z}[i] \to \boldsymbol{Z}$ given by $\delta(m + ni) = m^2 + n^2$ is a Euclidean valuation. Clearly (i) is satisfied.

To verify (ii), suppose that $a = m + ni, b = m' + n'i \in \boldsymbol{Z}[i]$.

Let $\frac{a}{b} = \frac{m + ni}{m' + n'i} = x + yi \in \boldsymbol{C}$.

Now let r be the closest integer to x and let s be the closest integer to y, so that $|x - r| \leq \frac{1}{2}$ and $|y - s| \leq \frac{1}{2}$, and let $q = r + si \in \boldsymbol{Z}[i]$.

We have

$$\begin{aligned}\delta(a - qb) &= \delta\left(\left(\frac{a}{b} - q\right)b\right) = \delta\left(\frac{a}{b} - q\right)\delta(b) \\ &= \delta((x - r) + (y - s)i)\delta(b) = ((x - r)^2 + (y - s)^2)\delta(b) \\ &\leq \left(\left(\frac{1}{2}\right)^2 + \left(\frac{1}{2}\right)^2\right)\delta(b) = \frac{1}{2}\delta(b) < \delta(b)\end{aligned}$$

To verify (iii), observe $\delta(a) \leq \delta(a)\delta(b) = \delta(ab)$.

19.26 Proposition If D is a Euclidean domain then D is a PID (and hence a UFD).

Proof: Let D be a Euclidean domain with valuation δ, and let I be an ideal in D.

Choose $b \in I$ such that $\delta(b)$ is as small as possible. Clearly $\langle b \rangle \subseteq I$.

Now suppose $a \in I$. There are q and r in D such that

$$a = qb + r \text{ and either } r = 0 \text{ or } \delta(r) < \delta(b).$$

Since $\delta(b)$ is minimal, we have $r = 0$ and hence $a = qb$, so that $a \in \langle b \rangle$. Hence $I \subseteq \langle b \rangle$, and so $I = \langle b \rangle$, and so is a principal ideal.

Examples (a) $F[x]$ is a Euclidean domain for each field F, and so is a PID and hence a UFD. We already knew this from 19.20.

(b) By example (b) above $\mathbf{Z}[i]$ is a Euclidean domain. By 19.26 $\mathbf{Z}[i]$ is a PID and hence a UFD.

The converse of 19.26 is false: there are PID's which fail to be Euclidean domains. For example, $\mathbf{Z}[\frac{1}{2}(1+\sqrt{-19}\,)]$ is a PID, but is not a Euclidean domain.

We conclude with an important theorem. The proof is rather long and so is omitted.

19.27 Theorem (Gauss) If D is a UFD then $D[x]$ is also a UFD.

Example Z is a UFD, and so $\mathbf{Z}[x]$ is also a UFD.

This concludes our study of rings. In the next chapter, our attention returns to vector spaces.

Exercise 19

1. Find all of the units in each of the following rings:

   (i) $\mathbb{Z}_{10}$ (ii) $\mathbb{Z}_{12}$ (iii) $\mathbb{Z}_{14}$ (iv) $\mathbb{Z}_{15}$

   For the remaining elements, state which are associates of each other.

2. Show that each of the following is a multiplicative norm:

   (i) $N : \mathbb{Z}[i] \to \mathbb{Z}$ given by $N(a+bi) = a^2 + b^2$

   (ii) $N : \mathbb{Z}[\sqrt{-3}\,] \to \mathbb{Z}$ given by $N(a+b\sqrt{-3}\,) = a^2 + 3b^2$

3. Decide which of the following are irreducible in $\mathbb{Z}[i]$:

   5 7 11 13

4. Show that in $\mathbb{Z}[\sqrt{-5}\,]$ the only units are 1 and -1.

   Show that $1 + \sqrt{-5}$ is irreducible in $\mathbb{Z}[\sqrt{-5}\,]$.

   Show that $1 + \sqrt{-5}$ is *not* prime in $\mathbb{Z}[\sqrt{-5}\,]$.

   Find two distinct factorisations of 6 into irreducibles in the ring

   $$\mathbb{Z}[\sqrt{-5}\,] = \{a + b\sqrt{-5} \;:\; a, b \in \mathbb{Z}\}.$$

   Is $\mathbb{Z}[\sqrt{-5}\,]$ a UFD?

5. Show that $\delta : \mathbb{Z}[\sqrt{-2}\,] \to \mathbb{Z}$ given by

   $$\delta(m + n\sqrt{-2}\,) = m^2 + 2n^2$$

   is a Euclidean valuation.

   Is $\mathbb{Z}[\sqrt{-2}\,]$ a UFD?

Chapter 20: Vector Spaces with Products

The material in this chapter and those that follow builds upon the linear algebra that has already been studied in chapters 3 through 7. We will define an *algebra* to be a vector space, which has the extra structure of a *product* or *multiplication*. We begin by reviewing some basic ideas about real vector spaces.

Suppose that V is a non-empty set, and that we have a binary operation $+$ on V (*vector addition*) and an operation which combines a real number λ with a vector $v \in V$ to give a vector $\lambda v \in V$ (*multiplication by a scalar*).

Recall from chapter 3 that V is a *real vector space* if $(V,+)$ is an abelian group and we have:

$$(\lambda+\mu)v = \lambda v + \mu v$$

$$\lambda(u+v) = \lambda u + \lambda v$$

$$\lambda(\mu v) = (\lambda\mu)v$$

$$1v = v$$

for all $u, v \in V$ and all $\lambda, \mu \in \boldsymbol{R}$.

Examples (a) $\boldsymbol{R}^2$ is a real vector space. Vector addition is defined by

$$\begin{pmatrix} x_1 \\ y_1 \end{pmatrix} + \begin{pmatrix} x_2 \\ y_2 \end{pmatrix} = \begin{pmatrix} x_1 + x_2 \\ y_1 + y_2 \end{pmatrix}$$

and multiplication by a scalar λ by

$$\lambda \begin{pmatrix} x \\ y \end{pmatrix} = \begin{pmatrix} \lambda x \\ \lambda y \end{pmatrix}$$

(b) We can view the set of complex numbers $\boldsymbol{C}$ as a real vector space. Vector addition is addition of complex numbers and multiplication by a scalar is defined by

$$\lambda(x+yi)=\lambda x+\lambda yi \text{ for all } \lambda\in\boldsymbol{R}.$$

Of course, this is essentially the same as example (a), via the correspondence

$$\begin{pmatrix} x \\ y \end{pmatrix} \leftrightarrow x+yi$$

We say that the real vector spaces $\boldsymbol{R}^2$ and $\boldsymbol{C}$ are *isomorphic.*

(c) The set of 2×2 matrices with entries in $\boldsymbol{R}$, written $Mat(2,\boldsymbol{R})$, is a real vector space under matrix addition, and multiplication by a scalar given by

$$\lambda\begin{pmatrix} a & b \\ c & d \end{pmatrix}=\begin{pmatrix} \lambda a & \lambda b \\ \lambda c & \lambda d \end{pmatrix}$$

This vector space is isomorphic to $\boldsymbol{R}^4$.

Similarly, the set of 2×2 matrices with entries in $\boldsymbol{C}$, written $Mat(2,\boldsymbol{C})$, is a complex vector space of dimension 4. We will prefer to view $Mat(2,\boldsymbol{C})$ as a *real* vector space of dimension 8.

(d) The set of polynomials with real coefficients, $\boldsymbol{R}[x]$, is a real vector space under polynomial addition and multiplication by a scalar.

Recall from chapter 4 that a subset U of a vector space V is a *subspace* of V if U is itself a real vector space under the same operations of vector addition and scalar multiplication of V.

Recall also that a non-empty subset U is a subspace if and only if

(i) if $u,v\in U$ then $u+v\in U$

(ii) if $v\in U$ and $\lambda\in\boldsymbol{R}$ then $\lambda v\in U$

20.3 Definition Suppose that V and W are real vector spaces. A mapping $f: V \to W$ is called a *linear transformation* if

(i) $f(u+v) = f(u) + f(v)$ for all $u, v \in V$

(ii) $f(\lambda v) = \lambda f(v)$ for all $\lambda \in \boldsymbol{R}$ and all $v \in V$.

Recall that for any linear transformation f we have

$$f(\mathbf{0}) = \mathbf{0} \text{ and } f(-v) = -f(v).$$

A linear transformation that is bijective is called an *isomorphism.*

If there is an isomorphism between a pair of vector spaces V and W then we say that V is *isomorphic* to W, and we write $V \cong W$.

Examples (a) $f: \boldsymbol{R}^2 \to \boldsymbol{R}^3$ given by

$$f\begin{pmatrix} x \\ y \end{pmatrix} = \begin{pmatrix} x \\ y \\ x+y \end{pmatrix}$$

is a linear transformation. You should check this.

(b) $f: \boldsymbol{R}^2 \to \boldsymbol{C}$ given by $f\begin{pmatrix} x \\ y \end{pmatrix} = x + yi$ is an isomorphism of real vector spaces. Again, you should check this.

A linear transformation $f: V \to V$ with the same vector space as both domain and codomain is called an *endomorphism* or *operator.*

Example We may define an endomorphism of $\boldsymbol{R}^2$ by

$$f\begin{pmatrix} x \\ y \end{pmatrix} = \begin{pmatrix} x+2y \\ 3x+4y \end{pmatrix} = \begin{pmatrix} 1 & 2 \\ 3 & 4 \end{pmatrix}\begin{pmatrix} x \\ y \end{pmatrix}$$

In this way, each endomorphism of $\boldsymbol{R}^n$ can be thought of as an $n \times n$ matrix.

Suppose that V is a real vector space.

Recall from chapter 5 that a set of vectors $B = \{v_1, v_2, \ldots, v_n\}$ is a *basis* for V if

(i) *B spans V*: any vector $v \in V$ may be written as

$$v = \lambda_1 v_1 + \lambda_2 v_2 + ... + \lambda_n v_n \text{ for some } \lambda_1, ..., \lambda_n \in \boldsymbol{R}.$$

(ii) *B* is *linearly independent*:

If $\lambda_1 v_1 + \lambda_2 v_2 + ... + \lambda_n v_n = \mathbf{0}$ then $\lambda_1 = \lambda_2 = ... = \lambda_n = 0$.

Examples (a) $\left\{ \begin{pmatrix} 1 \\ 0 \end{pmatrix}, \begin{pmatrix} 0 \\ 1 \end{pmatrix} \right\}$ is the *standard basis* for $\boldsymbol{R}^2$.

Another basis for $\boldsymbol{R}^2$ is $\left\{ \begin{pmatrix} 1 \\ 2 \end{pmatrix}, \begin{pmatrix} 3 \\ 7 \end{pmatrix} \right\}$.

(b) The standard basis for $\boldsymbol{C}$, viewed as a real vector space, is $\{1, i\}$.

(c) The standard basis for $Mat(2, \boldsymbol{R})$ is

$$\left\{ \begin{pmatrix} 1 & 0 \\ 0 & 0 \end{pmatrix}, \begin{pmatrix} 0 & 1 \\ 0 & 0 \end{pmatrix}, \begin{pmatrix} 0 & 0 \\ 1 & 0 \end{pmatrix}, \begin{pmatrix} 0 & 0 \\ 0 & 1 \end{pmatrix} \right\}$$

(d) The standard basis for $\boldsymbol{R}[x]$ is $\{1, x, x^2, x^3, ...\}$.

Suppose that $B = \{v_1, v_2, ..., v_n\}$ is a basis for a vector space V. By 5.7, any vector in V may be written uniquely as a linear combination of vectors in B. This is why bases are important.

If a vector space V has a basis with a finite number of vectors then *all bases for V have the same number of vectors*. The number of vectors in each basis of V is called the *dimension*, written dim V.

Examples For the examples above

$\dim \boldsymbol{R}^2 = 2$ $\dim \boldsymbol{C} = 2$ $\dim Mat(2, \boldsymbol{R}) = 4$

The basis for $\boldsymbol{R}[x]$ has infinitely many vectors. This is an example of a vector space of *infinite dimension*.

We are now ready to give the definition of an algebra:

20.1 Definition Suppose that V is a real vector space. V is a *real algebra* if there is a binary operation $\cdot$ on V, known as a *product*, which satisfies:

$$(u+v)\cdot w = u\cdot w + v\cdot w$$

$$u\cdot(v+w) = u\cdot v + u\cdot w$$

$$\lambda u\cdot v = u\cdot \lambda v = \lambda(u\cdot v)$$

for all $u, v, w \in V$ and all $\lambda \in \boldsymbol{R}$.

These three conditions may be summed up by saying that the product is *bilinear*, so that $u\cdot v$ is linear both in u and in v.

Examples (a) We can make the vector space $\boldsymbol{R}^2$ into an algebra by defining a product:

$$\begin{pmatrix} x_1 \\ y_1 \end{pmatrix} \cdot \begin{pmatrix} x_2 \\ y_2 \end{pmatrix} = \begin{pmatrix} x_1x_2 \\ y_1y_2 \end{pmatrix}$$

However, this algebra has no geometric significance and so is not particularly useful.

(b) We already mentioned that as real vector spaces $\boldsymbol{C}$ is isomorphic to $\boldsymbol{R}^2$. The field of complex numbers already has a bilinear product, and so is a real algebra. However, this real algebra is not isomorphic to that of example (a).

(c) The vector space $Mat(2, \boldsymbol{R})$ is a real algebra under matrix multiplication.

(d) The vector space $\boldsymbol{R}[x]$ is a real algebra under polynomial multiplication.

(e) $\boldsymbol{R}^3$ is a real algebra under the cross product:

$$\begin{pmatrix} x_1 \\ y_1 \\ z_1 \end{pmatrix} \times \begin{pmatrix} x_2 \\ y_2 \\ z_2 \end{pmatrix} = \begin{pmatrix} y_1z_2 - y_2z_1 \\ -x_1z_2 + x_2z_1 \\ x_1y_2 - x_2y_1 \end{pmatrix}$$

An algebra is *commutative* if $u \cdot v = v \cdot u$ for all $u, v \in V$.

Examples (a), (b) and (d) above are commutative, whereas (c) and (e) are not.

An algebra is *associative* if $(u \cdot v) \cdot w = u \cdot (v \cdot w)$ for all $u, v, w \in V$

Each of the examples (a) through (d) above is associative, whereas (e) is not.

An algebra has a *unity* if there is $1 \in V$ such that $1 \cdot v = v = v \cdot 1$ for all $v \in V$.

For example, the unity of $Mat(2, \boldsymbol{R})$ is $\begin{pmatrix} 1 & 0 \\ 0 & 1 \end{pmatrix}$, whereas the algebra of example (e) does not have a unity.

20.2 Definition Suppose that A is a real algebra and that $B \subseteq A$.

B is a *subalgebra* of A if B is itself a real algebra under the same operations as A.

Examples Let $B = \left\{ \begin{pmatrix} x & 0 \\ 0 & y \end{pmatrix} : x, y \in \boldsymbol{R} \right\}$

B is a subalgebra of $Mat(2, \boldsymbol{R})$.

20.3 Test for a Subalgebra

Let A be a real algebra. A non-empty subset B of A is a subalgebra if and only if:

(i) if $x, y \in B$ then $x + y \in B$

(ii) if $x \in B$ and $\lambda \in \boldsymbol{R}$ then $\lambda x \in B$

(iii) if $x, y \in B$ then $x \cdot y \in B$.

Conditions (i) and (ii) ensure that B is a subspace of A. Condition (iii) ensures that B is closed under the product.

As we did for rings, we shall often write $x \cdot y$ as xy.

Example For the example above:

$$\begin{pmatrix} x & 0 \\ 0 & y \end{pmatrix} + \begin{pmatrix} x' & 0 \\ 0 & y' \end{pmatrix} = \begin{pmatrix} x+x' & 0 \\ 0 & y+y' \end{pmatrix} \in B$$

$$\lambda \begin{pmatrix} x & 0 \\ 0 & y \end{pmatrix} = \begin{pmatrix} \lambda x & 0 \\ 0 & \lambda y \end{pmatrix} \in B$$

$$\begin{pmatrix} x & 0 \\ 0 & y \end{pmatrix} \begin{pmatrix} x' & 0 \\ 0 & y' \end{pmatrix} = \begin{pmatrix} xx' & 0 \\ 0 & yy' \end{pmatrix} \in B$$

20.4 Definition Suppose that A and B are real algebras.

A linear transformation $f: A \longrightarrow B$ is called an *algebra map* if

$$f(x \cdot y) = f(x) \cdot f(y)$$

for all $x, y \in A$.

Example The mapping $f: \boldsymbol{C} \rightarrow Mat(2, \boldsymbol{R})$ given by

$$f(x+yi) = \begin{pmatrix} x & y \\ -y & x \end{pmatrix}$$

is an algebra map. Let $z_1 = x_1 + y_1 i$ and $z_2 = x_2 + y_2 i$.

We show first that f is a linear transformation:

$$f(z_1 + z_2) = f((x_1 + x_2) + (y_1 + y_2)i) = \begin{pmatrix} x_1 + x_2 & y_1 + y_2 \\ -y_1 - y_2 & x_1 + x_2 \end{pmatrix}$$

$$= \begin{pmatrix} x_1 & y_1 \\ -y_1 & x_1 \end{pmatrix} + \begin{pmatrix} x_2 & y_2 \\ -y_2 & x_2 \end{pmatrix} = f(z_1) + f(z_2)$$

$$f(\lambda(x+yi)) = f(\lambda x + \lambda yi) = \begin{pmatrix} \lambda x & \lambda y \\ -\lambda y & \lambda x \end{pmatrix} = \lambda \begin{pmatrix} x & y \\ -y & x \end{pmatrix} = \lambda f(x+yi)$$

Finally, to show that f is an algebra map:

$$f(z_1z_2) = f((x_1+y_1i)(x_2+y_2i)) = f((x_1x_2-y_1y_2)+(x_1y_2+x_2y_1)i)$$

$$= \begin{pmatrix} x_1x_2-y_1y_2 & x_1y_2+x_2y_1 \\ -x_1y_2-x_2y_1 & x_1x_2-y_1y_2 \end{pmatrix} = \begin{pmatrix} x_1 & y_1 \\ -y_1 & x_1 \end{pmatrix}\begin{pmatrix} x_2 & y_2 \\ -y_2 & x_2 \end{pmatrix}$$

$$= f(z_1)f(z_2).$$

An algebra map that is bijective is called an *isomorphism.* If there is an isomorphism $A \to B$ then we say that the algebras A and B are *isomorphic*, and write $A \cong B$.

Examples (a) Define a product on $\boldsymbol{R}^2$ by

$$\begin{pmatrix} x \\ y \end{pmatrix} \times \begin{pmatrix} x' \\ y' \end{pmatrix} = \begin{pmatrix} xx'-yy' \\ xy'+x'y \end{pmatrix}$$

The mapping $f: \boldsymbol{R}^2 \to \boldsymbol{C}$ given by $f\begin{pmatrix} x \\ y \end{pmatrix} = x+yi$ is an isomorphism. You should check this.

(b) Although $\boldsymbol{R}^2$ and $\boldsymbol{C}$ are isomorphic as real vector spaces, the algebra on $\boldsymbol{R}^2$ with product

$$\begin{pmatrix} x_1 \\ y_1 \end{pmatrix} \cdot \begin{pmatrix} x_2 \\ y_2 \end{pmatrix} = \begin{pmatrix} x_1x_2 \\ y_1y_2 \end{pmatrix}$$

is not isomorphic to $\boldsymbol{C}$.

20.5 Definition Suppose that A and B are real algebras. We make

$$A \times B = \{(x,y) : x \in A, y \in B\}$$

into an algebra by defining

$$(x,y)+(x',y') = (x+x', y+y')$$

$$\lambda(x,y) = (\lambda x, \lambda y)$$

$$(x,y)\cdot(x',y') = (xx', yy')$$

This algebra is called the *direct sum*, and is denoted $A \oplus B$.

20.6 Proposition $\dim A \oplus B = \dim A + \dim B$

Proof: Suppose that A has basis $\{u_1, u_2, ..., u_m\}$ and B has basis $\{v_1, v_2, ..., v_n\}$.

Then $\{(u_1, 0), (u_2, 0), , ..., (u_m, 0), (0, v_1), (0, v_2), ..., (0, v_n)\}$ is a basis for $A \oplus B$.

Linear independence: Suppose

$$\begin{aligned} &\lambda_1(u_1, 0) + \lambda_2(u_2, 0) + ... + \lambda_m(u_m, 0) \\ &+ \mu_1(0, v_1) + \mu_2(0, v_2) + ... + \mu_n(0, v_n) \\ = &(0, 0) \end{aligned}$$

Then $(\lambda_1 u_1 + \lambda_2 u_2 + ... + \lambda_m u_m, \mu_1 v_1 + \mu_2 v_2 + ... + \mu_n v_n) = (0, 0)$

Hence $\lambda_i = 0$ for $1 \le i \le m$ and $\mu_j = 0$ for $1 \le j \le n$.

Spanning: Suppose $(u, v) \in A \oplus B$.

Then $u = \lambda_1 u_1 + \lambda_2 u_2 + ... + \lambda_m u_m$ for some scalars λ_i, and $v = \mu_1 v_1 + \mu_2 v_2 + ... + \mu_n v_n$ for some scalars μ_j. We have

$$\begin{aligned} (u, v) &= (u, 0) + (0, v) \\ &= \lambda_1(u_1, 0) + \lambda_2(u_2, 0) + ... + \lambda_m(u_m, 0) \\ &\quad + \mu_1(0, v_1) + \mu_2(0, v_2) + ... + \mu_n(0, v_n) \end{aligned}$$

Example

$\boldsymbol{R} \oplus \boldsymbol{C}$ is the algebra of ordered pairs (x, z) with $x \in \boldsymbol{R}$ and $z \in \boldsymbol{C}$.

Vector addition is given by $(x_1, z_1) + (x_2, z_2) = (x_1 + x_2, z_1 + z_2)$ and multiplication by a real scalar by $\lambda(x, z) = (\lambda x, \lambda z)$.

The product in this algebra is $(x_1, z_1) \cdot (x_2, z_2) = (x_1 x_2, z_1 z_2)$.

A basis for $\boldsymbol{R} \oplus \boldsymbol{C}$ as a real vector space is $\{(1, 0), (0, 1), (0, i)\}$ and hence

$$\dim \boldsymbol{R} \oplus \boldsymbol{C} = 1 + 2 = 3.$$

Exercises 20

1. Convince yourself that each of the following is a real vector space under the operations given in the chapter:

$\boldsymbol{R}^2$ $\quad$ $\boldsymbol{C}$

$Mat(2, \boldsymbol{R})$ $\quad$ $\boldsymbol{R}[x]$

2. State the dimension of each of the following real vector spaces:

(i) $\boldsymbol{R}^4$ $\quad$ (ii) $Mat(3, \boldsymbol{R})$

(iii) $Mat(2, \boldsymbol{C})$

Write down a basis for each of these real vector spaces.

3. Show that each of the following products is bilinear:

(i) $$\begin{pmatrix} x_1 \\ y_1 \end{pmatrix} \cdot \begin{pmatrix} x_2 \\ y_2 \end{pmatrix} = \begin{pmatrix} x_1 x_2 \\ y_1 y_2 \end{pmatrix} \text{ on } \boldsymbol{R}^2$$

(ii) $(x_1 + y_1 i)(x_2 + y_2 i) = (x_1 x_2 - y_1 y_2) + (x_1 y_2 + x_2 y_1)i$ on $\boldsymbol{C}$.

(iii) $$\begin{pmatrix} x_1 \\ y_1 \\ z_1 \end{pmatrix} \times \begin{pmatrix} x_2 \\ y_2 \\ z_2 \end{pmatrix} = \begin{pmatrix} y_1 z_2 - y_2 z_1 \\ -x_1 z_2 + x_2 z_1 \\ x_1 y_2 - x_2 y_1 \end{pmatrix} \text{ on } \boldsymbol{R}^3$$

4. Which of the following are subspaces of $Mat(2, \boldsymbol{R})$? Which are subalgebras?

(i) $$\left\{ \begin{pmatrix} x & 0 \\ 0 & y \end{pmatrix} : x, y \in \boldsymbol{R} \right\}$$ (ii) $$\left\{ \begin{pmatrix} 0 & x \\ y & 0 \end{pmatrix} : x, y \in \boldsymbol{R} \right\}$$

(iii) $$\left\{ \begin{pmatrix} x & y \\ -y & x \end{pmatrix} : x, y \in \boldsymbol{R} \right\}$$ (iv) $$\left\{ \begin{pmatrix} x & y \\ y & -x \end{pmatrix} : x, y \in \boldsymbol{R} \right\}$$

5. Show that

$$\left\{\begin{pmatrix} 1 & 0 \\ 0 & 1 \end{pmatrix}, \begin{pmatrix} 1 & 0 \\ 0 & -1 \end{pmatrix}, \begin{pmatrix} 0 & 1 \\ 1 & 0 \end{pmatrix}, \begin{pmatrix} 0 & 1 \\ -1 & 0 \end{pmatrix}\right\}$$

is a basis for $Mat(2, \boldsymbol{R})$.

6. Show that $f : \boldsymbol{R}^4 \to Mat(2, \boldsymbol{R})$ given by

$$f\begin{pmatrix} w \\ x \\ y \\ z \end{pmatrix} = \begin{pmatrix} w & x \\ y & z \end{pmatrix}$$

is an isomorphism of vector spaces.

7. Write down a basis for each of the following real vector spaces:

(i) $\boldsymbol{R} \oplus \boldsymbol{R}$ (ii) $\boldsymbol{C} \oplus \boldsymbol{C}$

(iii) $\boldsymbol{R}^2 \oplus \boldsymbol{R}^3$

Chapter 21: The Exterior Algebra of a Vector Space

We begin with an example. Consider the vector space $\boldsymbol{R}^3$ with basis

$$\{e_1, e_2, e_3\}.$$

We form a new real vector space of dimension 8, denoted $\Lambda\boldsymbol{R}^3$, with basis the set of products of basis vectors in $\boldsymbol{R}^3$:

$$\{1, e_1, e_2, e_3, e_1 \wedge e_2, e_1 \wedge e_3, e_2 \wedge e_3, e_1 \wedge e_2 \wedge e_3\}$$

The product $\wedge$ is called the *wedge product.* We think of the wedge product $u \wedge v$ geometrically as a parallelogram with u and v as two of the non-parallel sides.

Similarly, $u \wedge v \wedge w$ may be thought of as a parallelepiped. This product is assumed to be associative. The wedge product on vectors e_i is *anti-commutative*, so that for example

$$e_2 \wedge e_1 = -e_1 \wedge e_2$$

Notice also that we have $e_1 \wedge e_1 = -e_1 \wedge e_1$ and so $e_1 \wedge e_1 = \mathbf{0}$.

It is sometimes helpful to think of $\Lambda\boldsymbol{R}^3$ as the direct sum

$$\Lambda^0\boldsymbol{R}^3 \oplus \Lambda^1\boldsymbol{R}^3 \oplus \Lambda^3\boldsymbol{R}^3 \oplus \Lambda^3\boldsymbol{R}^3$$

where $\Lambda^0\boldsymbol{R}^3$ has basis $\{1\}, \Lambda^1\boldsymbol{R}^3$ has basis $\{e_1, e_2, e_3\}$, $\Lambda^2\boldsymbol{R}^3$ has basis $\{e_1 \wedge e_2, e_1 \wedge e_3, e_2 \wedge e_3\}$ and $\Lambda^3\boldsymbol{R}^3$ has basis $\{e_1 \wedge e_2 \wedge e_3\}$.

21.1 Definition Suppose that V is a real vector space with basis $\{e_1, e_2, ..., e_n\}$.

The *exterior algebra* of V, denoted ΛV, is a real vector space of dimension 2^n with basis the set of all products $e_{i_1} \wedge e_{i_2} \wedge ... \wedge e_{i_r}$ such that $i_1 < i_2 < ... < i_r$ and $r \leq n$. By convention, the product of zero e_i's is taken to be 1.

The vectors e_i satisfy the relations $e_i \wedge e_j = -e_j \wedge e_i$ for all i and j. As in the previous example, it follows that $e_i \wedge e_i = \mathbf{0}$ for each i.

As we did for $\Lambda \boldsymbol{R}^3$, we may think of ΛV as a direct sum

$$\Lambda^0 V \oplus \Lambda^1 V \oplus \Lambda^2 V \oplus \ldots \oplus \Lambda^n V$$

Notice that $\dim \Lambda^i V = \binom{n}{i}$ for each *i*.

In particular, if $\dim V = n$ then $\Lambda^n V$ is 1 dimensional.

21.2 Proposition For all $u, v \in V$ we have

(i) $v \wedge v = \mathbf{0}$

(ii) $v \wedge u = -u \wedge v$

The proof is left as an exercise.

21.3 Proposition A set of vectors $\{v_1, v_2, \ldots, v_n\}$ is linearly *dependent* if and only if

$$v_1 \wedge v_2 \wedge \ldots \wedge v_n = \mathbf{0}$$

Proof: "$\Rightarrow$" Suppose that $\{v_1, v_2, \ldots, v_n\}$ is linearly *dependent*, so that one of the v_i may be expressed as a linear combination of the others. We suppose, without loss of generality, that

$$v_1 = \lambda_2 v_2 + \lambda_3 v_3 + \ldots + \lambda_n v_n \text{ for some } \lambda_2, \lambda_3, \ldots, \lambda_n \in \boldsymbol{R}.$$

Then

$$v_1 \wedge v_2 \wedge v_3 \wedge \ldots \wedge v_n = (\lambda_2 v_2 + \lambda_3 v_3 + \ldots + \lambda_n v_n) \wedge v_2 \wedge v_3 \wedge \ldots \wedge v_n$$

$$= \lambda_2 v_2 \wedge v_2 \wedge v_3 \wedge \ldots \wedge v_n + \lambda_3 v_3 \wedge v_2 \wedge v_3 \wedge \ldots \wedge v_n + \ldots$$

$$+ \lambda_n v_n \wedge v_2 \wedge v_3 \wedge \ldots \wedge v_n$$

$$= \mathbf{0}$$

by 21.2(i), since each term contains a repetition of a v_i.

The proof of "$\Leftarrow$" is omitted.

Examples (a) Consider the set of vectors

$$\left\{ \begin{pmatrix} 1 \\ 0 \\ -2 \end{pmatrix}, \begin{pmatrix} 0 \\ 1 \\ 3 \end{pmatrix}, \begin{pmatrix} 1 \\ 1 \\ 1 \end{pmatrix} \right\}$$

$$\begin{aligned}
&(e_1 - 2e_3) \wedge (e_2 + 3e_3) \wedge (e_1 + e_2 + e_3) \\
&= (e_1 - 2e_3) \wedge (e_2 \wedge e_1 + e_2 \wedge e_3 + 3e_3 \wedge e_1 + 3e_3 \wedge e_2) \\
&= e_1 \wedge e_2 \wedge e_3 + 3e_1 \wedge e_3 \wedge e_2 - 2e_3 \wedge e_2 \wedge e_1 \\
&= (1 - 3 + 2)e_1 \wedge e_2 \wedge e_3 \\
&= \mathbf{0}
\end{aligned}$$

Hence the vectors are linearly *dependent.*

(b) Consider the set of vectors

$$\left\{ \begin{pmatrix} 1 \\ 0 \\ -2 \end{pmatrix}, \begin{pmatrix} 0 \\ 1 \\ 3 \end{pmatrix}, \begin{pmatrix} 1 \\ 1 \\ 0 \end{pmatrix} \right\}$$

$$\begin{aligned}
&(e_1 - 2e_3) \wedge (e_2 + 3e_3) \wedge (e_1 + e_2) \\
&= (e_1 - 2e_3) \wedge (e_2 \wedge e_1 + 3e_3 \wedge e_1 + 3e_3 \wedge e_2) \\
&= 3e_1 \wedge e_3 \wedge e_2 - 2e_3 \wedge e_2 \wedge e_1 \\
&= -e_1 \wedge e_2 \wedge e_3 \\
&\neq \mathbf{0}
\end{aligned}$$

Hence the vectors are linearly *independent.*

Endomorphisms of $\Lambda^n V$

Given an endomorphism $V \rightarrow V$ represented by a square matrix A there are several ways of letting A induce an endomorphism of $\Lambda^n V$. For example, each of the following gives an endomorphism of $\Lambda^3 \boldsymbol{R}^3$:

$$u \wedge v \wedge w \mapsto Au \wedge v \wedge w + u \wedge Av \wedge w + u \wedge v \wedge Aw$$

$$u \wedge v \wedge w \mapsto Au \wedge Av \wedge w + Au \wedge v \wedge Aw + u \wedge Av \wedge Aw$$

$$u \wedge v \wedge w \mapsto Au \wedge Av \wedge Aw$$

We denote these endomorphisms by $A^{(1)}, A^{(2)}$ and $A^{(3)}$ respectively.

Since if $\dim V = n$ then $\Lambda^n V$ is 1 dimensional, an endomorphism of $\Lambda^n V$ is simply multiplication by a scalar. This observation will be useful to us in the sections that follow. Notice also that Ae_i gives the *i*th column of the matrix A.

Determinants

The standard basis for $\boldsymbol{R}^2$ is $\left\{ e_1 = \begin{pmatrix} 1 \\ 0 \end{pmatrix}, e_2 = \begin{pmatrix} 0 \\ 1 \end{pmatrix} \right\}$.

Consider the matrix $A = \begin{pmatrix} a & b \\ c & d \end{pmatrix}$ in $Mat(2, \boldsymbol{R})$.

We have

$$Ae_1 \wedge Ae_2 = \begin{pmatrix} a \\ c \end{pmatrix} \wedge \begin{pmatrix} b \\ d \end{pmatrix} = (ae_1 + ce_2) \wedge (be_1 + de_2)$$

$$= abe_1 \wedge e_1 + ade_1 \wedge e_2 + cbe_2 \wedge e_1 + cde_2 \wedge e_2$$

$$= (ad - cb)e_1 \wedge e_2$$

You should recognise the quantity $ad - cb$ as the determinant of the matrix A. This suggests the following definition:

21.4 Definition Let A be an $n \times n$ real matrix.

$$A^{(n)}(e_1 \wedge e_2 \wedge ... \wedge e_n) = Ae_1 \wedge Ae_2 \wedge ... \wedge Ae_n = \lambda e_1 \wedge e_2 \wedge ... \wedge e_n$$

The scalar λ is called the *determinant* of the matrix A, denoted $\det A$.

Example We will use the exterior algebra to calculate the determinant of

$$A = \begin{pmatrix} 1 & 0 & 2 \\ 0 & 4 & 1 \\ 3 & -1 & 0 \end{pmatrix}$$

$$\begin{aligned}(\det A)e_1 \wedge e_2 \wedge e_3 &= Ae_1 \wedge Ae_2 \wedge Ae_3 \\ &= \begin{pmatrix} 1 \\ 0 \\ 3 \end{pmatrix} \wedge \begin{pmatrix} 0 \\ 4 \\ -1 \end{pmatrix} \wedge \begin{pmatrix} 2 \\ 1 \\ 0 \end{pmatrix} \\ &= (e_1 + 3e_3) \wedge (4e_2 - e_3) \wedge (2e_1 + e_2) \\ &= (e_1 + 3e_3) \wedge (8e_2 \wedge e_1 - 2e_3 \wedge e_1 - e_3 \wedge e_2) \\ &= -e_1 \wedge e_3 \wedge e_2 + 24e_3 \wedge e_2 \wedge e_1 \\ &= -23e_1 \wedge e_2 \wedge e_3\end{aligned}$$

Hence $\det A = -23$.

Of course, there are easier methods to calculate the determinant. But the definition via the exterior algebra gives us powerful means of proving results about determinants. In particular, notice that the determinant is the wedge product of the columns of the matrix. Here is a selection of results.

21.5 Proposition If two columns of a matrix are the same then the determinant of that matrix is 0.

Proof: If the *i*th and the *j*th columns of the matrix A are the same, then we have $Ae_i = Ae_j$.

Hence in the product $Ae_1 \wedge Ae_2 \wedge \ldots \wedge Ae_n$ two of the factors are the same. Hence by 21.2(i) the product, and hence the determinant, is zero.

21.6 Proposition Suppose that $\det A = \lambda$. If two columns of A are exchanged, the new matrix has determinant $-\lambda$.

Proof: If the *i*th and the *j*th columns of the matrix A are exchanged, then in the product $Ae_1 \wedge Ae_2 \wedge \ldots \wedge Ae_n$ the order of the factors Ae_i and Ae_j is exchanged.

By 21.2(ii), this changes the sign of the product, and hence the sign of the determinant.

21.7 Proposition Let A be an $n \times n$ real matrix. Let a_{ij} denote the entry at the intersection of the *i*th row and the *j*th column and let A_{ij} denote the $(n-1)\times(n-1)$ matrix obtained from A by removing the *i*th row and the *j*th column.

Then we have

$$\det A = a_{11}\det A_{11} - a_{21}\det A_{21} + \ldots + (-1)^{n+1} a_{n1}\det A_{n1}.$$

This is called a *Laplace expansion.*

Proof: Again we give the proof for the 3×3 case:

$$\det A(e_1 \wedge e_2 \wedge e_3)$$

$$= \begin{pmatrix} a_{11} \\ a_{21} \\ a_{31} \end{pmatrix} \wedge \begin{pmatrix} a_{12} \\ a_{22} \\ a_{32} \end{pmatrix} \wedge \begin{pmatrix} a_{13} \\ a_{23} \\ a_{33} \end{pmatrix}$$

$$= \begin{pmatrix} a_{11} \\ a_{21} \\ a_{31} \end{pmatrix} \wedge (a_{12}e_1 + a_{22}e_2 + a_{32}e_3) \wedge (a_{13}e_1 + a_{23}e_2 + a_{33}e_3)$$

$$= \begin{pmatrix} a_{11} \\ a_{21} \\ a_{31} \end{pmatrix} \wedge \{(a_{12}a_{23} - a_{22}a_{13})e_1 \wedge e_2$$

$$+ (a_{12}a_{33} - a_{32}a_{13})e_1 \wedge e_3 + (a_{22}a_{33} - a_{32}a_{23})e_2 \wedge e_3\}$$

$$= (a_{11}e_1 + a_{21}e_2 + a_{31}e_3)$$

$$\wedge \{\det A_{31} e_1 \wedge e_2 + \det A_{21} e_1 \wedge e_3 + \det A_{11} e_2 \wedge e_3\}$$

$$= (a_{11}\det A_{11} - a_{21}\det A_{21} + a_{31}\det A_{31}) e_1 \wedge e_2 \wedge e_3$$

21.8 Proposition Let A and B be $n \times n$ real matrices. We have

$$\det AB = \det A \det B.$$

Proof:

$$\begin{aligned}
\det AB(e_1 \wedge e_2 \wedge \ldots \wedge e_n) &= (AB)^{(n)}(e_1 \wedge e_2 \wedge \ldots \wedge e_n) \\
&\quad ABe_1 \wedge ABe_2 \wedge \ldots \wedge ABe_n \\
&= A^{(n)}(Be_1 \wedge Be_2 \wedge \ldots \wedge Be_n) \\
&= A^{(n)} \circ B^{(n)}(e_1 \wedge e_2 \wedge \ldots \wedge e_n) \\
&= \det A \det B(e_1 \wedge e_2 \wedge \ldots \wedge e_n)
\end{aligned}$$

The Trace of a Matrix

The *trace* of a matrix A is the sum of its diagonal entries.

For example $\begin{pmatrix} 1 & 2 \\ 3 & 4 \end{pmatrix}$ has trace $1+4=5$ and $\begin{pmatrix} 3 & 0 & 1 \\ 6 & 5 & 2 \\ 4 & 8 & 7 \end{pmatrix}$ has trace $3+5+7=15$.

For the matrix $\begin{pmatrix} a & b \\ c & d \end{pmatrix}$

$$\begin{aligned}
A^{(1)}(e_1 \wedge e_2) &= Ae_1 \wedge e_2 + e_1 \wedge Ae_2 \\
&= \begin{pmatrix} a \\ c \end{pmatrix} \wedge e_2 + e_1 \wedge \begin{pmatrix} b \\ d \end{pmatrix} \\
&= (ae_1 + ce_2) \wedge e_2 + e_1 \wedge (be_1 + de_2) \\
&= (a+d)e_1 \wedge e_2
\end{aligned}$$

This suggests the following definition:

21.9 Definition Let A be an $n \times n$ real matrix.

$$A^{(1)}(e_1 \wedge e_2 \wedge \ldots \wedge e_n) = Ae_1 \wedge e_2 \wedge \ldots \wedge e_n + e_1 \wedge Ae_2 \wedge \ldots \wedge e_n + \ldots$$
$$+ e_1 \wedge e_2 \wedge \ldots \wedge Ae_n$$
$$= \lambda e_1 \wedge e_2 \wedge \ldots \wedge e_n$$

The scalar λ is called the *trace* of the matrix A, denoted tr A.

Here are some properties of the trace:

21.10 Proposition Let A and B be $n \times n$ real matrices. Then

(i) $\text{tr}\,(A+B) = \text{tr}\,A + \text{tr}\,B$.

(ii) $\text{tr}\,A^T = \text{tr}\,A$.

(iii) $\text{tr}\,\lambda A = \lambda\,\text{tr}\,A$.

Proof:

(i) Using the subscript notation of 21.7

$$\text{tr}\,(A+B) = (a_{11} + b_{11}) + \ldots + (a_{nn} + b_{nn})$$
$$= (a_{11} + \ldots + a_{nn}) + (b_{11} + \ldots + b_{nn}) = \text{tr}\,A + \text{tr}\,B$$

(ii) Since transposing a matrix leaves the diagonal elements unchanged, the result is clear.

(iii) $\text{tr}\,\lambda A = (\lambda a_{11} + \ldots + \lambda a_{nn}) = \lambda(a_{11} + \ldots + a_{nn}) = \lambda\,\text{tr}\,A$.

We do *not* have $\text{tr}\,AB = \text{tr}A\,\text{tr}B$, however we do have the following useful result:

21.11 Proposition For a pair of $n \times n$ real matrices A and B we have

$$\text{tr}\,AB = \text{tr}\,BA.$$

Proof: We give the proof for the 3×3 case:

$$
\begin{aligned}
&\operatorname{tr} A \operatorname{tr} B(e_1 \wedge e_2 \wedge e_3) \\
&= A^{(1)} \circ B^{(1)}(e_1 \wedge e_2 \wedge e_3) \\
&= A^{(1)}(Be_1 \wedge e_2 \wedge e_3 + e_1 \wedge Be_2 \wedge e_3 + e_1 \wedge e_2 \wedge Be_3) \\
&= ABe_1 \wedge e_2 \wedge e_3 + Be_1 \wedge Ae_2 \wedge e_3 + Be_1 \wedge e_2 \wedge Ae_3 \\
&\quad + Ae_1 \wedge Be_2 \wedge e_3 + e_1 \wedge ABe_2 \wedge e_3 + e_1 \wedge Be_2 \wedge Ae_3 \\
&\quad + Ae_1 \wedge e_2 \wedge Be_3 + e_1 \wedge Ae_2 \wedge Be_3 + e_1 \wedge e_2 \wedge ABe_3 \\
&= \operatorname{tr} AB(e_1 \wedge e_2 \wedge e_3) + Be_1 \wedge Ae_2 \wedge e_3 + Be_1 \wedge e_2 \wedge Ae_3 \\
&\quad + Ae_1 \wedge Be_2 \wedge e_3 + e_1 \wedge Be_2 \wedge Ae_3 + Ae_1 \wedge e_2 \wedge Be_3 \\
&\quad + e_1 \wedge Ae_2 \wedge Be_3
\end{aligned}
$$

Similarly

$$
\begin{aligned}
&\operatorname{tr} B \operatorname{tr} A(e_1 \wedge e_2 \wedge e_3) \\
&= B^{(1)} \circ A^{(1)}(e_1 \wedge e_2 \wedge e_3) \\
&= \operatorname{tr} BA(e_1 \wedge e_2 \wedge e_3) + Be_1 \wedge Ae_2 \wedge e_3 + Be_1 \wedge e_2 \wedge Ae_3 \\
&\quad + Ae_1 \wedge Be_2 \wedge e_3 + e_1 \wedge Be_2 \wedge Ae_3 + Ae_1 \wedge e_2 \wedge Be_3 \\
&\quad + e_1 \wedge Ae_2 \wedge Be_3
\end{aligned}
$$

All except the first terms of these expansions are equal. Since operators on a one-dimensional vector space commute, we have

$$A^{(1)} \circ B^{(1)} = B^{(1)} \circ A^{(1)}$$

Hence $\operatorname{tr} AB = \operatorname{tr} BA$ as required.

The Characteristic Polynomial

Recall from chapter 6 that to find the eigenvalues of a square matrix A we calculated $\det(A - xI)$. This gave the *characteristic polynomial*, which we denote $\chi_A(x)$. The roots of $\chi_A(x)$ are the eigenvalues of A.

Example The matrix $\begin{pmatrix} 2 & 2 & -1 \\ 0 & 3 & 1 \\ 0 & 0 & 4 \end{pmatrix}$ has characteristic polynomial

$$-x^3 + 9x^2 - 26x + 24.$$

We notice that the coefficient of x^2 is the trace and the constant is the determinant of A. This suggests that we ought to be able to make sense of the other coefficients in terms of operators $A^{(i)}$.

21.12 Proposition The characteristic polynomial of an $n \times n$ matrix A is

$$\chi_A(x) = \sum_{i=0}^{n} A^{(i)}(-x)^{n-i}$$

Example For a 3×3 matrix A we have

$$\begin{aligned}
&\chi_A(x)(e_1 \wedge e_2 \wedge e_3) \\
&= (\det(A - xI))e_1 \wedge e_2 \wedge e_3 \\
&= (A - xI)e_1 \wedge (A - xI)e_2 \wedge (A - xI)e_3 \\
&= (Ae_1 - xe_1) \wedge (Ae_2 - xe_2) \wedge (Ae_3 - xe_3) \\
&= (Ae_1 - xe_1) \\
&\quad \wedge (Ae_2 \wedge Ae_3 - Ae_2 \wedge xe_3 - xe_2 \wedge Ae_3 + xe_2 \wedge xe_3) \\
&= Ae_1 \wedge Ae_2 \wedge Ae_3 - Ae_1 \wedge Ae_2 \wedge xe_3 - Ae_1 \wedge xe_2 \wedge Ae_3 \\
&\quad + Ae_1 \wedge xe_2 \wedge xe_3 - xe_1 \wedge Ae_2 \wedge Ae_3 + xe_1 \wedge Ae_2 \wedge xe_3 \\
&\quad + xe_1 \wedge xe_2 \wedge Ae_3 - xe_1 \wedge xe_2 \wedge xe_3 \\
&= Ae_1 \wedge Ae_2 \wedge Ae_3 \\
&\quad - x(Ae_1 \wedge Ae_2 \wedge e_3 + Ae_1 \wedge e_2 \wedge Ae_3 + e_1 \wedge Ae_2 \wedge Ae_3) \\
&\quad + x^2(Ae_1 \wedge e_2 \wedge e_3 + e_1 \wedge Ae_2 \wedge e_3 + e_1 \wedge e_2 \wedge Ae_3) \\
&\quad - x^3 e_1 \wedge e_2 \wedge e_3 \\
&= (A^{(3)} - xA^{(2)} + x^2 A^{(1)} - x^3)e_1 \wedge e_2 \wedge e_3
\end{aligned}$$

Exercises 21

1. Write down the basis for the exterior algebra $\Lambda \boldsymbol{R}^4$.

2. Simplify each of the following:

   (i) $8e_1 \wedge e_2 + 5e_2 \wedge e_1$

   (ii) $6e_1 \wedge e_3 \wedge e_2 + 7e_3 \wedge e_2 \wedge e_1 + 10e_2 \wedge e_3 \wedge e_1$

   (iii) $e_1 \wedge e_2 \wedge e_3 \wedge (e_1 + e_2 + e_3)$

3. Expand and simplify each of the following:

   (i) $(e_1 + 3e_2) \wedge (4e_1 - 5e_2)$

   (ii) $(2e_1 + e_2 + 4e_3) \wedge (3e_1 - 6e_2 + 2e_3)$

   (iii) $(2e_1 + 5e_2) \wedge (3e_1 - e_2) \wedge (4e_1 + 7e_2)$

   (iv) $(e_1 + e_2 \wedge e_3) \wedge (e_1 + e_2 \wedge e_3)$

   (v) $(e_1 \wedge e_3 + 3e_2 \wedge e_4) \wedge (2e_1 + 10e_2 \wedge e_3 - 5e_4)$

4. Show that for all $u, v \in \boldsymbol{R}^3$ we have

   (i) $v \wedge v = \mathbf{0}$ (ii) $v \wedge u = -u \wedge v$

   Hint: choose a basis $\{e_1, e_2, e_3\}$ and write $v = xe_1 + ye_2 + ze_3$ and $u = x'e_1 + y'e_2 + z'e_3$.

5. Using exterior products determine whether or not each of the following sets of vectors is linearly independent:

   (i) $$\left\{ \begin{pmatrix} 1 \\ 0 \\ 2 \\ 1 \end{pmatrix}, \begin{pmatrix} 1 \\ -1 \\ 0 \\ 0 \end{pmatrix}, \begin{pmatrix} 0 \\ 1 \\ 2 \\ 1 \end{pmatrix} \right\}$$

(ii) $$\left\{ \begin{pmatrix} 1 \\ 0 \\ 2 \\ 1 \end{pmatrix}, \begin{pmatrix} 1 \\ -1 \\ 0 \\ 0 \end{pmatrix}, \begin{pmatrix} 0 \\ 1 \\ 0 \\ 3 \end{pmatrix} \right\}$$

6. Using exterior products, find the determinant of each of the following matrices:

(i) $\begin{pmatrix} 3 & 5 \\ 4 & 8 \end{pmatrix}$ (ii) $\begin{pmatrix} 2 & 0 & 1 \\ 3 & 4 & -1 \\ 0 & -1 & 2 \end{pmatrix}$

(iii) $\begin{pmatrix} 2 & 0 & 2 \\ -3 & 4 & 1 \\ 0 & -1 & -1 \end{pmatrix}$

7. Show that for a 2×2 matrix A we have

$$\det A = \frac{1}{2}((\text{tr } A)^2 - \text{tr } A^2)$$

8. Consider the matrix $A = \begin{pmatrix} 2 & 1 & -1 \\ 0 & 2 & 0 \\ 0 & 0 & 3 \end{pmatrix}$.

(i) Write down $\text{tr } A$ and $\det A$.

(ii) Calculate $A^{(2)}(e_1 \wedge e_2 \wedge e_3)$

Hence write down the characteristic equation of A.

9. **HARD!** For a 3×3 matrix A show that

$$6 \det A = (\text{tr } A)^3 - 3\text{tr } A \text{ tr } A^2 + 2\text{tr } A^3$$

Chapter 22: Lie Algebras

22.1 Definition A *real Lie algebra* is a real vector space L with a bilinear product, written $[x,y]$, satisfying

(i) $[y,x] = -[x,y]$

(ii) $[[x,y],z] + [[y,z],x] + [[z,x],y] = \mathbf{0}$

for all $x, y, z \in L$.

(i) expresses the fact that Lie algebras are *anti-commutative.* Notice that an immediate consequence of (i) is that $[x,x] = \mathbf{0}$ for all $x \in L$.

(ii) is known as the *Jacobi identity.*

Lie algebras are not in general associative.

Examples (a) Consider $\boldsymbol{R}^3$ with basis $\{i,j,k\}$. The cross product is defined on the basis vectors by $i \times j = k$, $j \times k = i$ and $k \times i = j$, and also $i \times i = j \times j = k \times k = \mathbf{0}$. In column vector notation this leads to the product:

$$\begin{pmatrix} x_1 \\ y_1 \\ z_1 \end{pmatrix} \times \begin{pmatrix} x_2 \\ y_2 \\ z_2 \end{pmatrix} = \begin{pmatrix} y_1 z_2 - y_2 z_1 \\ -x_1 z_2 + x_2 z_1 \\ x_1 y_2 - x_2 y_1 \end{pmatrix}.$$

Under this product $\boldsymbol{R}^3$ is a Lie algebra. You should check that $\times$ is anti-commutative and satisfies the Jacobi identity.

(b) Define a new product on $Mat(n, \boldsymbol{R})$ by

$$[A,B] = AB - BA \text{ for all } A, B \in Mat(n, \boldsymbol{R})$$

This is called the *commutator* of A and B. It is anti-commutative since

$$[B,A] = BA - AB = -(AB - BA) = -[A,B]$$

You should also check that the Jacobi identity is satisfied.

This is known as the *general linear Lie algebra*, and is denoted $gl(n, \boldsymbol{R})$.

It is of dimension n^2.

By considering complex matrices we may also define the general linear Lie algebra $gl(n, \boldsymbol{C})$.

Considered as a *real* vector space it is of dimension $2n^2$.

(c) Let V be a real vector space. An *endomorphism* of V is a linear transformation $V \to V$.

We denote the set of endomorphisms of V by $End(V)$. We may make this into a vector space by defining vector addition by

$$(f+g)(v) = f(v) + g(v)$$

and multiplication by a scalar λ by

$$(\lambda f)(v) = \lambda f(v).$$

You should convince yourself that the vector space axioms are all satisfied. We may make $End(V)$ into a Lie algebra by defining a product by

$$[f, g] = f \circ g - g \circ f$$

where $f \circ g$ denotes the composite function of f and g.

22.2 Definition Let L be a Lie algebra. A subspace K of L is a *Lie subalgebra* if

$$[x, y] \in K \text{ for all } x, y \in K.$$

Examples The *special linear Lie algebra* $sl(n, \boldsymbol{R})$ comprises matrices with a trace of 0.

We check first that $sl(n, \boldsymbol{R})$ is a subspace of $gl(n, \boldsymbol{R})$: if $A, B \in sl(n, \boldsymbol{R})$ then

$$\text{tr}\,(A + B) = \text{tr}\,A + \text{tr}\,B = 0 + 0 = 0$$

$$\text{tr}\,\lambda A = \lambda\,\text{tr}\,A = \lambda \times 0 = 0.$$

Since by 21.11 $\text{tr}\,AB = \text{tr}\,BA$, we have

$$\text{tr}\,[A, B] = \text{tr}\,(AB - BA) = \text{tr}\,AB - \text{tr}\,BA = 0.$$

Hence $sl(n, \boldsymbol{R})$ is a Lie subalgebra of $gl(n, \boldsymbol{R})$.

Notice that for a matrix with trace 0, once $n-1$ diagonal entries are chosen the nth diagonal entry is already determined. Hence $sl(n,\boldsymbol{R})$ is of dimension n^2-1.

For example, a basis for $sl(2,\boldsymbol{R})$ is

$$\left\{\begin{pmatrix} 0 & 1 \\ 0 & 0 \end{pmatrix}, \begin{pmatrix} 0 & 0 \\ 1 & 0 \end{pmatrix}, \begin{pmatrix} 1 & 0 \\ 0 & -1 \end{pmatrix}\right\}.$$

Similarly, $sl(n,\boldsymbol{C})$ is a Lie subalgebra of $gl(n,\boldsymbol{C})$ of dimension $2(n^2-1)$.

22.3 Definition We say that a Lie algebra L is *abelian* if $[x,y]=0$ for all $x,y\in L$.

Example $\left\{\begin{pmatrix} x & 0 \\ 0 & y \end{pmatrix} : x,y\in\boldsymbol{R}\right\}$ is abelian.

To see this, observe that

$$\left[\begin{pmatrix} x & 0 \\ 0 & y \end{pmatrix}, \begin{pmatrix} x' & 0 \\ 0 & y' \end{pmatrix}\right] = \begin{pmatrix} xx' & 0 \\ 0 & yy' \end{pmatrix} - \begin{pmatrix} xx' & 0 \\ 0 & yy' \end{pmatrix} = \begin{pmatrix} 0 & 0 \\ 0 & 0 \end{pmatrix}.$$

22.4 Definition For any Lie algebra L, the *derived algebra* L' is the subspace of L spanned by the set $\{[x,y] : x,y\in L\}$.

Examples (a) If L is abelian, then the derived algebra is $L'=\{\mathbf{0}\}$, the trivial Lie algebra.

(b) Let L be the Lie algebra on $\boldsymbol{R}^3$ given by the cross product, with basis $\{i,j,k\}$.

We have $[i,j]=k$, $[j,k]=i$ and $[k,i]=j$. Hence the derived algebra contains i, j and k, so that $L'=L$.

22.5 Definition The *centre* of a Lie algebra L is

$$Z(L)=\{x\in L : [x,y]=\mathbf{0} \text{ for all } y\in L\}.$$

Examples (a) If L is abelian then $Z(L) = L$.

(b) Let L be the Lie algebra on $\boldsymbol{R}^3$ given by the cross product. Suppose $u \in Z(L)$. If $u \neq \mathbf{0}$ then there is $v \in \boldsymbol{R}^3$ such that $u \times v \neq \mathbf{0}$. Hence $u = \mathbf{0}$ and we have $Z(L) = \{\mathbf{0}\}$.

22.6 The Heisenberg Algebra

Let L be a Lie algebra of dimension 3.

Suppose $\dim L' = 1$ and $L' \subseteq Z(L)$.

We claim that there is only one such algebra.

Choose $e_1, e_2 \in L$ such that $[e_1, e_2] \neq \mathbf{0}$. We have $[e_1, e_2] \in L'$ and we set $e_3 = [e_1, e_2]$.

Notice that since $e_3 \in Z(L)$ we have $[e_1, e_3] = [e_2, e_3] = \mathbf{0}$.

Now suppose $\lambda_1 e_1 + \lambda_2 e_2 + \lambda_3 e_3 = \mathbf{0}$.

Then $\lambda_1[e_1, e_1] + \lambda_2[e_2, e_1] + \lambda_3[e_3, e_1] = \mathbf{0}$ and so $\lambda_2 = 0$.

Similarly $\lambda_1[e_1, e_2] + \lambda_2[e_2, e_2] + \lambda_3[e_3, e_2] = \mathbf{0}$ and so $\lambda_1 = 0$.

It follows that $\lambda_3 = 0$, so that $\{e_1, e_2, e_3\}$ is linearly independent and is a basis for L.

The Lie Algebras $o(n)$, $u(n)$ and $su(n)$

A real matrix is said to be *skew-symmetric* if $A^T = -A$. For example, the following matrices are skew-symmetric:

$$\begin{pmatrix} 0 & 1 \\ -1 & 0 \end{pmatrix} \qquad \begin{pmatrix} 0 & 2 & 3 \\ -2 & 0 & -5 \\ -3 & 5 & 0 \end{pmatrix}$$

Notice that in any skew-symmetric matrix the diagonal entries are all zero.

22.7 Proposition The set of $n \times n$ real skew-symmteric matrices is a Lie subalgebra of $gl(n, \boldsymbol{R})$.

Proof: If A and B are skew-symmetric then

$$(A+B)^T = A^T + B^T = (-A) + (-B) = -(A+B)$$

so that $A+B$ is skew-symmetric.

$(\lambda A)^T = \lambda A^T = \lambda(-A) = -\lambda A$ so that λA is skew-symmetric.

Finally,

$$\begin{aligned}[A,B]^T &= (AB-BA)^T = (AB)^T - (BA)^T \\ &= B^T A^T - A^T B^T = (-B)(-A) - (-A)(-B) \\ &= BA - AB = -(AB - BA) \\ &= -[A,B]\end{aligned}$$

so that $[A,B]$ is skew-symmetric. This Lie algebra is denoted $o(n)$.

A matrix in $gl(n, \boldsymbol{R})$ has n^2 entries, but if the matrix is skew-symmetric then the n diagonal entries are 0, leaving $n^2 - n$ entries. To obtain the dimension on $o(n)$ we divide this number by 2 to account for the skew-symmetry, giving

$$\dim o(n) = \frac{n^2 - n}{2} = \tfrac{1}{2} n(n-1).$$

The *conjugate transpose* of a complex matrix A is given by

$$A^* = \left(\overline{A}\right)^T = \overline{(A^T)}.$$

For example for $A = \begin{pmatrix} 2+i & 3-2i \\ -1+4i & 5 \end{pmatrix}$ we have

$$A^* = \begin{pmatrix} 2-i & -1-4i \\ 3+2i & 5 \end{pmatrix}.$$

A complex matrix is said to be *skew-hermitian* if $A^* = -A$. For example, the following matrices are skew-hermitian:

$$\begin{pmatrix} i & 2+3i \\ -2+3i & 2i \end{pmatrix} \qquad \begin{pmatrix} 2i & i & 3 \\ i & -5i & 5+2i \\ -3 & -5+2i & 4i \end{pmatrix}$$

Notice that in any skew-hermitian matrix the diagonal entries have real part zero. The set of $n \times n$ complex skew-hermitian matrices forms a subspace of $gl(n, \boldsymbol{C})$, denoted $u(n)$.

It is left as an exercise to show that $u(n)$ is a Lie subalgebra of $gl(n, \boldsymbol{C})$.

A matrix in $gl(n, \boldsymbol{C})$ has $2n^2$ *real* entries (counting each complex entry as two reals). Excluding the diagonal, there are $2n^2 - 2n$ entries, and we divide this number by 2 to account for the skew-symmetry. There are n pure imaginary entries on the diagonal, giving

$$\dim u(n) = \frac{2n^2 - 2n}{2} + n = n^2$$

as a real vector space.

Finally, $su(n)$ comprises $n \times n$ complex skew-hermitian matrices with trace 0. It is a Lie subalgebra of $gl(n, \boldsymbol{C})$. Since when $n-1$ of the diagonal entries are chosen, the nth is determined by the condition that the trace be 0, we have

$$\dim su(n) = n^2 - 1.$$

Exercises 22

1. Verify that the cross product on $\boldsymbol{R}^3$ satisfies:

   (i) $u \times v = -v \times u$

   (ii) $(u \times v) \times w + (v \times w) \times u + (w \times u) \times v = \mathbf{0}$

   Hint: You may find it useful to use the identity $(u \times v) \times w = (u.w)v - (v.w)u$ *where the quantities in brackets on the right hand side are scalar products.*

2. Let L be a real algebra, with product written [,].

   Show that if $[u,u] = \mathbf{0}$ for all $u \in L$ then $[x,y] = -[y,x]$.

   Hint: consider $[x+y, x+y]$.

3. Check that the product on $Mat(n, \boldsymbol{R})$ given by

   $$[A,B] = AB - BA$$

   satisfies the Jacobi identity.

4. Show that the centre $Z(L)$ is a Lie subalgebra of L.

5. Let $L = gl(n, \boldsymbol{R})$. Show that the derived algebra satisfies

   $$L' \subseteq sl(n, \boldsymbol{R}).$$

6. Show that $u(n)$ is a Lie subalgebra of $gl(n, \boldsymbol{C})$.

7. Show that

   $$\left\{ \frac{1}{2}\begin{pmatrix} 0 & 1 \\ -1 & 0 \end{pmatrix}, \frac{1}{2}\begin{pmatrix} 0 & i \\ i & 0 \end{pmatrix}, \frac{1}{2}\begin{pmatrix} i & 0 \\ 0 & -i \end{pmatrix} \right\}$$

   is a basis for $su(2)$. Hence show that $su(2)$ is isomorphic to the Lie algebra on $\boldsymbol{R}^3$ given by the cross product.

8. Show that

   $$\left\{ \begin{pmatrix} 0 & 1 & 0 \\ -1 & 0 & 0 \\ 0 & 0 & 0 \end{pmatrix}, \begin{pmatrix} 0 & 0 & 1 \\ 0 & 0 & 0 \\ -1 & 0 & 0 \end{pmatrix}, \begin{pmatrix} 0 & 0 & 0 \\ 0 & 0 & 1 \\ 0 & -1 & 0 \end{pmatrix} \right\}$$

is a basis for $o(3)$. Hence show that $o(3)$ is isomorphic to the Lie algebra on $\boldsymbol{R}^3$ given by the cross product.

9. A matrix is said to be *upper triangular* if all of the entries below the leading diagonal are zero. For example, these matrices are upper triangular:

$$\begin{pmatrix} 1 & 2 \\ 0 & 3 \end{pmatrix} \qquad \begin{pmatrix} 2 & 5 & 3 \\ 0 & 4 & -2 \\ 0 & 0 & 6 \end{pmatrix}$$

Let $ut(n, \boldsymbol{R})$ denote the set of $n \times n$ upper triangular matrices.

Show that $ut(3, \boldsymbol{R})$ is a subspace of $gl(3, \boldsymbol{R})$.

Show that $ut(3, \boldsymbol{R})$ is a Lie subalgebra of $gl(3, \boldsymbol{R})$.

What is the dimension of $ut(n, \boldsymbol{R})$?

What is the centre $Z(ut(n, \boldsymbol{R}))$?

10. A matrix is said to be *strictly upper triangular* if only the entries above the leading diagonal are non-zero. For example, these matrices are strictly upper triangular:

$$\begin{pmatrix} 0 & 2 \\ 0 & 0 \end{pmatrix} \qquad \begin{pmatrix} 0 & 5 & 3 \\ 0 & 0 & -2 \\ 0 & 0 & 0 \end{pmatrix}$$

Let $sut(n, \boldsymbol{R})$ denote the set of $n \times n$ strictly upper triangular matrices.

Show that $sut(3, \boldsymbol{R})$ is a Lie subalgebra of $sl(3, \boldsymbol{R})$.

Show that $sut(3, \boldsymbol{R})$ is isomorphic to the Heisenberg Lie algebra.

What is the dimension of $sut(n, \boldsymbol{R})$?

11. Show that for $L = ut(3, \boldsymbol{R})$ the derived algebra is

$$L' = sut(3, \boldsymbol{R}).$$

What is the derived subalgebra of $sut(3, \boldsymbol{R})$?

Chapter 23: Matrix Groups and their Tangent Spaces

Recall from chapter 1 that a *group* is a set G with a binary operation $*$ on G satisfying:

Associativity: $(a * b) * c = a * (b * c)$ for all $a, b, c \in G$

Identity: There is an element $e \in G$ such that

$e * a = a = a * e$ for all $a \in G$

Inverses: For each $a \in G$ there is an element $a^{-1} \in G$ such that $a * a^{-1} = e = a^{-1} * a$

These three statements are known as the *group axioms.* In this chapter we study groups of matrices of real and complex numbers.

Examples The invertible matrices in $Mat(n, \boldsymbol{R})$ form a group known as the *general linear group*, denoted $GL(n, \boldsymbol{R})$. These are the matrices in $Mat(n, \boldsymbol{R})$ with non-zero determinant.

Similarly, the invertible matrices in $Mat(n, \boldsymbol{C})$ form a group denoted $GL(n, \boldsymbol{C})$.

Recall from chapter 2 that if G is a group, then a subset $H \subseteq G$ is a *subgroup* if H is itself a group under the same operation. There is an easy test to determine whether a non-empty subset is a subgroup: $H \subseteq G$ is a subgroup if and only if

(i) if $a, b \in H$ then $ab \in H$

(ii) if $a \in H$ then $a^{-1} \in H$.

Examples Consider the matrices in $Mat(n, \boldsymbol{R})$ with determinant 1.

Since by 21.8 $\det AB = \det A \det B$, it follows that if A and B have determinant 1 then so do AB and A^{-1}. Hence these matrices form a subgroup of $GL(n, \boldsymbol{R})$. It is known as the *special linear group*, and is denoted $SL(n, \boldsymbol{R})$.

Similarly, complex matrices with determinant 1 form a subgroup of $GL(n, \boldsymbol{C})$, denoted $SL(n, \boldsymbol{C})$.

Recall from 7.11 that a real matrix A that has transpose equal to its inverse, so that $A^T = A^{-1}$, is called an *orthogonal matrix*. This is equivalent to

$$AA^T = I = A^T A.$$

Examples (a) $A = \begin{pmatrix} \frac{1}{\sqrt{2}} & -\frac{1}{\sqrt{2}} \\ \frac{1}{\sqrt{2}} & \frac{1}{\sqrt{2}} \end{pmatrix}$ is orthogonal, since we have

$AA^T = \begin{pmatrix} \frac{1}{\sqrt{2}} & -\frac{1}{\sqrt{2}} \\ \frac{1}{\sqrt{2}} & \frac{1}{\sqrt{2}} \end{pmatrix}\begin{pmatrix} \frac{1}{\sqrt{2}} & \frac{1}{\sqrt{2}} \\ -\frac{1}{\sqrt{2}} & \frac{1}{\sqrt{2}} \end{pmatrix} = \begin{pmatrix} 1 & 0 \\ 0 & 1 \end{pmatrix}$ and similarly $A^T A = I$.

(b) $A = \begin{pmatrix} \frac{1}{\sqrt{3}} & \frac{1}{\sqrt{6}} & \frac{1}{\sqrt{2}} \\ \frac{1}{\sqrt{3}} & \frac{1}{\sqrt{6}} & -\frac{1}{\sqrt{2}} \\ \frac{1}{\sqrt{3}} & -\frac{2}{\sqrt{6}} & 0 \end{pmatrix}$ is an orthogonal matrix.

To see this, calculate AA^T and $A^T A$ and observe that each is I.

Suppose that A and B are orthogonal matrices. Then

$$(AB)^T AB = B^T A^T AB = B^T IB = B^T B = I$$

$$AB(AB)^T = ABB^T A^T = AIA^T = AA^T = I$$

Hence AB is also orthogonal. Also, A^{-1} is orthogonal. Hence the set of $n \times n$ orthogonal matrices form a subgroup of $GL(n, \boldsymbol{R})$. It is known as the *orthogonal group*, and denoted $O(n)$.

Of particular interest is the subgroup of $O(n)$ consisting of all $n \times n$ orthogonal matrices with determinant 1. This is called the *special orthogonal group*, and denoted $SO(n)$. The matrices in $SO(n)$ represent rotations about the origin in $\boldsymbol{R}^n$.

A complex matrix A such that $A^* = A^{-1}$, where A^* is the conjugate transpose, is said to be *unitary*. Equivalently, $AA^* = I = A^*A$.

Examples $\frac{1}{\sqrt{2}}\begin{pmatrix} 1 & i \\ i & 1 \end{pmatrix}$ and $\frac{1}{3}\begin{pmatrix} \frac{4}{\sqrt{5}} + \frac{2}{\sqrt{5}}i & 2+i \\ -\sqrt{5} & 2 \end{pmatrix}$ are unitary.

The set of unitary matrices form a subgroup of $GL(n, \boldsymbol{C})$. It is known the *unitary group*, and denoted $U(n)$.

We also consider the subgroup of $U(n)$ consisting of all $n \times n$ unitary matrices with determinant 1. This is called the *special unitary group*, and denoted $SU(n)$.

Let G and H be groups. Recall from chapter 2 that a mapping $f: G \to H$ is a *group homomorphism* if

$$f(a \cdot b) = f(a) \cdot f(b) \text{ for all } a, b \in G.$$

A bijective homomorphism is called an *isomorphism*.

Example We define a mapping $f: U(1) \to SO(2)$ by

$$f(x+yi) = \begin{pmatrix} x & y \\ -y & x \end{pmatrix}$$

It is left as an exercise to show that f is an isomorphism. Hence $U(1) \cong SO(2)$.

Our goal for the rest of this chapter is to show that, apart from the isomorphism of the last example, all of the matrix groups $SO(n)$, $U(n)$ and $SU(n)$ are distinct. It turns out to be easy to distinguish the groups $O(n)$ from the others: each group $O(n)$ comes in two 'pieces' or *components*, one consisting of matrices with determinant 1, the other of matrices of determinant -1. We say that $O(n)$ is not *path-connected.* By contrast, each of the groups $SO(n)$, $U(n)$ and $SU(n)$ is path-connected.

For $a, b \in \boldsymbol{R}$ with $a < b$, the *open interval* is

$$(a, b) = \{x \in \boldsymbol{R} : a < x < b\}.$$

Let G be a matrix group. A continuous mapping $\gamma : (a,b) \to G$ is called a *curve* in G.

The *derivative* of γ at t is

$$\gamma'(t) = \lim_{\delta t \to 0} \frac{\gamma(t+\delta t) - \gamma(t)}{\delta t}.$$

Note that $\gamma'(t)$ is a matrix. We say that γ is a *differentiable curve* if the limit $\gamma'(t)$ exists for all $t \in (a,b)$.

Example $\gamma(t) = \begin{pmatrix} 1 & t^2 \\ 0 & 1 \end{pmatrix}$ defines a curve in $SL(2,\boldsymbol{R})$.

We have $\gamma'(t) = \lim_{\delta t \to 0} \frac{1}{\delta t}\left(\begin{pmatrix} 1 & (t+\delta t)^2 \\ 0 & 1 \end{pmatrix} - \begin{pmatrix} 1 & t^2 \\ 0 & 1 \end{pmatrix}\right) = \begin{pmatrix} 0 & 2t \\ 0 & 0 \end{pmatrix}$,

so $\gamma(t)$ is a differentiable curve.

23.1 Definition Let G be a group of matrices and $a > 0$ be a real number. Consider the set S of differentiable curves $\gamma : (-a,a) \to G$ such that $\gamma(0) = I$.

The *tangent space* of G is

$$TG = \{\gamma'(0) : \gamma \in S\}.$$

23.2 Proposition Let G be a group of real $n \times n$ matrices.

The tangent space TG is a subspace of $Mat(n,\boldsymbol{R})$.

Proof: Suppose $\gamma_1'(0), \gamma_2'(0) \in TG$.

Then $\gamma_1\gamma_2(0) = \gamma_1(0)\gamma_2(0) = I \cdot I = I$, so $(\gamma_1\gamma_2)'(0) \in TG$.

By the product rule we have $(\gamma_1\gamma_2)'(t) = \gamma_1'(t)\gamma_2(t) + \gamma_1(t)\gamma_2'(t)$.

$$\begin{aligned} \text{So } (\gamma_1\gamma_2)'(0) &= \gamma_1'(0)\gamma_2(0) + \gamma_1(0)\gamma_2'(0) \\ &= \gamma_1'(0) \cdot I + I \cdot \gamma_2'(0) = \gamma_1'(0) + \gamma_2'(0) \end{aligned}$$

and hence $\gamma_1'(0) + \gamma_2'(0) \in TG$.

Now suppose $\gamma'(0) \in TG$ and $\lambda \in \boldsymbol{R}$, and let $\delta(t) = \gamma(\lambda t)$.

Then $\delta(0) = \gamma(\lambda \cdot 0) = I$, so $\delta'(0) \in TG$.

By the chain rule we have $\delta'(t) = \lambda\gamma'(\lambda t)$ and so $\delta'(0) = \lambda\gamma'(0)$.

Hence $\lambda\gamma'(0) \in TG$.

Similarly, if G is a group of complex $n \times n$ matrices then TG is a subspace of $Mat(n, \boldsymbol{C})$.

23.3 Proposition Let G be a group of real or complex $n \times n$ matrices.

The tangent space TG is a Lie subalgebra of $gl(n, \boldsymbol{R})$ or $gl(n, \boldsymbol{C})$ respectively under the product

$$[\gamma_1'(0), \gamma_2'(0)] = \gamma_1'(0)\gamma_2'(0) - \gamma_2'(0)\gamma_1'(0)$$

The proof is omitted.

Example Consider differentiable curves $\gamma : (-a, a) \to O(n)$ with $\gamma(0) = I$.

We have $\gamma(t)\gamma(t)^T = I$. Differentiating with respect to t we get

$$\frac{d}{dt}(\gamma(t)\gamma(t)^T) = \frac{dI}{dt} = \mathbf{0}.$$

By the product rule we have $\gamma'(t)\gamma(t)^T + \gamma(t)\gamma'(t)^T = \mathbf{0}$.

At $t = 0$ we get $\gamma'(0)\gamma(0)^T + \gamma(0)\gamma'(0)^T = \gamma'(0) + \gamma'(0)^T = \mathbf{0}$.

Hence $\gamma'(0)$ is skew-symmetric. It follows that $TO(n) \subseteq o(n)$. Shortly, we will show that this is an equality.

Recall that for a real number x the exponential function is defined by the power series

$$e^x = 1 + x + \tfrac{x^2}{2!} + \tfrac{x^3}{3!} + \ldots$$

For a real or complex square matrix A we define the exponential by

$$\exp A = I + A + \tfrac{1}{2!}A^2 + \tfrac{1}{3!}A^3 + \ldots$$

It can be shown that this series converges for all square matrices A. Notice that for the zero matrix we have $\exp \mathbf{0} = I$.

For real numbers x and y we have $e^{x+y} = e^x \cdot e^y$. Because matrix multiplication is not commutative, we need to be more careful in the case of matrices.

23.4 Proposition We have $\exp(A+B) = \exp A \cdot \exp B$

provided that the matrices A and B commute.

Proof:

$$\begin{aligned}
\exp A \cdot \exp B &= \left(I + A + \frac{1}{2!}A^2 + \ldots\right)\left(I + B + \frac{1}{2!}B^2 + \ldots\right) \\
&= I + A + B + \frac{1}{2!}A^2 + AB + \frac{1}{2!}B^2 + \ldots \\
&= I + A + B + \frac{1}{2!}(A+B)^2 + \ldots \\
&= \exp(A+B)
\end{aligned}$$

Notice how the middle step depends upon $AB = BA$.

The exponential of a matrix is always invertible: since A and $-A$ commute we have

$$\exp A \cdot \exp(-A) = \exp(A - A) = \exp \mathbf{0} = I.$$

Note that $\exp(A^T) = (\exp A)^T$, so that we can write $\exp A^T$ without ambiguity.

For each matrix $A \in gl(n, \boldsymbol{R})$ we define a differentiable curve by $\gamma(t) = \exp tA$.

Since the exponential of a matrix is invertible, such curves are in $GL(n, \boldsymbol{R})$. Notice that we have $\gamma(0) = \exp(0 \cdot A) = I$.

Differentiating gives $\gamma'(t) = A \exp tA$, and so $\gamma'(0) = A \exp \mathbf{0} = A$.

Because any matrix A belongs to $TGL(n, \boldsymbol{R})$ it follows that $TGL(n, \boldsymbol{R}) = gl(n, \boldsymbol{R})$.

Similarly, $TGL(n, \boldsymbol{C}) = gl(n, \boldsymbol{C})$.

23.5 Proposition $TO(n) = o(n)$

Proof: We have already shown that $TO(n) \subseteq o(n)$. Now suppose $A \in o(n)$, so that $A + A^T = \mathbf{0}$.

For $t \in \boldsymbol{R}$ we have $tA + tA^T = t(A + A^T) = \mathbf{0}$.

Then $\exp tA \cdot \exp tA^T = \exp(tA + tA^T) = I$, so that $\exp tA \in O(n)$.

So $\gamma(t) = \exp tA$ defines a differentiable curve in $O(n)$. We have $\gamma'(t) = A \exp tA$, so $\gamma'(0) = A$.

Hence $A \in TO(n)$.

23.6 Proposition $TU(n) = u(n)$

The proof is similar to 23.5 and is left as an exercise.

23.7 Lemma Let P be an invertible matrix, and let A be any square matrix of the same size. We have

(i) $(P^{-1}AP)^n = P^{-1}A^nP$ for each natural number n.

(ii) $\exp(P^{-1}AP) = P^{-1}(\exp A)P$

Proof: (i) $(P^{-1}AP)^n = \overbrace{(P^{-1}AP)(P^{-1}AP)...(P^{-1}AP)}^{n} = P^{-1}A^nP$

The proof of (ii) follows from (i) and is left as an exercise.

23.8 Theorem For any square matrix A we have

$$\det \exp A = e^{\operatorname{tr} A}$$

Sketch proof: We shall only prove the result for the case where the matrix A is diagonalisable, although it is true for all matrices. Recall that a matrix A is diagonalisable if there is an invertible matrix P such that $P^{-1}AP = \Lambda$, where Λ is a diagonal matrix:

$$\Lambda = \begin{pmatrix} \lambda_1 & 0 & \cdot & \cdot & \cdot & 0 \\ 0 & \lambda_2 & & & & \cdot \\ \cdot & & \cdot & & & \cdot \\ \cdot & & & \cdot & & \cdot \\ \cdot & & & & \cdot & 0 \\ 0 & \cdot & \cdot & \cdot & 0 & \lambda_n \end{pmatrix}$$

The λ_i on the leading diagonal are the eigenvalues. We have

$$\exp \Lambda = \begin{pmatrix} e^{\lambda_1} & 0 & \cdot & \cdot & \cdot & 0 \\ 0 & e^{\lambda_2} & & & & \cdot \\ \cdot & & \cdot & & & \cdot \\ \cdot & & & \cdot & & \cdot \\ \cdot & & & & \cdot & 0 \\ 0 & \cdot & \cdot & \cdot & 0 & e^{\lambda_n} \end{pmatrix}$$

We have $A = P\Lambda P^{-1}$ and so

$$\begin{aligned} \det \exp A &= \det \exp(P\Lambda P^{-1}) \\ &= \det(P(\exp \Lambda)P^{-1}) \\ &= \det P(\det \exp \Lambda) \det P^{-1} = \det \exp \Lambda = e^{\lambda_1} e^{\lambda_2} \ldots e^{\lambda_n} \\ &= e^{\lambda_1 + \lambda_2 + \ldots + \lambda_n} = e^{\operatorname{tr} \Lambda} = e^{\operatorname{tr} P^{-1}AP} = e^{\operatorname{tr} A}. \end{aligned}$$

23.9 Proposition $\quad TSL(n, \boldsymbol{R}) = sl(n, \boldsymbol{R})$

Proof: Suppose that $A \in TSL(n, \boldsymbol{R})$. Then there is a curve $\gamma(t) = \exp(tA)$ in $SL(n, \boldsymbol{R})$.

Since $\det \exp A = 1$ we have $e^{\operatorname{tr} A} = 1$ so that $\operatorname{tr} A = 0$. Hence $A \in sl(n, \boldsymbol{R})$.

Conversely, suppose $A \in sl(n, \boldsymbol{R})$ so that $\operatorname{tr} A = 0$. For $t \in \boldsymbol{R}$ we have $\operatorname{tr} tA = 0$, so $e^{\operatorname{tr} tA} = 1$.

Hence $\det \exp tA = 1$ so that $\gamma(t) = \exp(tA)$ defines a curve in $SL(n, \boldsymbol{R})$ with $\gamma'(0) = A$.

Hence $A \in TSL(n, \boldsymbol{R})$.

23.10 Proposition $TSO(n) = o(n)$

Proof: Since any curve in $SO(n)$ is in $O(n)$ we have $TSO(n) \subseteq o(n)$.

Now suppose $A \in o(n)$. Since the diagonal entries of a skew-symmetric matrix are 0 we have $\operatorname{tr} A = 0$. For $t \in \boldsymbol{R}$ we have $e^{\operatorname{tr} tA} = 1$ so that by 23.8 $\det \exp tA = 1$.

Hence $\gamma(t) = \exp tA$ defines a curve in $SO(n)$ with $\gamma'(0) = A$, so that $A \in TSO(n)$.

23.11 Theorem Isomorphic groups have isomorphic tangent spaces: If $G \cong H$ then $TG \cong TH$.

Proof: Suppose that $\phi : G \to H$ is a group homomorphism.

Suppose $\gamma'(0) \in TG$, where $\gamma(t)$ is a differentiable curve in G with $\gamma(0) = I$.

Then $\phi \circ \gamma(t)$ is a differentiable curve in H with $\phi \circ \gamma(0) = \phi(I) = I$.

Define $T_\phi : TG \to TH$ by $T_\phi(\gamma'(0)) = (\phi \circ \gamma)'(0)$.

We shall show that T_ϕ is a linear transformation:

First observe that

$$\phi \circ \gamma_1\gamma_2(t) = \phi(\gamma_1(t) \cdot \gamma_2(t)) = \phi(\gamma_1(t)) \cdot \phi(\gamma_2(t)) = \phi \circ \gamma_1(t) \cdot \phi \circ \gamma_2(t)$$

so that $\phi \circ \gamma_1\gamma_2 = (\phi \circ \gamma_1) \cdot (\phi \circ \gamma_2)$. We have

$$\begin{aligned} T_\phi(\gamma_1'(0) + \gamma_2'(0)) &= T_\phi((\gamma_1\gamma_2)'(0)) \\ &= (\phi \circ \gamma_1\gamma_2)'(0) = ((\phi \circ \gamma_1)(\phi \circ \gamma_2))'(0) \\ &= (\phi \circ \gamma_1)'(0) + (\phi \circ \gamma_2)'(0) \\ &= T_\phi(\gamma_1'(0)) + T_\phi(\gamma_2'(0)) \end{aligned}$$

For $\lambda \in \boldsymbol{R}$, if $\delta(t) = \gamma(\lambda t)$ then $\delta'(t) = \lambda\gamma'(\lambda t)$. We have

$$T_\phi(\lambda\gamma'(0)) = T_\phi(\delta'(0)) = (\phi \circ \delta)'(0) = \lambda(\phi \circ \gamma)'(0) = \lambda T_\phi(\gamma'(0))$$

If G and H are isomorphic groups of matrices, then there is an *invertible* homomorphism $\phi : G \to H$. Then T_ϕ is also invertible with inverse $T_{\phi^{-1}}$, and so is an isomorphism.

Notice that the converse of this result is *false*: for example $O(n)$ and $SO(n)$ have isomorphic tangent spaces, but are *not* isomorphic as groups of matrices.

We define the *dimension* of a matrix group to be the dimension of its tangent space. Notice that an immediate corollary of theorem 23.11 is that *isomorphic matrix groups have the same dimension.* It follows that if a pair of matrix groups have different dimensions then they are not isomorphic.

We saw in chapter 22 that $\dim o(n) = \frac{1}{2}n(n-1), \dim u(n) = n^2$ and $\dim su(n) = n^2 - 1$.

Hence these are the dimensions of the corresponding families of matrix groups $SO(n)$, $U(n)$ and $SU(n)$. We tabulate the dimensions of the first nine matrix groups in each family:

n	$SO(n)$	$U(n)$	$SU(n)$
1	0	1	0
2	1	4	**3**
3	**3**	9	8
4	6	16	**15**
5	10	25	24
6	**15**	***36***	35
7	21	49	48
8	28	64	63
9	***36***	81	80

We notice that $SO(2)$ and $U(1)$ have the same dimension. This is to be expected, since these groups are isomorphic. We also notice that $SO(3)$ and $SU(2)$ have the same dimension, and that $SO(6)$ and $SU(4)$ have the same dimension. This leads us to wonder whether these pairs of groups are isomorphic.

23.12 Definition The *centre* of a group G, written $Z(G)$, is the set of elements of G that commute with all other elements of G:

$$Z(G) = \{g \in G : gx = xg \text{ for all } x \in G\}$$

A *scalar matrix* is a matrix of the form

$$\lambda I = \begin{pmatrix} \lambda & 0\ldots & 0 \\ \vdots & \ddots & \vdots \\ 0 & \ldots 0 & \lambda \end{pmatrix}$$

Clearly a scalar matrix commutes with any other matrix, so the scalar matrices always belong to the centre of a matrix group. Indeed, they are often the only matrices in the centre.

We make good use of the following principle:

Isomorphic groups have isomorphic centres.

That is, if $G \cong H$ then $Z(G) \cong Z(H)$.

Examples (a) We calculate the centre of $SO(3)$. Since a matrix in the centre of $SO(3)$ must commute with every matrix in $SO(3)$, in particular it must commute with

$$\begin{pmatrix} 1 & 0 & 0 \\ 0 & 0 & 1 \\ 0 & -1 & 0 \end{pmatrix} \text{ and } \begin{pmatrix} 0 & 1 & 0 \\ -1 & 0 & 0 \\ 0 & 0 & 1 \end{pmatrix}.$$

Brute force calculation gives

$$\begin{pmatrix} a & b & c \\ d & e & f \\ g & h & i \end{pmatrix}\begin{pmatrix} 1 & 0 & 0 \\ 0 & 0 & 1 \\ 0 & -1 & 0 \end{pmatrix} = \begin{pmatrix} 1 & 0 & 0 \\ 0 & 0 & 1 \\ 0 & -1 & 0 \end{pmatrix}\begin{pmatrix} a & b & c \\ d & e & f \\ g & h & i \end{pmatrix}$$

$$\begin{pmatrix} a & -c & b \\ d & -f & e \\ g & -i & h \end{pmatrix} = \begin{pmatrix} a & b & c \\ g & h & i \\ -d & -e & -f \end{pmatrix}$$

Hence $b = c = 0, d = g = 0$ and $e = i$.

$$\begin{pmatrix} a & b & c \\ d & e & f \\ g & h & i \end{pmatrix}\begin{pmatrix} 0 & 1 & 0 \\ -1 & 0 & 0 \\ 0 & 0 & 1 \end{pmatrix} = \begin{pmatrix} 0 & 1 & 0 \\ -1 & 0 & 0 \\ 0 & 0 & 1 \end{pmatrix}\begin{pmatrix} a & b & c \\ d & e & f \\ g & h & i \end{pmatrix}$$

$$\begin{pmatrix} -b & a & c \\ -e & d & f \\ -h & g & i \end{pmatrix} = \begin{pmatrix} d & e & f \\ -a & -b & -c \\ g & h & i \end{pmatrix}$$

Hence $g = h = 0, c = f = 0$ and $a = e$.

Hence every matrix in the centre is of the form $\begin{pmatrix} a & 0 & 0 \\ 0 & a & 0 \\ 0 & 0 & a \end{pmatrix}$.

Since the determinant is 1, we have $a^3 = 1$ and so $a = 1$. Hence $Z(SO(3)) = \{I\}$.

(b) We calculate the centre of $SU(2)$. Since a matrix in the centre of $SU(2)$ must commute with every matrix in $SU(2)$, in particular it must commute with

$$\begin{pmatrix} 0 & 1 \\ -1 & 0 \end{pmatrix} \text{and} \begin{pmatrix} i & 0 \\ 0 & -i \end{pmatrix}.$$

$$\begin{pmatrix} a & b \\ c & d \end{pmatrix}\begin{pmatrix} 0 & 1 \\ -1 & 0 \end{pmatrix} = \begin{pmatrix} 0 & 1 \\ -1 & 0 \end{pmatrix}\begin{pmatrix} a & b \\ c & d \end{pmatrix}$$

$$\begin{pmatrix} -b & a \\ -d & c \end{pmatrix} = \begin{pmatrix} c & d \\ -a & -b \end{pmatrix}$$

Hence $a = d$ and $b = -c$.

$$\begin{pmatrix} a & b \\ c & d \end{pmatrix}\begin{pmatrix} i & 0 \\ 0 & -i \end{pmatrix} = \begin{pmatrix} i & 0 \\ 0 & -i \end{pmatrix}\begin{pmatrix} a & b \\ c & d \end{pmatrix}$$

$$\begin{pmatrix} ai & -bi \\ ci & -di \end{pmatrix} = \begin{pmatrix} ai & bi \\ -ci & -di \end{pmatrix}$$

Hence $b = c = 0$, so every matrix in the centre is of the form

$$\begin{pmatrix} a & 0 \\ 0 & a \end{pmatrix}.$$

Since the determinant is 1, we have $a^2 = 1$ and so $a = \pm 1$.

Hence $Z(SU(2)) = \{I, -I\}$.

Since the groups $SO(3)$ and $SU(2)$ have different centres they are not isomorphic, despite being of the same dimension. A similar thing can be shown for $SO(6)$ and $SU(4)$.

Exercises 23

1. Show that if A is an orthogonal matrix then $\det A = \pm 1$.

   Show that if A is a unitary matrix then $|\det A| = 1$.

2. Show that if A and B are unitary matrices then AB and A^{-1} are also unitary.

3. Show that $f: U(1) \to SO(2)$ given by

$$f(x + yi) = \begin{pmatrix} x & y \\ -y & x \end{pmatrix}$$

   is an isomorphism of groups.

4. (i) By considering differentiable curves in $U(n)$ satisfying $\gamma(0) = I$, show that $TU(n) \subseteq u(n)$.

   (ii) Using the exponential map show that $u(n) \subseteq TU(n)$.

5. Show that $TSU(n) = su(n)$.

6. Find the centres of the groups $SO(6)$ and $SU(4)$.

   Is $SO(6) \cong SU(4)$?

   Repeat this exercise for $SO(9)$ and $U(6)$.

   You may assume that the centres contain only scalar matrices.

Chapter 24: Normed Real Algebras

Suppose that V is a real vector space. Recall from chapter 7 that a mapping $V \times V \to \boldsymbol{R}$, written $\langle u, v\rangle$, is called a *real inner product* if it satisfies:

(i) $\langle u, v\rangle = \langle v, u\rangle$

(ii) $\langle u+v, w\rangle = \langle u, w\rangle + \langle v, w\rangle$

(iii) $\langle \lambda u, v\rangle = \lambda\langle u, v\rangle$

(iv) $\langle v, v\rangle \geq 0$, and $\langle v, v\rangle = 0$ if and only if $v = \mathbf{0}$

for all $u, v, w \in V$ and all $\lambda \in \boldsymbol{R}$.

A real vector space equipped with a real inner product is called a *real inner product space*

If V is finite dimensional with basis $\{e_1, e_2, ..., e_n\}$ then we can define an inner product by

$$\langle e_i, e_j\rangle = \begin{cases} 1 \text{ if } i = j \\ 0 \text{ if } i \neq j \end{cases}$$

This is the usual scalar product. For example, in $\boldsymbol{R}^2$ we have

$$\begin{aligned} &\langle xe_1 + ye_2, x'e_1 + y'e_2\rangle \\ &= xx'\langle e_1, e_1\rangle + xy'\langle e_1, e_2\rangle + x'y\langle e_2, e_1\rangle + yy'\langle e_2, e_2\rangle \\ &= xx' + yy'. \end{aligned}$$

Given a real inner product space V we can define a *norm* on V by

$$\| v \| = \sqrt{\langle v, v\rangle}$$

We call v a *unit vector* if $\| v \| = 1$.

Example In $\boldsymbol{R}^2$ we define the norm by

$$\left\| \begin{pmatrix} x \\ y \end{pmatrix} \right\| = \sqrt{\left\langle \begin{pmatrix} x \\ y \end{pmatrix}, \begin{pmatrix} x \\ y \end{pmatrix} \right\rangle} = \sqrt{x^2 + y^2}.$$

24.1 Definition Suppose that A is a real algebra with unity, with inner product and norm defined as above. We call A a *normed real algebra* if

$$\|uv\| = \|u\| \cdot \|v\| \text{ for all } u, v \in A,$$

so that the norm of a product of vectors is the product of their individual norms.

Examples (a) $\boldsymbol{R}$ is a normed real algebra: the norm is the modulus, which satisfies:

$$|xy| = |x| \cdot |y|$$

The unit vectors are 1 and -1.

(b) $\boldsymbol{C}$ is a normed real algebra: the norm is $\|x + yi\| = \sqrt{x^2 + y^2}$.

The unit vectors lie on a circle of radius 1 centred on the origin of the Argand diagram.

24.2 Definition A real algebra A with unity is called a *real division algebra* if for any non-zero $x \in A$ there is $x^{-1} \in A$ such that

$$x \cdot x^{-1} = 1 = x^{-1} \cdot x$$

Examples Both $\boldsymbol{R}$ and $\boldsymbol{C}$ are real division algebras.

In fact, every normed real algebra is a real division algebra. Our first objective is to prove this.

Suppose that A is a finite dimensional real algebra with basis $\{e_1, e_2, ..., e_n\}$, where $e_1 = 1$.

We write $\boldsymbol{R}^{\perp}$ for the subspace of A spanned by $\{e_2, ..., e_n\}$. Each element of A can be written uniquely as a sum $\alpha + \beta$, where $\alpha \in \boldsymbol{R}$ and $\beta \in \boldsymbol{R}^{\perp}$. We read $\boldsymbol{R}^{\perp}$ as "R perpendicular".

Notice that if $\alpha \in \boldsymbol{R}$ and $\beta \in \boldsymbol{R}^{\perp}$ then $\langle \alpha, \beta \rangle = 0$.

The *conjugate* of an element $x = \alpha + \beta$ is $\overline{x} = \alpha - \beta$.

Notice that if $x \in \boldsymbol{R}$ then $\overline{x} = x$ and if $x \in \boldsymbol{R}^{\perp}$ then $\overline{x} = -x$.

Example C has basis $\{1, i\}$ and a complex number $z = x + yi$ has conjugate $\overline{z} = x - yi$.

24.3 Proposition Let A be a normed real algebra. We have

(i) $\langle xz, yz \rangle = \langle x, y \rangle \langle z, z \rangle$

(ii) $\langle zx, zy \rangle = \langle z, z \rangle \langle x, y \rangle$

for all $x, y, z \in A$.

Proof: We prove (i). The proof of (ii) is left as an exercise.

First, we prove the easy case where $x = y$. Observe that

$$\|xz\|^2 = \|xz\| \cdot \|xz\| = \|x\|^2 \cdot \|z\|^2$$

so that $\langle xz, xz \rangle = \langle x, x \rangle \langle z, z \rangle$

Now replace x by $x + y$:

$$\langle (x+y)z, (x+y)z \rangle = \langle x+y, x+y \rangle \langle z, z \rangle$$

Expanding gives

$$\langle xz, xz \rangle + 2\langle xz, yz \rangle + \langle yz, yz \rangle = \{\langle x, x \rangle + 2\langle x, y \rangle + \langle y, y \rangle\} \langle z, z \rangle$$

and so

$$\langle x, x \rangle \langle z, z \rangle + 2\langle xz, yz \rangle + \langle y, y \rangle \langle z, z \rangle = \langle x, x \rangle \langle z, z \rangle + 2\langle x, y \rangle \langle z, z \rangle + \langle y, y \rangle \langle z, z \rangle$$

Hence by cancellation $\langle xz, yz \rangle = \langle x, y \rangle \langle z, z \rangle$.

24.4 Proposition Let A be a normed real algebra. We have

(i) $\langle xz, y \rangle = \langle x, y\overline{z} \rangle$

(ii) $\langle zx, y \rangle = \langle x, \overline{z}y \rangle$

for all $x, y, z \in A$.

Proof: We prove (i). The proof of (ii) is left as an exercise.

Let $z = a + \beta$, where $a \in \boldsymbol{R}$ and $\beta \in \boldsymbol{R}^{\perp}$. We have

$$\langle x(1+\beta), y(1+\beta)\rangle = \langle x, y\rangle\langle 1+\beta, 1+\beta\rangle \text{ by 24.3.}$$

Expanding gives

$$\langle x,y\rangle + \langle x, y\beta\rangle + \langle x\beta, y\rangle + \langle x\beta, y\beta\rangle = \langle x,y\rangle\{\langle 1,1\rangle + \langle 1,\beta\rangle + \langle \beta,1\rangle + \langle \beta,\beta\rangle\}$$

and so

$$\langle x,y\rangle + \langle x, y\beta\rangle + \langle x\beta, y\rangle + \langle x,y\rangle\langle \beta,\beta\rangle = \langle x,y\rangle + \langle x,y\rangle\langle \beta,\beta\rangle$$

since $\langle 1,\beta\rangle = \langle \beta,1\rangle = 0$.

Hence by cancellation

$$\langle x, y\beta\rangle + \langle x\beta, y\rangle = 0, \text{ and so } \langle x\beta, y\rangle = -\langle x, y\beta\rangle = \langle x, -y\beta\rangle.$$

Finally,

$$\begin{aligned}\langle xz, y\rangle &= \langle x(a+\beta), y\rangle = \langle xa, y\rangle + \langle x\beta, y\rangle \\ &= \langle x, ya\rangle + \langle x, -y\beta\rangle = \langle x, y(a-\beta)\rangle = \langle x, y\bar{z}\rangle\end{aligned}$$

24.5 Proposition Let A be a normed real algebra. We have $\overline{xy} = \bar{y}\,\bar{x}$ for all $x, y \in A$.

Proof: Three applications of 24.4 give

$$\langle \overline{xy}, z\rangle = \langle \overline{xy}1, z\rangle = \langle 1, xyz\rangle = \langle \bar{x}, yz\rangle = \langle \bar{y}\,\bar{x}, z\rangle$$

Since this is true for all z we have $\overline{xy} = \bar{y}\,\bar{x}$.

In the complex numbers the basis vector i squares to give -1. A similar thing is true of the basis vectors, except $e_1 = 1$, in any normed real algebra:

24.6 Proposition Let A be a normed real algebra with basis $\{e_1, e_2, ..., e_n\}$, where $e_1 = 1$.

For $i \geq 2$ we have $e_i^2 = -1$.

Proof: $\|\overline{e_i}\|^2 = \langle \overline{e_i}, \overline{e_i} \rangle = \langle -e_i, -e_i \rangle = \langle e_i, e_i \rangle = 1$.

Hence both e_i and $\overline{e_i}$ are unit vectors. It follows that $e_i\overline{e_i}$ is also a unit vector.

But by 24.5 $\overline{e_i\overline{e_i}} = \overline{\overline{e_i}}\ \overline{e_i} = e_i\overline{e_i}$, so that $e_i\overline{e_i}$ is real. Hence $e_i\overline{e_i} = \pm 1$.

For $i \geq 2$ we have $\overline{e_i} = -e_i$ and so $e_i^2 = \pm 1$.

Now if $e_i^2 = 1$ then $e_i^2 - 1 = (e_i - 1)(e_i + 1) = 0$, so that $e_i = \pm 1$.

But e_i is not real for $i \geq 2$, and so we must have $e_i^2 = -1$.

24.7 Proposition Let A be a normed real algebra with basis $\{e_1, e_2, ..., e_n\}$, where $e_1 = 1$.

For $i,j \geq 2$ with $i \neq j$ we have $e_ie_j = -e_je_i$.

Proof: First observe that for $i \neq j$ we have

$$\langle e_je_i, 1 \rangle = \langle e_j, \overline{e_i} \rangle = \langle e_j, -e_i \rangle = -\langle e_j, e_i \rangle = 0.$$

Hence for $i \neq j$ we have $e_je_i \in \boldsymbol{R}^{\perp}$.

Now $e_ie_j = (-e_i)(-e_j) = \overline{e_i}\ \overline{e_j} = \overline{e_je_i} = -e_je_i$.

24.8 Corollary For a normed real algebra A we have

$$x\overline{x} = \|x\|^2 \text{ for all } x \in A.$$

Proof: Suppose A has basis $\{e_1, e_2, ..., e_n\}$, where $e_1 = 1$.

If $x = x_1e_1 + x_2e_2 + ... + x_ne_n$ then $\overline{x} = x_1e_1 - x_2e_2 - ... - x_ne_n$.

Hence by 24.6 and 24.7 we have $x\overline{x} = x_1^2 + x_2^2 + ... + x_n^2 = \|x\|^2$.

We are now ready to prove the promised theorem:

24.9 Theorem Every normed real algebra is a real division algebra.

Proof: Suppose that A is a normed real algebra and $x \in A$ is non-zero. Then

$$x\bar{x} = \|x\|^2 = \bar{x}x$$

Since $\|x\| \neq 0$ we have $x \cdot \frac{1}{\|x\|^2}\bar{x} = 1 = \frac{1}{\|x\|^2}\bar{x} \cdot x$.

Hence $x^{-1} = \frac{1}{\|x\|^2}\bar{x}$.

24.10 Proposition There is no normed real algebra of dimension 3.

Proof: Suppose that A is a normed real algebra of dimension 3 with basis $\{e_1, e_2, e_3\}$ where $e_1 = 1$.

We have $\langle e_2e_3, 1\rangle = \langle e_2, -e_3\rangle = -\langle e_2, e_3\rangle = 0$.

Also $\langle e_2e_3, e_2\rangle = \langle e_3, -e_2^2\rangle = \langle e_3, 1\rangle = 0$ and $\langle e_2e_3, e_3\rangle = \langle e_2, -e_3^2\rangle = \langle e_2, 1\rangle = 0$.

Hence the set $\{1, e_2, e_3, e_2e_3\}$ is orthogonal and therefore linearly independent, by 7.8.

It follows that $\dim A \geq 4$, and so there is no normed real algebra of dimension 3.

24.11 Dimension 4: The Quaternions

The proof that there is no normed real algebra of dimension 3 suggests that we should examine dimension 4 more closely. Suppose that we take the basis $\{1, e_2, e_3, e_2e_3\}$.

By 24.6 we have $e_2^2 = e_3^2 = -1$. Also

$$(e_1e_2)^2 = e_1e_2e_1e_2 = -e_1^2e_2^2 = -(-1 \times -1) = -1.$$

Now we take $i = e_2, j = e_3$ and $k = e_2e_3$.

By definition we have $ij = k$. But also $jk - e_3e_2e_3 = -e_3^2e_2 = i$ and $ki = e_2e_3e_2 = -e_2^2e_3 = j$.

It follows that

$$ji = -k, \quad kj = -i, \quad \text{and} \quad ik = -j.$$

For example, $ji = (ki)i = ki^2 = k \times -1 = -k$.

The set of *quaternions* is

$$\boldsymbol{H} = \{w + xi + yj + zk : w, x, y, z \in \boldsymbol{R}\}.$$

Addition is like addition of vectors in $\boldsymbol{R}^4$:

$$\begin{aligned}&(w_1 + x_1 i + y_1 j + z_1 k) + (w_2 + x_2 i + y_2 j + z_2 k) \\ &= (w_1 + w_2) + (x_1 + x_2)i + (y_1 + y_2)j + (z_1 + z_2)k.\end{aligned}$$

Multiplication follows from the relations between *i*, *j* and *k* introduced above.

Example

$$\begin{aligned}(1 - i + 2j + k) \cdot (2 + i - j + 2k) &= 2 + i - j + 2k - 2i + 1 + k + 2j \\ &\quad + 4j - 2k + 2 + 4i + 2k + j + i - 2 \\ &= 3 + 4i + 6j + 3k.\end{aligned}$$

For a quaternion $q = w + xi + yj + zk$ we may define a norm by

$$\|q\| = \sqrt{w^2 + x^2 + y^2 + z^2}.$$

It is not difficult to show that for $q_1, q_2 \in \boldsymbol{H}$ we have

$$\|q_1 q_2\| = \|q_1\| \cdot \|q_2\|$$

so that $\boldsymbol{H}$ is a normed real algebra. By 24.9 it is a real division algebra. The conjugate is defined by $\overline{q} = w - xi - yj - zk$, so that multiplicative inverses are given by

$$q^{-1} = \frac{1}{\|q\|^2}\overline{q}.$$

A feature of quaternions that will be useful to us in chapter 25 is that they correspond to endomorphisms of $\boldsymbol{R}^4$.

We may regard a quaternion as vector in $\boldsymbol{R}^4$ via the correspondence

$$w + xi + yj + zk \leftrightarrow \begin{pmatrix} w \\ x \\ y \\ z \end{pmatrix}$$

Multiplying by a quaternion carries such a vector of $\boldsymbol{R}^4$ to another vector of $\boldsymbol{R}^4$.

Example Let $q = 1 + 2i + 3j + 4k$ and $v = w + xi + yj + zk.$

Then

$$\begin{aligned} qv &= (1 + 2i + 3j + 4k)(w + xi + yj + zk) \\ &= (w - 2x - 3y - 4z) + (2w + x - 4y + 3z)i \\ &\quad + (3w + 4x + y - 2z)j + (4w - 3x + 2y + z)k \end{aligned}$$

The quaternion qv corresponds to the vector

$$\begin{pmatrix} w - 2x - 3y - 4z \\ 2w + x - 4y + 3z \\ 3w + 4x + y - 2z \\ 4w - 3x + 2y + z \end{pmatrix}$$

and so the corresponding endomorphism of $\boldsymbol{R}^4$ is represented by the matrix

$$\begin{pmatrix} 1 & -2 & -3 & -4 \\ 2 & 1 & -4 & 3 \\ 3 & 4 & 1 & -2 \\ 4 & -3 & 2 & 1 \end{pmatrix}$$

24.12 Theorem There is no *associative* normed real algebra of dimension 5 or greater.

Proof: Suppose that A is an associative normed real algebra in which the set $\{e_1, e_2, e_3, e_4, e_5\}$ is linearly independent and $e_1 = 1$. Define an inner product in such a way that this set is orthonormal.

We have

$$\begin{aligned}(e_2e_3e_4)^2 &= e_2e_3e_4e_2e_3e_4 = e_2e_2e_3e_4e_3e_4 \\ &= -e_3e_4e_3e_4 = e_3e_3e_4e_4 = -1\times-1 = 1\end{aligned}$$

Also,

$$\begin{aligned}\overline{e_2e_3e_4} &= \overline{e_4}\ \overline{e_3}\ \overline{e_2} = (-e_4)(-e_3)(-e_2) \\ &= -e_4e_3e_2 = e_2e_3e_4,\end{aligned}$$

so that $e_2e_3e_4$ is real. Hence $e_2e_3e_4 = \pm 1$.

By a similar argument $e_2e_3e_5 = \pm 1$.

But then we have $e_2e_3e_4 = \pm e_2e_3e_5$.

Since A is a division algebra by 24.9, we can cancel to get $e_4 = \pm e_5$.

This cannot be so since e_4 and e_5 are linearly independent.

Hence there is no such algebra A.

We have shown that the only associative normed real algebras with unity are $\boldsymbol{R}$, $\boldsymbol{C}$ and $\boldsymbol{H}$. There is a fourth normed real algebra with unity known as the *octonions* $\mathbf{O}$, which is of dimension 8. However, this algebra is not associative.

Exercises 24

1. Show that $\|z_1 z_2\| = \|z_1\| \cdot \|z_2\|$ for all $z_1, z_2 \in \boldsymbol{C}$.

2. Prove 24.3(ii): in a normed real algebra A we have

$$\langle zx, zy \rangle = \langle z, z \rangle \langle x, y \rangle \text{ for all } x, y, z \in A.$$

3. Prove 24.4(ii): in a normed real algebra A we have

$$\langle zx, y \rangle = \langle x, \bar{z} y \rangle \text{ for all } x, y, z \in A.$$

4. For the quaternions

   $q_1 = 3 - i + 2j + k$ and $q_2 = 1 + 4i - 3j + 2k$ calculate:

   (i) $q_1 + q_2$ (ii) $q_1 - q_2$

   (iii) $q_1 q_2$ (iv) $q_2 q_1$

   (v) $\|q_1\|$ (vi) q_1^{-1}

   (vii) $\|q_2\|$ (viii) q_2^{-1}

   Show that for any quaternion q we have $q\bar{q} = \|q\|^2$.

5. Find the matrix representing the endomorphism of $\boldsymbol{R}^4$ given by $f(v) = qv$ for each of the following quaternions:

   (i) $q = i$

   (ii) $q = 2j + 3k$

   (iii) $q = 4 + 3i + 2j + k$

   Repeat this exercise for the endomorphisms given by

$$f(v) = vq.$$

6. Make $\boldsymbol{H}$ into a Lie algebra by defining $[q_1, q_2] = q_1 q_2 - q_2 q_1$.

   What is the centre of this algebra?

   What is the derived algebra?

Chapter 25: Tensor Products and Clifford Algebras

Suppose that U and V are real vector spaces. Consider the set of ordered pairs

$$U \times V = \{(u, v) : u \in U, v \in V\}$$

We define a new real vector space of formal linear combinations of ordered pairs in $U \times V$.

A typical element is of the form

$$\lambda_1(u_1, v_1) + \lambda_2(u_2, v_2) + \ldots + \lambda_n(u_n, v_n)$$

where $u_i \in U, v_i \in V$ and $\lambda_i \in \boldsymbol{R}$ for each i.

We write the ordered pair (u, v) as $u \otimes v$. We want the ordered pairs to behave like multiplication, so we introduce the relations

$$u \otimes (v_1 + v_2) = u \otimes v_1 + u \otimes v_2$$

$$(u_1 + u_2) \otimes v = u_1 \otimes v + u_2 \otimes v$$

$$\lambda u \otimes v = u \otimes \lambda v = \lambda(u \otimes v)$$

The vector space of formal linear combinations of pairs $u \otimes v$, subject to these relations, is called the *tensor product* of U and V. It is denoted $U \otimes V$.

25.1 Proposition If U and V are real vector spaces, where U has basis $\{u_1, u_2, \ldots, u_m\}$ and V has basis $\{v_1, v_2, \ldots, v_n\}$, then

$$\{u_i \otimes v_j : 1 \leq i \leq m, 1 \leq j \leq n\}$$

is a basis for $U \otimes V$. It follows that

$$\dim U \otimes V = \dim U \times \dim V.$$

The proof is somewhat technical and is omitted.

Example Suppose $\boldsymbol{R}^2$ has basis $\{u_1, u_2\}$ and $\boldsymbol{R}^3$ has basis $\{v_1, v_2, v_3\}$.

Then $\{u_1 \otimes v_1, u_1 \otimes v_2, u_1 \otimes v_3, u_2 \otimes v_1, u_2 \otimes v_2, u_2 \otimes v_3\}$ is a basis for $\boldsymbol{R}^2 \otimes \boldsymbol{R}^3$.

Hence $\boldsymbol{R}^2 \otimes \boldsymbol{R}^3 \cong \boldsymbol{R}^6$.

More generally, $\boldsymbol{R}^m \otimes \boldsymbol{R}^n \cong \boldsymbol{R}^{mn}$.

We extend the tensor product construction from real vector spaces to real algebras by defining a suitable product:

25.2 Definition Let A and B be real algebras. Form the tensor product $A \otimes B$ of A and B as vector spaces. Then $A \otimes B$ is a real algebra under the product

$$(x \otimes y) \cdot (x' \otimes y') = xx' \otimes yy'$$

Example $\boldsymbol{C} \otimes \boldsymbol{C}$ is a real algebra with basis

$$\{1 \otimes 1, 1 \otimes i, i \otimes 1, i \otimes i\}.$$

$\boldsymbol{C} \otimes \boldsymbol{H}$ has basis $\{1 \otimes 1, 1 \otimes i, 1 \otimes j, 1 \otimes k, i \otimes 1, i \otimes i, i \otimes j, i \otimes k\}$.

25.3 Proposition For all real algebras A, B and C we have

(i) $\boldsymbol{R} \otimes A \cong A$

(ii) $A \otimes B \cong B \otimes A$

(iii) $A \otimes (B \oplus C) \cong (A \otimes B) \oplus (A \otimes C)$

Proof: We define algebra maps as follows:

(i) $f : \boldsymbol{R} \otimes A \to A$ given by $f(x \otimes a) = xa$

(ii) $f : A \otimes B \to B \otimes A$ given by $f(a \otimes b) = b \otimes a$

(iii) $f : A \otimes (B \oplus C) \to (A \otimes B) \oplus (A \otimes C)$ given by

$$f(a \otimes (b, c)) = (a \otimes b, a \otimes c)$$

Each of these is a bijection, so the three isomorphisms follow.

25.4 Lemma Suppose that A is a real algebra with subalgebras B and C satisfying:

(i) $xy = yx$ for all $x \in B$ and $y \in C$

(ii) $A = BC = \{\sum xy : x \in B, y \in C\}$

(iii) $\dim A = \dim B \times \dim C$

Then we have $B \otimes C \cong A$.

Proof: Define $\phi : B \otimes C \to A$ by $\phi(x \otimes y) = xy$.

ϕ is an algebra map since

$$\begin{aligned}\phi((x \otimes y) \cdot (x' \otimes y')) &= \phi(xx' \otimes yy') \\ &= xx'yy' \\ &= xyx'y' \text{ by (i)} \\ &= \phi(x \otimes y) \cdot \phi(x' \otimes y')\end{aligned}$$

By (ii) we see that ϕ is surjective. By (iii) and the rank theorem 5.10 the mapping ϕ is injective.

Hence ϕ is an isomorphism.

25.5 Proposition Suppose that A is a real algebra. We have

$$Mat(n, \mathbf{R}) \otimes A \cong Mat(n, A)$$

Proof: Let $B = Mat(n, \mathbf{R})$ and $C = \{aI : a \in A\} \cong A$. The result follows by lemma 25.4 since:

(i) Scalar matrices of the form aI, where $a \in A$, commute with real matrices.

(ii) Take a basis for $Mat(n, \mathbf{R})$ comprising matrices E_{ij} having 1 for the (i, j)th entry and 0 for all other entries. Multiplying the E_{ij} by matrices aI gives matrices with a for the (i, j)th entry and 0 for all other entries. Adding such matrices together generates $Mat(n, A)$.

(iii) $\dim Mat(n, A) = n^2 \times \dim A = \dim Mat(n, \mathbf{R}) \times \dim A$.

Examples $Mat(2,\boldsymbol{R}) \otimes \boldsymbol{C} \cong Mat(2,\boldsymbol{C})$

$$Mat(3,\boldsymbol{R}) \otimes \boldsymbol{H} \cong Mat(3,\boldsymbol{H})$$

25.6 Corollary $Mat(m,\boldsymbol{R}) \otimes Mat(n,\boldsymbol{R}) \cong Mat(mn,\boldsymbol{R})$

Proof: By 25.5 we have

$$Mat(m,\boldsymbol{R}) \otimes Mat(n,\boldsymbol{R}) \cong Mat(m, Mat(n,\boldsymbol{R})) \cong Mat(mn,\boldsymbol{R}).$$

Example The elements of

$$Mat(2,\boldsymbol{R}) \otimes Mat(2,\boldsymbol{R}) \cong Mat(2, Mat(2,\boldsymbol{R}))$$

are

$$\begin{pmatrix} \begin{pmatrix} a_1 & b_1 \\ c_1 & d_1 \end{pmatrix} & \begin{pmatrix} a_2 & b_2 \\ c_2 & d_2 \end{pmatrix} \\ \begin{pmatrix} a_3 & b_3 \\ c_3 & d_3 \end{pmatrix} & \begin{pmatrix} a_4 & b_4 \\ c_4 & d_4 \end{pmatrix} \end{pmatrix}$$

By ignoring the internal brackets we see that such matrices are in $Mat(4,\boldsymbol{R})$.

25.7 Proposition $\boldsymbol{H} \otimes \boldsymbol{H} \cong Mat(4,\boldsymbol{R})$

Proof: Let B be the subalgebra of $Mat(4,\boldsymbol{R})$ comprising endomorphisms of $\boldsymbol{R}^4$ arising as *left* multiplication by a quaternion:

$$f_q : \boldsymbol{R}^4 \to \boldsymbol{R}^4 \text{ given by } f_q(v) = qv$$

Let C be the subalgebra of $Mat(4,\boldsymbol{R})$ comprising endomorphisms of $\boldsymbol{R}^4$ arising as *right* multiplication by a quaternion:

$$g_q : \boldsymbol{R}^4 \to \boldsymbol{R}^4 \text{ given by } g_q(v) = vq$$

Each of B and C is isomorphic to $\boldsymbol{H}$.

The result follows by lemma 25.4 since:

(i) $$f_{q_1} \circ g_{q_2}(v) = f_{q_1}(vq_2) = q_1(vq_2) = (q_1 v)q_2$$
$$= g_{q_2}(q_1 v) = g_{2_1} \circ f_{q_1}(v)$$

(ii) The kernel of the algebra map $\boldsymbol{H} \otimes \boldsymbol{H} \longrightarrow Mat(4, \boldsymbol{R})$ given by $q_1 \otimes q_2 \mapsto f_{q_1} \circ g_{q_2}$ is trivial, and hence the mapping is injective.

(iii) $\dim \boldsymbol{H} \times \dim \boldsymbol{H} = 4 \times 4 = \dim Mat(4, \boldsymbol{R})$, and so by (ii) and the rank theorem 5.10 the algebra map is also surjective.

25.8 Proposition $\boldsymbol{H} \otimes \boldsymbol{C} \cong Mat(2, \boldsymbol{C})$

Proof: We define an isomorphism $\boldsymbol{H} \otimes \boldsymbol{C} \longrightarrow Mat(2, \boldsymbol{C})$ of real vector spaces by specifying a correspondence between basis elements:

$$1 \otimes 1 \leftrightarrow \begin{pmatrix} 1 & 0 \\ 0 & 1 \end{pmatrix} \qquad 1 \otimes i \leftrightarrow \begin{pmatrix} -i & 0 \\ 0 & -i \end{pmatrix}$$

$$i \otimes 1 \leftrightarrow \begin{pmatrix} i & 0 \\ 0 & -i \end{pmatrix} \qquad i \otimes i \leftrightarrow \begin{pmatrix} 1 & 0 \\ 0 & -1 \end{pmatrix}$$

$$j \otimes 1 \leftrightarrow \begin{pmatrix} 0 & 1 \\ -1 & 0 \end{pmatrix} \qquad j \otimes i \leftrightarrow \begin{pmatrix} 0 & -i \\ i & 0 \end{pmatrix}$$

$$k \otimes 1 \leftrightarrow \begin{pmatrix} 0 & i \\ i & 0 \end{pmatrix} \qquad k \otimes i \leftrightarrow \begin{pmatrix} 0 & 1 \\ 1 & 0 \end{pmatrix}$$

By inspection, this is an algebra map.

25.9 Definition Suppose that V is a real vector space with basis $\{e_1, e_2, \dots, e_n\}$.

The Clifford algebra of V, denoted $Cl(V)$, is a vector space of dimension 2^n with basis the set of all products $e_{i_1} e_{i_2} \dots e_{i_r}$ such that $i_1 < i_2 < \dots < i_r$ and $0 \leq r \leq n$.

The product of zero e_i's is taken to be 1.

By definition the vectors e_i satisfy the relations $e_i^2 = -1$ for each i and $e_i e_j = -e_j e_i$ for $i \neq j$.

Examples (a) We may regard $\boldsymbol{R}$ as a one-dimensional real vector space with basis $\{e_1\}$.

The Clifford algebra $Cl(\boldsymbol{R})$ has basis $\{1, e_1\}$. Since $e_1^2 = -1$, we see that $Cl(\boldsymbol{R}) = \boldsymbol{C}$.

(b) The Clifford algebra $Cl(\boldsymbol{R}^2)$ has basis $\{1, e_1, e_2, e_1e_2\}$.

By taking $e_1 = i, e_2 = j$ and $e_1e_2 = k$ we see that $Cl(\boldsymbol{R}^2) = \boldsymbol{H}$

(c) The Clifford algebra $Cl(\boldsymbol{R}^3)$ has basis

$$\{1, e_1, e_2, e_3, e_1e_2, e_1e_3, e_2e_3, e_1e_2e_3\}.$$

Before we can determine the structure of this algebra, we need to develop more theory.

From now on we shall write $Cl(\boldsymbol{R}^n)$ as $Cl(n)$.

We define another family of algebras, denoted $Cl'(n)$. Each of these has the same basis as the corresponding $Cl(n)$, with basis vectors satisfying $e_ie_j = -e_je_i$ for $i \neq j$ as before, but this time $e_i^2 = 1$ for each i.

In the examples and proofs that follow, we write the basis of $Cl'(n)$ as products of vectors e_i'.

Examples (a) The algebra $Cl'(1)$ has basis $\{1, e_1'\}$.

We have $Cl'(1) \cong \boldsymbol{R} \oplus \boldsymbol{R}$ via the isomorphism

$$1 \leftrightarrow (1, 1) \text{ and } e_1' \leftrightarrow (1, -1).$$

(b) The algebra $Cl'(2)$ has basis $\{1, e_1', e_2', e_1'e_2'\}$.

We have $Cl'(2) \cong Mat(2, \boldsymbol{R})$ via the isomorphism

$$1 \leftrightarrow \begin{pmatrix} 1 & 0 \\ 0 & 1 \end{pmatrix}, e_1' \leftrightarrow \begin{pmatrix} 1 & 0 \\ 0 & -1 \end{pmatrix}, e_2' \leftrightarrow \begin{pmatrix} 0 & 1 \\ 1 & 0 \end{pmatrix}, e_1'e_2' \leftrightarrow \begin{pmatrix} 0 & 1 \\ -1 & 0 \end{pmatrix}$$

The following theorem allows us to determine the structure of further algebras $Cl(n)$ and $Cl'(n)$:

25.10 Theorem For $n \geq 1$ we have the following isomorphisms of algebras:

(i) $Cl'(n+2) \cong Cl(n) \otimes Cl'(2)$

(ii) $Cl(n+2) \cong Cl'(n) \otimes Cl(2)$

Proof: We prove (i). The proof of (ii) is similar and is an exercise.

Define $\phi : Cl'(n+2) \longrightarrow Cl(n) \otimes Cl'(2)$ by

$$\phi(e_i') = \begin{cases} 1 \otimes e_i' & \text{if } i = 1,2 \\ e_{i-2} \otimes e_1' e_2' & \text{if } 3 \leq i \leq n+2 \end{cases}$$

Since this mapping carries basis elements of $Cl'(n+2)$ to basis elements of $Cl(n) \otimes Cl'(2)$ it is an isomorphism of vectors spaces. We need to check that it is also an algebra map.

For $i = 1,2$ we have $\phi((e_i')^2) = \phi(1) = 1$ and

$\phi(e_i') \cdot \phi(e_i') = (1 \otimes e_i') \cdot (1 \otimes e_i') = 1 \otimes (e_i')^2 = 1 \otimes 1 = 1.$

For $i \geq 3$ we have $\phi((e_i')^2) = \phi(1) = 1$ and

$$\begin{aligned} \phi(e_i') \cdot \phi(e_i') &= (e_{i-2} \otimes e_1' e_2') \cdot (e_{i-2} \otimes e_1' e_2') \\ &= e_{i-2}^2 \otimes e_1' e_2' e_1' e_2' = -1 \otimes -1 = 1 \end{aligned}$$

We have $\phi(e_1' e_2') = \phi(e_1')\phi(e_2') = (1 \otimes e_1') \cdot (1 \otimes e_2') = 1 \otimes e_1' e_2'$ and

$\phi(e_2' e_1') = \phi(e_2')\phi(e_1') = (1 \otimes e_2') \cdot (1 \otimes e_1') = 1 \otimes e_2' e_1' = -(1 \otimes e_1' e_2')$

For $i = 1,2$ and $j \geq 3$ we have $\phi\left(e_i' e_j'\right) = \phi(e_i')\phi\left(e_j'\right)$

$$\begin{aligned} &= (1 \otimes e_i') \cdot (e_{j-2} \otimes e_1' e_2') \\ &= e_{j-2} \otimes e_i' e_1' e_2' = \begin{cases} e_{j-2} \otimes e_2' & \text{if } i = 1 \\ -e_{j-2} \otimes e_2' & \text{if } i = 2 \end{cases} \end{aligned}$$

Similarly $\phi\left(e_j' e_i'\right) = \phi\left(e_j'\right)\phi(e_i') = (e_{j-2} \otimes e_1' e_2') \cdot (1 \otimes e_i')$

$$= e_{j-2} \otimes e_1' e_2' e_i' = \begin{cases} -e_{j-2} \otimes e_2' & \text{if } i = 1 \\ e_{j-2} \otimes e_2' & \text{if } i = 2 \end{cases}$$

Finally, for $i,j \geq 3$ with $i \neq j$ we have

$$\begin{aligned}\phi(e_i'e_j') &= \phi(e_i')\phi(e_j') \\ &= (e_{i-2} \otimes e_1'e_2')(e_{j-2} \otimes e_1'e_2') \\ &= e_{i-2}e_{j-2} \otimes e_1'e_2'e_1'e_2' = -e_{i-2}e_{j-2} \otimes 1\end{aligned}$$

and

$$\begin{aligned}\phi(e_j'e_i') &= \phi(e_j')\phi(e_i') \\ &= (e_{j-2} \otimes e_1'e_2')(e_{i-2} \otimes e_1'e_2') \\ &= e_{j-2}e_{i-2} \otimes e_1'e_2'e_1'e_2' = e_{i-2}e_{j-2} \otimes 1\end{aligned}$$

This completes the proof.

We have already found that

$$Cl(1) = \boldsymbol{C} \qquad Cl'(1) = \boldsymbol{R} \oplus \boldsymbol{R}$$

$$Cl(2) = \boldsymbol{H} \qquad Cl'(2) = Mat(2, \boldsymbol{R})$$

Using Theorem 25.10 we may calculate further Clifford algebras.

$$Cl(3) = Cl'(1) \otimes Cl(2) = (\boldsymbol{R} \oplus \boldsymbol{R}) \otimes \boldsymbol{H} = \boldsymbol{H} \oplus \boldsymbol{H}$$

$$Cl'(3) = Cl(1) \otimes Cl'(2) = \boldsymbol{C} \otimes Mat(2, \boldsymbol{R}) = Mat(2, \boldsymbol{C})$$

$$Cl(4) = Cl'(2) \otimes Cl(2) = Mat(2, \boldsymbol{R}) \otimes \boldsymbol{H} = Mat(2, \boldsymbol{H})$$

$$Cl'(4) = Cl(2) \otimes Cl'(2) = \boldsymbol{H} \otimes Mat(2, \boldsymbol{R}) = Mat(2, \boldsymbol{H})$$

$$\begin{aligned}Cl(5) &= Cl'(3) \otimes Cl(2) = Mat(2, \boldsymbol{C}) \otimes \boldsymbol{H} \\ &= Mat(2, \boldsymbol{R}) \otimes \boldsymbol{C} \otimes \boldsymbol{H} = Mat(2, \boldsymbol{R}) \otimes Mat(2, \boldsymbol{C}) \\ &= Mat(2, \boldsymbol{R}) \otimes Mat(2, \boldsymbol{R}) \otimes \boldsymbol{C} = Mat(4, \boldsymbol{R}) \otimes \boldsymbol{C} \\ &= Mat(4, \boldsymbol{C})\end{aligned}$$

$$\begin{aligned}Cl'(5) &= Cl(3) \otimes Cl'(2) = (\boldsymbol{H} \oplus \boldsymbol{H}) \otimes Mat(2, \boldsymbol{R}) \\ &= \boldsymbol{H} \otimes Mat(2, \boldsymbol{R}) \oplus \boldsymbol{H} \otimes Mat(2, \boldsymbol{R}) \\ &= Mat(2, \boldsymbol{H}) \oplus Mat(2, \boldsymbol{H})\end{aligned}$$

$$
\begin{aligned}
Cl(6) &= Cl'(4) \otimes Cl(2) = Mat(2, \boldsymbol{H}) \otimes \boldsymbol{H} \\
&= Mat(2, \boldsymbol{R}) \otimes \boldsymbol{H} \otimes \boldsymbol{H} = Mat(2, \boldsymbol{R}) \otimes Mat(4, \boldsymbol{R}) \\
&= Mat(8, \boldsymbol{R})
\end{aligned}
$$

$$
\begin{aligned}
Cl'(6) &= Cl(4) \otimes Cl'(2) = Mat(2, \boldsymbol{H}) \otimes Mat(2, \boldsymbol{R}) \\
&= \boldsymbol{H} \otimes Mat(2, \boldsymbol{R}) \otimes Mat(2, \boldsymbol{R}) = \boldsymbol{H} \otimes Mat(4, \boldsymbol{R}) \\
&= Mat(4, \boldsymbol{H})
\end{aligned}
$$

You are invited to calculate a few more in the exercises.

We summarise the calculations of $Cl(n)$ and $Cl'(n)$ for n from 0 to 8 in a table:

n	$Cl(n)$	$Cl'(n)$	dim
0	$\boldsymbol{R}$	$\boldsymbol{R}$	1
1	$\boldsymbol{C}$	$\boldsymbol{R} \oplus \boldsymbol{R}$	2
2	$\boldsymbol{H}$	$Mat(2, \boldsymbol{R})$	4
3	$\boldsymbol{H} \oplus \boldsymbol{H}$	$Mat(2, \boldsymbol{C})$	8
4	$Mat(2, \boldsymbol{H})$	$Mat(2, \boldsymbol{H})$	16
5	$Mat(4, \boldsymbol{C})$	$Mat(2, \boldsymbol{H}) \oplus Mat(2, \boldsymbol{H})$	32
6	$Mat(8, \boldsymbol{R})$	$Mat(4, \boldsymbol{H})$	64
7	$Mat(8, \boldsymbol{R}) \oplus Mat(8, \boldsymbol{R})$	$Mat(8, \boldsymbol{C})$	128
8	$Mat(16, \boldsymbol{R})$	$Mat(16, \boldsymbol{R})$	256

We conclude with an interesting periodicity phenomenon exhibited by the real Clifford algebras: when n increases by 8, the new algebra $Cl(n+8)$ comprises 16×16 matrices with entries in the original algebra $Cl(n)$.

The algebras $Cl'(n)$ exhibit similar periodicity.

25.11 Proposition (i) $Cl(n+8) = Mat(16, Cl(n))$

(ii) $Cl'(n+8) = Mat(16, Cl'(n))$

Proof: We will prove (i).

$$
\begin{aligned}
Cl(n+8) &= Cl'(n+6) \otimes \boldsymbol{H} \\
&= Cl(n+4) \otimes Mat(2, \boldsymbol{R}) \otimes \boldsymbol{H} \\
&= Cl'(n+2) \otimes Mat(2, \boldsymbol{R}) \otimes \boldsymbol{H} \otimes \boldsymbol{H} \\
&= Cl(n) \otimes Mat(2, \boldsymbol{R}) \otimes Mat(2, \boldsymbol{R}) \otimes \boldsymbol{H} \otimes \boldsymbol{H} \\
&= Cl(n) \otimes Mat(4, \boldsymbol{R}) \otimes Mat(4, \boldsymbol{R}) \\
&= Cl(n) \otimes Mat(16, \boldsymbol{R}) \\
&= Mat(16, Cl(n))
\end{aligned}
$$

The proof of (ii) is very similar, and is left as an exercise.

Examples We see already from the table that

$$Cl(8) = Mat(16, Cl(0)) = Mat(16, \boldsymbol{R})$$

Similarly we have $Cl(9) = Mat(16, Cl(1)) = Mat(16, \boldsymbol{C})$

and $Cl(10) = Mat(16, Cl(2)) = Mat(16, \boldsymbol{H})$

Similarly, for the algebras $Cl'(n)$:

$$
\begin{aligned}
Cl'(9) &= Mat(16, Cl'(1)) = Mat(16, \boldsymbol{R} \oplus \boldsymbol{R}) \\
&= Mat(16, \boldsymbol{R}) \oplus Mat(16, \boldsymbol{R}).
\end{aligned}
$$

$$
\begin{aligned}
Cl'(10) &= Mat(16, Cl'(2)) = Mat(16, Mat(2, \boldsymbol{R})) \\
&= Mat(16, \boldsymbol{R}) \otimes Mat(2, \boldsymbol{R}) = Mat(32, \boldsymbol{R}).
\end{aligned}
$$

Exercises 25

1. State the dimension of each of the following real vector spaces:

   $\boldsymbol{R}^3 \otimes \boldsymbol{R}^4$ $\qquad$ $\boldsymbol{C} \otimes \boldsymbol{C}$

   $Mat(2, \boldsymbol{R}) \otimes Mat(3, \boldsymbol{R})$ $\qquad$ $Mat(2, \boldsymbol{H}) \otimes Mat(3, \boldsymbol{C})$

2. Show that

$$(2u_1 + 3u_2) \otimes (4v_1 - v_2)$$
$$= 8u_1 \otimes v_1 - 2u_1 \otimes v_2 + 12u_2 \otimes v_1 - 3u_2 \otimes v_2$$

3. If A has basis $\{u_1, u_2, u_3\}$ and B has basis $\{v_1, v_2\}$, write down a basis for $A \otimes B$.

4. **HARD!** By choosing suitable bases prove the following isomorphism of real algebras:

$$\boldsymbol{C} \otimes \boldsymbol{C} \cong \boldsymbol{C} \oplus \boldsymbol{C}$$

5. Using theorem 25.10 determine the structure of each of the following algebras:

   (i) $Cl(7)$ $\qquad$ (ii) $Cl'(7)$

   (iii) $Cl(8)$ $\qquad$ (iv) $Cl'(8)$

6. Using proposition 25.11 determine the structure of each of the following algebras:

   (i) $Cl(24)$ $\qquad$ (ii) $Cl(25)$

   (iii) $Cl(26)$ $\qquad$ (iv) $Cl(27)$

7. Prove 25.11(ii): $\qquad Cl'(n+8) = Mat(16, Cl'(n))$

Solutions to the Exercises

Chapter 1

1. Suppose $b * a = c * a$. By G3, the element a has an inverse a^{-1} which we may combine on the right of both sides of the equation

$$(b * a) * a^{-1} = (c * a) * a^{-1}$$

Then by G1

$$b * (a * a^{-1}) = c * (a * a^{-1})$$

so that $b * e = c * e$ by G3 and finally $b = c$ by G2.

2. Suppose that the same entry appears at the intersection of the row corresponding to element a and the columns corresponding to elements b_1 and b_2, so that $a * b_1 = a * b_2$. By cancellation $b_1 = b_2$, and so each element can appear only once in a row. By counting, each element appears exactly once. The argument is similar for each column.

3. $\begin{pmatrix} 1 & 2 & 3 \\ 1 & 2 & 3 \end{pmatrix} = e$ is the identity $\qquad \begin{pmatrix} 1 & 2 & 3 \\ 2 & 1 & 3 \end{pmatrix} = (12)$

$$\begin{pmatrix} 1 & 2 & 3 \\ 1 & 3 & 2 \end{pmatrix} = (23) \qquad \begin{pmatrix} 1 & 2 & 3 \\ 3 & 2 & 1 \end{pmatrix} = (13)$$

$$\begin{pmatrix} 1 & 2 & 3 \\ 3 & 1 & 2 \end{pmatrix} = (132) \qquad \begin{pmatrix} 1 & 2 & 3 \\ 2 & 3 & 1 \end{pmatrix} = (123)$$

4. $\sigma \circ \pi = \begin{pmatrix} 1 & 2 & 3 & 4 & 5 \\ 1 & 3 & 5 & 4 & 2 \end{pmatrix} \qquad \pi \circ \sigma = \begin{pmatrix} 1 & 2 & 3 & 4 & 5 \\ 1 & 5 & 3 & 2 & 4 \end{pmatrix}$

$$\sigma^{-1} = \begin{pmatrix} 1 & 2 & 3 & 4 & 5 \\ 4 & 2 & 1 & 5 & 3 \end{pmatrix} \qquad \pi^2 = \begin{pmatrix} 1 & 2 & 3 & 4 & 5 \\ 2 & 1 & 4 & 3 & 5 \end{pmatrix}$$

5. (1 3 5 7)(2 9)(4 8 12)(6 11 10)

6. $\begin{pmatrix} 1 & 2 & 3 & 4 & 5 & 6 & 7 & 8 & 9 & 10 \\ 4 & 3 & 9 & 6 & 7 & 1 & 5 & 2 & 8 & 10 \end{pmatrix}$

7.

(i) $\sigma = \begin{pmatrix} 1 & 2 & 3 & 4 & 5 & 6 \\ 3 & 1 & 5 & 4 & 2 & 6 \end{pmatrix}\begin{pmatrix} 1 & 2 & 3 & 4 & 5 & 6 \\ 6 & 1 & 2 & 5 & 4 & 3 \end{pmatrix} = \begin{pmatrix} 1 & 2 & 3 & 4 & 5 & 6 \\ 2 & 6 & 4 & 5 & 1 & 3 \end{pmatrix}$

(ii) $\sigma = \begin{pmatrix} 1 & 2 & 3 & 4 & 5 & 6 \\ 6 & 5 & 4 & 3 & 2 & 1 \end{pmatrix}\begin{pmatrix} 1 & 2 & 3 & 4 & 5 & 6 \\ 3 & 2 & 4 & 1 & 6 & 5 \end{pmatrix} = \begin{pmatrix} 1 & 2 & 3 & 4 & 5 & 6 \\ 5 & 6 & 1 & 4 & 2 & 3 \end{pmatrix}$

8. Cayley table for symmetries of a rectangle:

$\circ$	I	R	M_1	M_2
I	I	R	M_1	M_2
R	R	I	M_2	M_1
M_1	M_1	M_2	I	R
M_2	M_2	M_1	R	I

Cayley table for D_4:

$\circ$	I	R	R^2	R^3	M_1	M_2	M_3	M_4
I	I	R	R^2	R^3	M_1	M_2	M_3	M_4
R	R	R^2	R^3	I	M_4	M_1	M_2	M_3
R^2	R^2	R^3	I	R	M_3	M_4	M_1	M_2
R^3	R^3	I	R	R^2	M_2	M_3	M_4	M_1
M_1	M_1	M_2	M_3	M_4	I	R	R^2	R^3
M_2	M_2	M_3	M_4	M_1	R^3	I	R	R^2
M_3	M_3	M_4	M_1	M_2	R^2	R^3	I	R
M_4	M_4	M_1	M_2	M_3	R	R^2	R^3	I

9.
$$\{\ldots,-5,0,5,10,15,\ldots\}$$
$$\{\ldots,-4,1,6,11,16,\ldots\}$$
$$\{\ldots,-3,2,7,12,17,\ldots\}$$
$$\{\ldots,-2,3,8,13,18,\ldots\}$$
$$\{\ldots,-1,4,9,14,19,\ldots\}.$$

10.

$(\boldsymbol{Z}_5,+)$

+	0	1	2	3	4
0	0	1	2	3	4
1	1	2	3	4	0
2	2	3	4	0	1
3	3	4	0	1	2
4	4	0	1	2	3

$(\boldsymbol{Z}_6,+)$

+	0	1	2	3	4	5
0	0	1	2	3	4	5
1	1	2	3	4	5	0
2	2	3	4	5	0	1
3	3	4	5	0	1	2
4	4	5	0	1	2	3
5	5	0	1	2	3	4

$(\boldsymbol{Z}_7^*,\cdot)$

·	1	2	3	4	5	6
1	1	2	3	4	5	6
2	2	4	6	1	3	5
3	3	6	2	5	1	4
4	4	1	5	2	6	3
5	5	3	1	6	4	2
6	6	5	4	3	2	1

Chapter 2

1. 0° is of order 1. 180° is of order 2. 90° and 270° are of order 4, and the remaining elements are of order 8.

2. Square: $\{0^\circ, 180^\circ\}$

   Hexagon: $\{0^\circ, 180^\circ\}, \{0^\circ, 120^\circ, 240^\circ\}$

   Octagon: $\{0^\circ, 180^\circ\}, \{0^\circ, 90^\circ, 180^\circ, 270^\circ\}$

   Pentagon: There are none.

3. (i) The sum of two integers is an integer; the negative of an integer is an integer.

(ii) The product of two positive reals is positive; the reciprocal of a positive real is positive.

(iii) Suppose $A, B \in SL(2, \boldsymbol{R})$ so that $\det A = 1, \det B = 1$.

Then $\det AB = 1 \times 1 = 1$ so that $AB \in SL(2, \boldsymbol{R})$;

$\det A^{-1} = 1^{-1} = 1$ so that $A^{-1} \in SL(2, \boldsymbol{R})$.

4. Rectangle: $\{I, R\}, \{I, M_1\}, \{I, M_2\}$

D_4: $\{I, R^2\}, \{I, R, R^2, R^3\}, \{I, M_i\}$ for $i = 1, 2, 3, 4$.

$\{I, R^2, M_1, M_3\}, \{I, R^2, M_2, M_4\}$

D_5: $\{I, R, R^2, R^3, R^4\}, \{I, M_i\}$ for $i = 1, 2, 3, 4, 5$.

5.

$$\begin{pmatrix} a_1 & b_1 \\ -b_1 & a_1 + b_1 \end{pmatrix}\begin{pmatrix} a_2 & b_2 \\ -b_2 & a_2 + b_2 \end{pmatrix}$$

$$= \begin{pmatrix} a_1a_2 - b_1b_2 & a_1b_2 + a_2b_1 + b_1b_2 \\ -a_2b_1 - a_1b_2 - b_1b_2 & a_1a_2 + a_1b_2 + a_2b_1 \end{pmatrix} \in H$$

$$\begin{pmatrix} a & b \\ -b & a + b \end{pmatrix}^{-1} = \frac{1}{a^2 + ab + b^2}\begin{pmatrix} a + b & -b \\ b & a \end{pmatrix} \in H$$

Hence H is a subgroup of $GL(2, \boldsymbol{Q})$.

6. D_4: R^2 and each of the reflections is of order 2; R and R^3 are of order 4; I is of order 1.

Q_8: The identity is of order 1; -1 is of order 2; the remaining six elements are of order 4.

$\boldsymbol{Z}_8$: 0 is of order 1; 4 is of order 2; 2 and 6 are of order 4; the remaining four elements are of order 8.

7. Suppose $G = \langle a \rangle$ is a cyclic group, and $x, y \in G$.

Then $x = a^p$ and $y = a^q$ for some $p, q \in \boldsymbol{Z}$.

$$xy = a^p a^q = a^{p+q} = a^{q+p} = a^q a^p = yx$$

Hence G is abelian.

The converse is false - for example $(\boldsymbol{Q}, +)$ is abelian but not cyclic.

8. (i) $$\begin{pmatrix} 1 & 1 \\ 0 & 1 \end{pmatrix}\begin{pmatrix} a & b \\ c & d \end{pmatrix} = \begin{pmatrix} a & b \\ c & d \end{pmatrix}\begin{pmatrix} 1 & 1 \\ 0 & 1 \end{pmatrix}$$

$$\begin{pmatrix} a+c & b+d \\ c & d \end{pmatrix} = \begin{pmatrix} a & a+b \\ c & c+d \end{pmatrix}.$$

Top left: $a + c = a$ and so $c = 0$

Top right: $b + d = a + b$ and so $a = d$.

Hence $$C\begin{pmatrix} 1 & 1 \\ 0 & 1 \end{pmatrix} = \left\{\begin{pmatrix} a & b \\ 0 & a \end{pmatrix} : a, b \in \boldsymbol{Q}, a \neq 0\right\}$$

(ii) $$\begin{pmatrix} 1 & 2 \\ 3 & 4 \end{pmatrix}\begin{pmatrix} a & b \\ c & d \end{pmatrix} = \begin{pmatrix} a & b \\ c & d \end{pmatrix}\begin{pmatrix} 1 & 2 \\ 3 & 4 \end{pmatrix}$$

$$\begin{pmatrix} a+2c & b+2d \\ 3a+4c & 3b+4d \end{pmatrix} = \begin{pmatrix} a+3b & 2a+4b \\ c+3d & 2c+4d \end{pmatrix}$$

Top left: $a + 2c = a + 3b$ and so $2c = 3b, c = \frac{3}{2}b$

Top right: $b + 2d = 2a + 4b$ and so $2d = 2a + 3b, d = a + \frac{3}{2}b$

Hence $$C\begin{pmatrix} 1 & 2 \\ 3 & 4 \end{pmatrix} = \left\{\begin{pmatrix} a & b \\ \frac{3}{2}b & a + \frac{3}{2}b \end{pmatrix} : a, b \in \boldsymbol{Q}, a^2 + \frac{3}{2}ab - \frac{3}{3}b^2 \neq 0\right\}$$

9. (a) $f(2 \times 2) = f(4) = 8$ whereas $f(2) \times f(2) = 4 \times 4 = 16$

So f is *not* a homomorphism, and so not an epimorphism, monomorphism or isomorphism.

(b) $f(m+n) = 4(m+n) = 4m + 4n = f(m) + f(n)$

so f is a homomorphism. f is injective, since if $f(m) = f(n)$ so that $4m = 4n$ then $m = n$.

f is *not* surjective, since for example $6 \in 2\boldsymbol{Z}$ but there is no $n \in \boldsymbol{Z}$ such that $f(n) = 6$. Hence f is a monomorphism, but is not an epimorphism or an isomorphism.

(c) $f(MN) = \det MN = \det M \cdot \det N = f(M) \cdot f(N)$

so f is a homomorphism. f is surjective, since for any $x \in \boldsymbol{Q}^*$

$$M = \begin{pmatrix} x & 0 \\ 0 & 1 \end{pmatrix} \in GL(2, \boldsymbol{Q})$$

has $\det M = x \cdot 1 - 0 \cdot 0 = x$. However, f is not injective, for example for

$$M = \begin{pmatrix} 2 & 1 \\ 1 & 2 \end{pmatrix} \text{ and } N = \begin{pmatrix} 3 & 0 \\ 0 & 1 \end{pmatrix}$$

we have $\det M = 3$ and $\det N = 3$.

Hence f is an epimorphism, but is *not* a monomorphism or an isomorphism.

(d) $f(x+y) = 3(x+y) = 3x + 3y = f(x) + f(y)$

so f is a homomorphism. f is injective, since if $f(x) = f(y)$ then $3x = 3y$ so that $x = y$.

Also f is surjective since for any $x \in \boldsymbol{R}$ we have $f(\frac{x}{3}) = x$.

Hence f is a monomorphism, an epimorphism and an isomorphism.

10. $(\boldsymbol{Z}_6, +)$: 0 is of order 1 3 is of order 2

2 and 4 are of order 3 1 and 5 are of order 6.

$(\boldsymbol{Z}_5, +)$: 0 is of order 1 1, 2, 3 and 4 are of order 5.

Proper non-trivial subgroups of $(\boldsymbol{Z}_6, +)$ are $\{0, 3\}$ and $\{0, 2, 4\}$.

$(\boldsymbol{Z}_5, +)$ has no proper non-trivial subgroups.

11.

$\times$	1	i	-1	$-i$
1	1	i	-1	$-i$
i	i	-1	$-i$	1
-1	-1	$-i$	1	i
$-i$	$-i$	1	i	-1

Isomorphism:

G	1	i	-1	$-i$
$\mathbf{Z}_4$	0	1	2	3

12. One isomorphism $f: (\mathbf{Z}_6, +) \to (\mathbf{Z}_7^*, \cdot)$ is given by

$$f(0) = 1,\ f(1) = 3,\ f(2) = 2,\ f(3) = 6,\ f(4) = 4,\ f(5) = 5.$$

There are other isomorphisms.

13. *Associativity:*

$$\begin{aligned}(g_1, h_1)\cdot((g_2, h_2)\cdot(g_3, h_3)) &= (g_1, h_1)\cdot(g_2g_3, h_2h_3)\\ &= (g_1(g_2g_3), h_1(h_2h_3))\\ &= ((g_1g_2)g_3, (h_1h_2)h_3)\\ &= (g_1g_2, h_1h_2)\cdot(g_3, h_3)\\ &= ((g_1, h_1)\cdot(g_2, h_2))\cdot(g_3, h_3)\end{aligned}$$

Identity: $(g, h)\cdot(e_G, e_H) = (ge_G, he_H) = (g, h)$

$(e_G, e_H)\cdot(g, h) = (e_Gg, e_Hh) = (g, h)$

Inverses: $(g, h)\cdot(g^{-1}, h^{-1}) = (gg^{-1}, hh^{-1}) = (e_G, e_H)$

$(g^{-1}, h^{-1})\cdot(g, h) = (g^{-1}g, h^{-1}h) = (e_G, e_H)$

14. Isomorphism:

V	e	a	b	c
$\mathbf{Z}_2 \times \mathbf{Z}_2$	$(0,0)$	$(0,1)$	$(1,0)$	$(1,1)$

15. An isomorphism for the first part:

$Z_2 \times Z_5$	$(0,0)$	$(1,1)$	$(0,2)$	$(1,3)$	$(0,4)$	$(1,0)$	$(0,1)$	$(1,2)$	$(0,3)$	$(1,4)$
Z_{10}	0	1	2	3	4	5	6	7	8	9

The second part is similar: $Z_3 \times Z_5$ is cyclic and generated by $(1,1)$

16. The groups are *not* isomorphic: in Z_{18} the element 1 is of order 18; in $Z_3 \times Z_6$ all non-identity elements are of order 2, 3 or 6.

Chapter 3

1. (i) $\begin{pmatrix} 6 \\ -8 \end{pmatrix}$ (ii) $\begin{pmatrix} 5 \\ -10 \\ 15 \end{pmatrix}$ (iii) $\begin{pmatrix} 11 \\ 8 \end{pmatrix}$

(iv) $\begin{pmatrix} 6 \\ 7 \end{pmatrix}$ (v) $\begin{pmatrix} 5 \\ 3 \\ -3 \end{pmatrix}$ (vi) $\begin{pmatrix} -1 \\ 3 \\ -5 \end{pmatrix}$

2. (i) $\sqrt{89}$ (ii) $\sqrt{40}$

(iii) $\sqrt{90}$ (iv) $\sqrt{194}$

3. (i) 6 (ii) -7

(iii) 19 (iv) 20

(i) 49° (ii) 121°

(iii) 60° (iv) 48°

4. First we verify the abelian group axioms:

Commutativity:

$$\begin{pmatrix} x_1 \\ y_1 \end{pmatrix} + \begin{pmatrix} x_2 \\ y_2 \end{pmatrix} = \begin{pmatrix} x_1 + x_2 \\ y_1 + y_2 \end{pmatrix} = \begin{pmatrix} x_2 + x_1 \\ y_2 + y_1 \end{pmatrix} = \begin{pmatrix} x_2 \\ y_2 \end{pmatrix} + \begin{pmatrix} x_1 \\ y_1 \end{pmatrix}$$

Associativity:

$$\left\{\begin{pmatrix} x_1 \\ y_1 \end{pmatrix}+\begin{pmatrix} x_2 \\ y_2 \end{pmatrix}\right\}+\begin{pmatrix} x_3 \\ y_3 \end{pmatrix}=\begin{pmatrix} x_1+x_2 \\ y_1+y_2 \end{pmatrix}+\begin{pmatrix} x_3 \\ y_3 \end{pmatrix}=\begin{pmatrix} (x_1+x_2)+x_3 \\ (y_1+y_2)+y_3 \end{pmatrix}$$

$$=\begin{pmatrix} x_1+(x_2+x_3) \\ y_1+(y_2+y_3) \end{pmatrix}=\begin{pmatrix} x_1 \\ y_1 \end{pmatrix}+\begin{pmatrix} x_2+x_3 \\ y_2+y_3 \end{pmatrix}$$

$$=\begin{pmatrix} x_1 \\ y_1 \end{pmatrix}+\left\{\begin{pmatrix} x_2 \\ y_2 \end{pmatrix}+\begin{pmatrix} x_3 \\ y_3 \end{pmatrix}\right\}$$

Identity: $$\begin{pmatrix} x \\ y \end{pmatrix}+\begin{pmatrix} 0 \\ 0 \end{pmatrix}=\begin{pmatrix} x+0 \\ y+0 \end{pmatrix}=\begin{pmatrix} x \\ y \end{pmatrix}$$

Inverses: $$\begin{pmatrix} x \\ y \end{pmatrix}+\begin{pmatrix} -x \\ -y \end{pmatrix}=\begin{pmatrix} x+(-x) \\ y+(-y) \end{pmatrix}=\begin{pmatrix} 0 \\ 0 \end{pmatrix}$$

Next, we verify the properties relating to multiplication by a scalar:

V1:

$$(\lambda+\mu)\begin{pmatrix} x \\ y \end{pmatrix}=\begin{pmatrix} (\lambda+\mu)x \\ (\lambda+\mu)y \end{pmatrix}=\begin{pmatrix} \lambda x+\mu x \\ \lambda y+\mu y \end{pmatrix}$$

$$=\begin{pmatrix} \lambda x \\ \lambda y \end{pmatrix}+\begin{pmatrix} \mu x \\ \mu y \end{pmatrix}=\lambda\begin{pmatrix} x \\ y \end{pmatrix}+\mu\begin{pmatrix} x \\ y \end{pmatrix}$$

V2:

$$\lambda\left\{\begin{pmatrix} x_1 \\ y_1 \end{pmatrix}+\begin{pmatrix} x_2 \\ y_2 \end{pmatrix}\right\}=\lambda\begin{pmatrix} x_1+x_2 \\ y_1+y_2 \end{pmatrix}=\begin{pmatrix} \lambda(x_1+x_2) \\ \lambda(y_1+y_2) \end{pmatrix}$$

$$=\begin{pmatrix} \lambda x_1+\lambda x_2 \\ \lambda y_1+\lambda y_2 \end{pmatrix}=\begin{pmatrix} \lambda x_1 \\ \lambda y_1 \end{pmatrix}+\begin{pmatrix} \lambda x_2 \\ \lambda y_2 \end{pmatrix}$$

$$=\lambda\begin{pmatrix} x_1 \\ y_1 \end{pmatrix}+\lambda\begin{pmatrix} x_2 \\ y_2 \end{pmatrix}$$

V3: $$\lambda\left(\mu\begin{pmatrix}x\\y\end{pmatrix}\right)=\lambda\begin{pmatrix}\mu x\\\mu y\end{pmatrix}=\begin{pmatrix}\lambda(\mu x)\\\lambda(\mu y)\end{pmatrix}=\begin{pmatrix}(\lambda\mu)x\\(\lambda\mu)y\end{pmatrix}=(\lambda\mu)\begin{pmatrix}x\\y\end{pmatrix}$$

V4: $$1\begin{pmatrix}x\\y\end{pmatrix}=\begin{pmatrix}1x\\1y\end{pmatrix}=\begin{pmatrix}x\\y\end{pmatrix}$$

5.

V1: $((\lambda+\mu)f)(x)=(\lambda+\mu)f(x)=\lambda f(x)+\mu f(x)=(\lambda f+\mu f)(x)$

V2: $(\lambda(f+g))(x)=\lambda(f+g)(x)=\lambda f(x)+\lambda g(x)=(\lambda f)(x)+(\lambda g)(x)$

V3: $(\lambda(\mu f))(x)=\lambda(\mu f)(x)=\lambda\mu f(x)=((\lambda\mu)f)(x)$

Chapter 4

1. Suppose that $\begin{pmatrix}x_1\\y_1\end{pmatrix},\begin{pmatrix}x_2\\y_2\end{pmatrix}\in U$.

Then $2x_1+3y_1=0$ and $2x_2+3y_2=0$.

We have $\begin{pmatrix}x_1\\y_1\end{pmatrix}+\begin{pmatrix}x_2\\y_2\end{pmatrix}=\begin{pmatrix}x_1+x_2\\y_1+y_2\end{pmatrix}\in U$,

because $2(x_1+x_2)+3(y_1+y_2)=2x_1+3y_1+2x_2+3y_2=0+0=0$.

Similarly, $\lambda\begin{pmatrix}x_1\\y_1\end{pmatrix}=\begin{pmatrix}\lambda x_1\\\lambda y_1\end{pmatrix}\in U$ because $2\lambda x_1+3\lambda y_1=\lambda.0=0$.

Hence U is a subspace of $\boldsymbol{R}^2$.

3. Suppose that $f,g\in U$. Then $f(0)=0$ and $g(0)=0$.

Then $(f+g)(0)=f(0)+g(0)=0+0=0$ so that $f+g\in U$.

Also $(\lambda f)(0)=\lambda f(0)=\lambda.0=0$ so that $\lambda f\in U$.

Hence U is a subspace of $[\boldsymbol{R},\boldsymbol{R}]$.

4. Suppose that $u,v\in U_1\cap U_2$.

Then $u,v\in U_1$, and since U_1 is a subspace $u+v\in U_1$.

Similarly $u,v\in U_2$, so $u+v\in U_2$, and hence $u+v\in U_1\cap U_2$.

Also, $\lambda v \in U_1$ and $\lambda v \in U_2$ since U_1 and U_2 are subspaces.

Hence $\lambda v \in U_1 \cap U_2$, so that $U_1 \cap U_2$ is a subspace by 4.2.

For the union, consider the following subspaces of $\boldsymbol{R}^2$:

$$U_1 = \left\{ \begin{pmatrix} x \\ 0 \end{pmatrix} : x \in \boldsymbol{R} \right\} \text{ and } U_2 = \left\{ \begin{pmatrix} 0 \\ y \end{pmatrix} : y \in \boldsymbol{R} \right\}$$

Then $\begin{pmatrix} 1 \\ 0 \end{pmatrix}, \begin{pmatrix} 0 \\ 1 \end{pmatrix} \in U_1 \cup U_2$ but $\begin{pmatrix} 1 \\ 0 \end{pmatrix} + \begin{pmatrix} 0 \\ 1 \end{pmatrix} = \begin{pmatrix} 1 \\ 1 \end{pmatrix} \notin U_1 \cup U_2$,

and so $U_1 \cup U_2$ is not a subspace of $\boldsymbol{R}^2$.

5. Suppose that $u_1 + u_2, v_1 + v_2 \in U_1 + U_2$, with $u_1, v_1 \in U_1$ and $u_2, v_2 \in U_2$

Then $(u_1 + u_2) + (v_1 + v_2) = (u_1 + v_1) + (u_2 + v_2) \in U_1 + U_2$

and $\lambda(v_1 + v_2) = \lambda v_1 + \lambda v_2 \in U_1 + U_2$.

Hence $U_1 + U_2$ is a subspace by 4.2.

6.

$$f\left\{ \begin{pmatrix} x_1 \\ y_1 \\ z_1 \end{pmatrix} + \begin{pmatrix} x_2 \\ y_2 \\ z_2 \end{pmatrix} \right\} = f\begin{pmatrix} x_1 + x_2 \\ y_1 + y_2 \\ z_1 + z_2 \end{pmatrix} = \begin{pmatrix} (x_1 + x_2) + (y_1 + y_2) \\ (y_1 + y_2) + (z_1 + z_2) \end{pmatrix}$$

$$= \begin{pmatrix} x_1 + y_1 \\ y_1 + z_1 \end{pmatrix} + \begin{pmatrix} x_2 + y_2 \\ y_2 + z_2 \end{pmatrix} = f\begin{pmatrix} x_1 \\ y_1 \\ z_1 \end{pmatrix} + f\begin{pmatrix} x_2 \\ y_2 \\ z_2 \end{pmatrix}.$$

$$f\left(\lambda \begin{pmatrix} x \\ y \\ z \end{pmatrix} \right) = f\begin{pmatrix} \lambda x \\ \lambda y \\ \lambda z \end{pmatrix} = \begin{pmatrix} \lambda x + \lambda y \\ \lambda y + \lambda z \end{pmatrix} = \lambda \begin{pmatrix} x + y \\ y + z \end{pmatrix} = \lambda f \begin{pmatrix} x \\ y \\ z \end{pmatrix}.$$

8. Consider the mapping given in the chapter:

$$f\begin{pmatrix} w+xi \\ y+zi \end{pmatrix} = \begin{pmatrix} w \\ x \\ y \\ z \end{pmatrix}$$

We show first that this is a linear transformation:

$$f\left\{\begin{pmatrix} w_1+x_1 i \\ y_1+z_1 i \end{pmatrix} + \begin{pmatrix} w_2+x_2 i \\ y_2+z_2 i \end{pmatrix}\right\} = f\begin{pmatrix} (w_1+w_2)+(x_1+x_2)i \\ (y_1+y_2)+(z_1+z_2)i \end{pmatrix}$$

$$= \begin{pmatrix} w_1+w_2 \\ x_1+x_2 \\ y_1+y_2 \\ z_1+z_2 \end{pmatrix} = \begin{pmatrix} w_1 \\ x_1 \\ y_1 \\ z_1 \end{pmatrix} + \begin{pmatrix} w_2 \\ x_2 \\ y_2 \\ z_2 \end{pmatrix}$$

$$= f\begin{pmatrix} w_1+x_1 i \\ y_1+z_1 i \end{pmatrix} + f\begin{pmatrix} w_2+x_2 i \\ y_2+z_2 i \end{pmatrix}$$

$$f\left\{\lambda\begin{pmatrix} w+xi \\ y+zi \end{pmatrix}\right\} = f\begin{pmatrix} \lambda w+\lambda xi \\ \lambda y+\lambda zi \end{pmatrix} = \begin{pmatrix} \lambda w \\ \lambda x \\ \lambda y \\ \lambda z \end{pmatrix} = \lambda\begin{pmatrix} w \\ x \\ y \\ z \end{pmatrix} = \lambda f\begin{pmatrix} w+xi \\ y+zi \end{pmatrix}$$

Next we show that f is injective: suppose

$$f\begin{pmatrix} w_1+x_1 i \\ y_1+z_1 i \end{pmatrix} = f\begin{pmatrix} w_2+x_2 i \\ y_2+z_2 i \end{pmatrix}.$$

Then $\begin{pmatrix} w_1 \\ x_1 \\ y_1 \\ z_1 \end{pmatrix} = \begin{pmatrix} w_2 \\ x_2 \\ y_2 \\ z_2 \end{pmatrix}$ so that $w_1 = w_2, x_1 = x_2, y_1 = y_2$ and $z_1 = z_2$.

It follows that

$$\begin{pmatrix} w_1 + x_1 i \\ y_1 + z_1 i \end{pmatrix} = \begin{pmatrix} w_2 + x_2 i \\ y_2 + z_2 i \end{pmatrix}$$

Finally we show that f is surjective:

For any $v = \begin{pmatrix} w \\ x \\ y \\ z \end{pmatrix} \in \boldsymbol{R}^4$ we have $v = f\begin{pmatrix} w + xi \\ y + zi \end{pmatrix}$.

9. $$\begin{aligned} g \circ f(u + v) &= g(f(u + v)) = g(f(u) + f(v)) \\ &= g(f(u)) + g(f(v)) = g \circ f(u) + g \circ f(v) \end{aligned}$$

$$g \circ f(\lambda v) = g(f(\lambda v)) = g(\lambda f(v)) = \lambda g(f(v)) = \lambda g \circ f(v)$$

10. $$\begin{pmatrix} 1 & 2 & 3 & 4 \\ 1 & -1 & 1 & -1 \\ 1 & 0 & 1 & 0 \\ 0 & 1 & 0 & -1 \\ 3 & -5 & 1 & -6 \end{pmatrix}$$

11. Suppose that $v = u_1 + u_2$ and $v = v_1 + v_2$.

Then $u_1 + u_2 = v_1 + v_2$ and so $u_1 - v_1 = v_2 - u_2 \in U_1 \cap U_2 = \{\mathbf{0}\}$.

Hence $u_1 - v_1 = v_2 - u_2 = 0$, so that $u_1 = v_1$ and $u_2 = v_2$.

Chapter 5

1. (i) Suppose that

$$\lambda_1 \begin{pmatrix} 2 \\ 1 \end{pmatrix} + \lambda_2 \begin{pmatrix} 1 \\ -1 \end{pmatrix} = \begin{pmatrix} 0 \\ 0 \end{pmatrix}$$

We obtain the simultaneous equations

$$2\lambda_1 + \lambda_2 = 0$$

$$\lambda_1 - \lambda_2 = 0$$

with solution $\lambda_1 = 0,\ \lambda_2 = 0$. Hence the vectors are linearly independent.

(ii) Suppose that we wish to express a vector in $\boldsymbol{R}^2$ as a linear combination of the two vectors:

$$\begin{pmatrix} x \\ y \end{pmatrix} = \lambda_1 \begin{pmatrix} 2 \\ 1 \end{pmatrix} + \lambda_2 \begin{pmatrix} 1 \\ -1 \end{pmatrix}$$

We obtain the simultaneous equations

$$x = 2\lambda_1 + \lambda_2$$

$$y = \lambda_1 - \lambda_2$$

with solution $\lambda_1 = \frac{1}{3}x + \frac{1}{3}y, \lambda_2 = \frac{1}{3}x - \frac{2}{3}y$.

Hence the vectors span $\boldsymbol{R}^2$.

2. (i) $\det\begin{pmatrix} 1 & 1 & 2 \\ 3 & -2 & 1 \\ 0 & -1 & 1 \end{pmatrix} = -10$, so the vectors form a basis.

(ii) $\det\begin{pmatrix} 1 & 1 & 2 \\ 3 & -2 & 1 \\ 0 & -1 & -1 \end{pmatrix} = 0$, so the vectors do *not* form a basis.

4. Define $f: P \to \boldsymbol{R}^3$ as follows:

$$f(ax^2 + bx + c) = \begin{pmatrix} a \\ b \\ c \end{pmatrix}.$$

We show that f is a linear transformation:

$$f((a_1x^2+b_1x+c_1)+(a_2x^2+b_2x+c_2))$$
$$=f((a_1+a_2)x^2+(b_1+b_2)x+(c_1+c_2))$$
$$=\begin{pmatrix} a_1+a_2 \\ b_1+b_2 \\ c_1+c_2 \end{pmatrix}=\begin{pmatrix} a_1 \\ b_1 \\ c_1 \end{pmatrix}+\begin{pmatrix} a_2 \\ b_2 \\ c_2 \end{pmatrix}$$
$$=f(a_1x^2+b_1x+c_1)+f(a_2x^2+b_2x+c_2)$$

$$f(\lambda(ax^2+bx+c))=f(\lambda ax^2+\lambda bx+\lambda c)=\begin{pmatrix} \lambda a \\ \lambda b \\ \lambda c \end{pmatrix}$$

$$=\lambda\begin{pmatrix} a \\ b \\ c \end{pmatrix}=\lambda f(ax^2+bx+c)$$

5. (i) $P=\begin{pmatrix} 7 & 3 \\ 2 & 1 \end{pmatrix}$ (ii) $\begin{pmatrix} 7 & 3 \\ 2 & 1 \end{pmatrix}\begin{pmatrix} 2 \\ 3 \end{pmatrix}=\begin{pmatrix} 23 \\ 7 \end{pmatrix}$

(iii) $\begin{pmatrix} 1 & -3 \\ -2 & 7 \end{pmatrix}\begin{pmatrix} 5 \\ 1 \end{pmatrix}=\begin{pmatrix} 2 \\ -3 \end{pmatrix}$

6. $M=\begin{pmatrix} 1 & 3 \\ 2 & -1 \\ 4 & 1 \end{pmatrix}$ $P=\begin{pmatrix} 1 & 1 \\ 1 & 3 \end{pmatrix}$ $Q=\begin{pmatrix} 1 & 1 & 0 \\ 1 & 0 & 1 \\ 0 & 1 & 1 \end{pmatrix}$ $N=\begin{pmatrix} 0 & 1 \\ 4 & 8 \\ 1 & -2 \end{pmatrix}$

7. (i) is linearly independent, whereas (ii) and (iii) are not.

Chapter 6

1. (i) $\lambda^2-7\lambda-30=0$ from which we obtain $\lambda=10, \lambda=-3$.

The respective eigenvectors are $\begin{pmatrix} 1 \\ 1 \end{pmatrix}$ and $\begin{pmatrix} 7 \\ -6 \end{pmatrix}$.

(ii) $\lambda^2 - 8\lambda + 15 = 0$ from which we obtain $\lambda = 3, \lambda = 5$.

The respective eigenvectors are $\begin{pmatrix} 5 \\ -3 \end{pmatrix}$ and $\begin{pmatrix} 2 \\ -1 \end{pmatrix}$.

(iii) $\lambda = 1, \lambda = 2, \lambda = -2$.

The respective eigenvectors are $\begin{pmatrix} 2 \\ 0 \\ 1 \end{pmatrix}, \begin{pmatrix} 3 \\ 1 \\ 0 \end{pmatrix}, \begin{pmatrix} 0 \\ 1 \\ -2 \end{pmatrix}$.

(iv) $(\lambda - 2)^2(\lambda - 6) = 0$, hence $\lambda = 2$ with algebraic multiplicity 2, and $\lambda = 6$ with algebraic multiplicity 1.

For $\lambda = 2$ the eigenspace is spanned by $\begin{pmatrix} 1 \\ -1 \\ 0 \end{pmatrix}$ and $\begin{pmatrix} 1 \\ 0 \\ -1 \end{pmatrix}$.

For $\lambda = 6$ the eigenvector is $\begin{pmatrix} 1 \\ 2 \\ 1 \end{pmatrix}$.

2. Eigenvalues are $\lambda = 2, \lambda = -5$ with eigenvectors

$$\begin{pmatrix} 1 \\ 1 \end{pmatrix} \text{ and } \begin{pmatrix} 1 \\ 2 \end{pmatrix}.$$

Matrix for f with respect to eigenvector basis is $\begin{pmatrix} 2 & 0 \\ 0 & -5 \end{pmatrix}$.

3. Eigenvalues are $\lambda = 3$ with algebraic multiplicity 2 and $\lambda = 5$ with multiplicity 1.

For $\lambda = 5$ the eigenvector is $\begin{pmatrix} 1 \\ 2 \\ 1 \end{pmatrix}$.

For $\lambda = 3$ the eigenspace is spanned by $\begin{pmatrix} 0 \\ 1 \\ 0 \end{pmatrix}$ and $\begin{pmatrix} 1 \\ 0 \\ -1 \end{pmatrix}$.

$$N = \begin{pmatrix} 5 & 0 & 0 \\ 0 & 3 & 0 \\ 0 & 0 & 3 \end{pmatrix}$$

4. $De^{ax} = ae^{ax}$, eigenvalue a.

$D^2 \sin ax = Da \cos ax = -a^2 \sin ax$, eigenvalue $-a^2$.

Chapter 7

1. (i) $\begin{pmatrix} x_1 \\ y_1 \end{pmatrix} \cdot \begin{pmatrix} x_2 \\ y_2 \end{pmatrix} = x_1 x_2 + y_1 y_2 = \begin{pmatrix} x_2 \\ y_2 \end{pmatrix} \cdot \begin{pmatrix} x_1 \\ y_1 \end{pmatrix}$

(ii) $$\left\{\begin{pmatrix} x_1 \\ y_1 \end{pmatrix} + \begin{pmatrix} x_2 \\ y_2 \end{pmatrix}\right\} \cdot \begin{pmatrix} x_3 \\ y_3 \end{pmatrix} = \begin{pmatrix} x_1 + x_2 \\ y_1 + y_2 \end{pmatrix} \cdot \begin{pmatrix} x_3 \\ y_3 \end{pmatrix}$$
$$= (x_1 + x_2)x_3 + (y_1 + y_2)y_3$$
$$= (x_1 x_3 + y_1 y_3) + (x_2 x_3 + y_2 y_3)$$
$$= \begin{pmatrix} x_1 \\ y_1 \end{pmatrix} \cdot \begin{pmatrix} x_3 \\ y_3 \end{pmatrix} + \begin{pmatrix} x_2 \\ y_2 \end{pmatrix} \cdot \begin{pmatrix} x_3 \\ y_3 \end{pmatrix}$$

(iii) $$\left\{\lambda \begin{pmatrix} x_1 \\ y_1 \end{pmatrix}\right\} \cdot \begin{pmatrix} x_2 \\ y_2 \end{pmatrix} = \begin{pmatrix} \lambda x_1 \\ \lambda y_1 \end{pmatrix} \cdot \begin{pmatrix} x_2 \\ y_2 \end{pmatrix} = \lambda x_1 x_2 + \lambda y_1 y_2$$
$$= \lambda(x_1 x_2 + y_1 y_2) = \lambda \left\{\begin{pmatrix} x_1 \\ y_1 \end{pmatrix} \cdot \begin{pmatrix} x_2 \\ y_2 \end{pmatrix}\right\}$$

(iv) $\begin{pmatrix} x \\ y \end{pmatrix} \cdot \begin{pmatrix} x \\ y \end{pmatrix} = x^2 + y^2 \geq 0,$

and $x^2 + y^2 = 0$ if and only if $\begin{pmatrix} x \\ y \end{pmatrix} = \begin{pmatrix} 0 \\ 0 \end{pmatrix}$.

2. $\langle u, \lambda v \rangle = \langle \lambda v, u \rangle = \lambda \langle v, u \rangle = \lambda \langle u, v \rangle$

$\langle \mathbf{0}, v \rangle = \langle 0v, v \rangle = 0\langle v, v \rangle = 0$

4. (i) is an inner product, (ii), (iii) and (iv) are not.

You should also think about the *reasons.*

5. (i)

$$\|u+v\|^2 = \langle u+v, u+v\rangle = \langle u,u\rangle + 2\langle u,v\rangle + \langle v,v\rangle$$
$$= \|u\|^2 + 2\langle u,v\rangle + \|v\|^2$$

(ii) is similar.

(iii)

$$\frac{1}{4}\{\|u+v\|^2 - \|u-v\|^2\}$$
$$= \frac{1}{4}\{\|u\|^2 + 2\langle u,v\rangle + \|v\|^2 - \|u\|^2 + 2\langle u,v\rangle - \|v\|^2\}$$
$$= \langle u,v\rangle$$

6. (i) $$\langle f,g\rangle = \int_0^1 f(t)g(t)dt = \int_0^1 g(t)f(t)dt = \langle g,f\rangle$$

(ii)

$$\langle f+g,h\rangle = \int_0^1 (f(t)+g(t))h(t)dt$$
$$= \int_0^1 f(t)h(t)dt + \int_0^1 g(t)h(t)dt = \langle f,h\rangle + \langle g,h\rangle$$

(iii) $$\langle \lambda f,g\rangle = \int_0^1 \lambda f(t)g(t)dt = \lambda \int_0^1 f(t)g(t)dt = \lambda\langle f,g\rangle$$

(iv) $\langle f,f\rangle = \int_0^1 f(t)^2 dt \geq 0$ because $f(t)^2 \geq 0$ and $\langle f,f\rangle = 0$ if and only if $f(t) = 0$ for all t because the area bounded by the zero function is 0.

$$\langle x^2, 4x-3\rangle = \int_0^1 x^2(4x-3)dx = \int_0^1 4x^3 - 3x^2.dx = [x^4 - x^3]_0^1 = 0$$

Integrating by parts:

$$\langle \sin \pi x, \cos \pi x \rangle = \int_0^1 \sin \pi x \cos \pi x dx = [\sin^2 \pi x]_0^1 - \int_0^1 \cos \pi x \sin \pi x dx dx$$

$$= 0 - \langle \sin \pi x, \cos \pi x \rangle.$$

Hence $\langle \sin \pi x, \cos \pi x \rangle = 0$.

7. (i) $P = \begin{pmatrix} \frac{1}{\sqrt{2}} & -\frac{1}{\sqrt{2}} \\ \frac{1}{\sqrt{2}} & \frac{1}{\sqrt{2}} \end{pmatrix}$ $N - \begin{pmatrix} 3 & 0 \\ 0 & -1 \end{pmatrix}$

(ii) $P = \begin{pmatrix} \frac{1}{\sqrt{2}} & 0 & \frac{1}{\sqrt{2}} \\ 0 & 1 & 0 \\ -\frac{1}{\sqrt{2}} & 0 & \frac{1}{\sqrt{2}} \end{pmatrix}$ $N = \begin{pmatrix} 0 & 0 & 0 \\ 0 & 1 & 0 \\ 0 & 0 & 2 \end{pmatrix}$

8. $\frac{1}{\sqrt{5}}\begin{pmatrix} 1 \\ 2 \end{pmatrix}$ $\frac{1}{\sqrt{14}}\begin{pmatrix} 1 \\ 2 \\ 3 \end{pmatrix}$ $\frac{1}{\sqrt{3}}\begin{pmatrix} 1 \\ -1 \\ 1 \end{pmatrix}$

$\frac{1}{\sqrt{30}}\begin{pmatrix} 1 \\ -2 \\ 3 \\ -4 \end{pmatrix}$

9. Suppose $u_1, u_2 \in v^{\perp}$ so that $\langle u_1, v \rangle = 0$ and $\langle u_2, v \rangle = 0$.

Then $\langle u_1 + u_2, v \rangle = \langle u_1, v \rangle + \langle u_2, v \rangle = 0 + 0 = 0$ so that $u_1 + u_2 \in v^{\perp}$.

Also $\langle \lambda u_1, v \rangle = \lambda \langle u_1, v \rangle = \lambda \times 0 = 0$ so that $\lambda u_1 \in v^{\perp}$.

Hence $v^{\perp}$ is a subspace of V.

11. $$\langle u, v + w \rangle = \overline{\langle v + w, u \rangle} = \overline{\langle v, u \rangle + \langle w, u \rangle}$$

$$= \overline{\langle v, u \rangle} + \overline{\langle w, u \rangle} = \langle u, v \rangle + \langle u, w \rangle$$

Chapter 8

1. $0+3\boldsymbol{Z}=\{\ldots,-3,0,3,6,9,\ldots\}$

   $1+3\boldsymbol{Z}=\{\ldots,-2,1,4,7,10,\ldots\}$

   $2+3\boldsymbol{Z}=\{\ldots,-1,2,5,8,11,\ldots\}$

2. (i) $\{I,R^2\},\{R,R^3\},\{M_1,M_3\},\{M_2,M_4\}$

   (ii) $\{I,R,R^2,R^3\},\{M_1,M_2,M_3,M_4\}$

   (iii) Left cosets $\{I,M_1\},\{R,M_4\},\{R^2,M_3\},\{R^3,M_2\}$

   Right cosets $\{I,M_1\},\{R,M_2\},\{R^2,M_3\},\{R^3,M_4\}$

   In (i) and (ii), the left and right cosets are the same.

3. We show first that ϕ_g is a homomorphism:

$$\phi_g(ab)=gabg^{-1}=gag^{-1}gbg^{-1}=\phi_g(a)\phi_g(b).$$

   The inverse mapping $\phi_g^{-1}:G\to G$ is given by $\phi_g^{-1}(a)=g^{-1}ag$ Hence ϕ_g^{-1} is bijective and so is an isomorphism.

4. (i) $\{0,6\},\{0,4,8\},\{0,3,6,9\},\{0,2,4,6,8,10\}$.

   (ii) $\{I,R^2\},\{I,R,R^2,R^3\},\{I,M_i\}$ for $i=1,2,3,4$,

   $\{I,R^2,M_1,M_3\}$ and $\{I,R^2,M_2,M_4\}$.

   (iii) $\{e,a\},\{e,b\}$ and $\{e,c\}$.

   (iv) $\{1,-1\},\{1,i,-1,-i\},\{1,j,-1,-j\}$ and $\{1,k,-1,-k\}$.

   All subgroups are normal, except $\{I,M_i\}$ for $i=1,2,3,4$ in D_4.

5. (i) Define $\phi:\boldsymbol{Z}\to\boldsymbol{Z}_3$ by $\phi(a)=[a]$.

ϕ is readily seen to be an epimorphism with $\ker\phi=3\boldsymbol{Z}$.

By the first isomorphism theorem 8.13 we have $\boldsymbol{Z}/3\boldsymbol{Z}\cong\boldsymbol{Z}_3$.

(ii) Define $\phi:D_4\to\boldsymbol{Z}_2$ by $\phi(a)=0$ for $a=I,R,R^2,R^3$ and $\phi(a)=1$ for $a=M_1,M_2,M_3,M_4$

ϕ is readily seen to be an epimorphism with $\ker\phi = \{I, R, R^2, R^3\}$.

By the first isomorphism theorem we have $D_4/\{I, R, R^2, R^3\} \cong \boldsymbol{Z}_2$.

(iii) Define $\phi : D_4 \to V$ by $\phi(I) = \phi(R^2) = e, \phi(R) = \phi(R^3) = a$ $\phi(M_1) = \phi(M_3) = b$ and $\phi(M_2) = \phi(M_4) = c$.

It is easily checked that ϕ is an epimorphism with $\ker\phi = \{I, R^2\}$.

By the first isomorphism theorem we have $D_4/\{I, R^2\} \cong V$.

6. Suppose that G is abelian. For $a, b \in G$ we have

$$aN \cdot bN = abN = baN = bN \cdot aN$$

and hence G/N is abelian. The converse is false. In question 5(ii) above we saw that $D_4/\{I, R, R^2, R^3\} \cong \boldsymbol{Z}_2$ is abelian whilst D_4 itself is non-abelian.

Chapter 9

1. (i) $e \cdot x = exe^{-1} = exe = x$ for all $x \in X$.

(ii) For all $g_1, g_2 \in G$ and all $x \in X$ we have

$$\begin{aligned} g_1 g_2 \cdot x &= (g_1 g_2)x(g_1 g_2)^{-1} \\ &= g_1 g_2 x g_2^{-1} g_1^{-1} \\ &= g_1 \cdot (g_2 x g_2^{-1}) \\ &= g_1 \cdot (g_2 \cdot x). \end{aligned}$$

2. (a) (i) Orbit is $\{x + y + z\}$ and stabiliser is D_3.

(ii) Orbit is is $\{x^2 + xyz, y^2 + xyz, z^2 + xyz\}$ and stabiliser is $\{I, M_1\}$

(iii) Stabiliser: $\{I\}$ Orbit:

$\{x + 2y + 3z, x + 2z + 3y, y + 2x + 3z, y + 2z + 3x, z + 2x + 3y, z + 2y + 3x\}$

(iv) Orbit: $\{(x - y)(y - z)(z - x), -(x - y)(y - z)(z - x)\}$

Stabiliser: $\{I, R, R^2\}$

(b) (i) $\{6,7,8,9\}$ $\{I\}$

(ii) $\{12,13,14,15\}$ $\{I\}$

(iii) $\{1\}$ $\{I,R,R^2,R^3\}$

(c) (i) Orbit is $\left\{\begin{pmatrix}0\\0\end{pmatrix}\right\}$ and stabiliser is the whole group $GL(2,\boldsymbol{Q})$.

(ii) Orbit is all non-zero vectors in $\boldsymbol{R}^2$ since for any non-zero vector $\begin{pmatrix}x\\y\end{pmatrix}$ we may take either $\begin{pmatrix}x&1\\y&0\end{pmatrix}$ or $\begin{pmatrix}x&0\\y&1\end{pmatrix}$ (at least one of which is invertible) so that

$$\begin{pmatrix}x&1\\y&0\end{pmatrix}\begin{pmatrix}1\\0\end{pmatrix}=\begin{pmatrix}x\\y\end{pmatrix} \text{ or } \begin{pmatrix}x&0\\y&1\end{pmatrix}\begin{pmatrix}1\\0\end{pmatrix}=\begin{pmatrix}x\\y\end{pmatrix}.$$

The stabiliser consists of all of the matrices satisfying

$\begin{pmatrix}a&b\\c&d\end{pmatrix}\begin{pmatrix}1\\0\end{pmatrix}=\begin{pmatrix}1\\0\end{pmatrix}$ so that $a=1, c=0$, that is

$$\left\{\begin{pmatrix}1&b\\0&d\end{pmatrix}: b,d\in\boldsymbol{Q}\right\}$$

3. Suppose that $g_1,g_2\in G_x$ so that $g_1\cdot x=x$ and $g_2\cdot x=x$. Then $g_1g_2\cdot x=g_1\cdot(g_2\cdot x)=g_1\cdot x=x$, and so $g_1g_2\in G_x$.

Since $x=g_1\cdot x$ we have $g_1^{-1}\cdot x=g_1^{-1}\cdot(g_1\cdot x)=g_1^{-1}g_1\cdot x=e\cdot x=x$ so that $g_1^{-1}\in G_x$.

Hence G_x is a subgroup of G.

4. There are $3^4=81$ colourings. If the number of orbits under the action of $\{I,R,R^2,R^3\}$ is m then

$$81+3+9+3=4m \text{ and so } m=24.$$

If the number of orbits under the action of D_4 is m then

$$81+3+9+3+9+27+9+27=8m \text{ and so } m=21.$$

5. For D_4 the orbits are the conjugacy classes

$$\{I\}, \{R, R^3\}, \{R^2\}, \{M_1, M_3\}, \{M_2, M_4\}$$

The corresponding stabilisers are

$$D_4, \{I, R, R^2, R^3\}, D_4, \{I, R^2, M_1, M_3\}, \{I, R^2, M_2, M_4\}$$

For Q_8 the orbits are $\{1\}, \{-1\}, \{i, -i\}, \{j, -j\}, \{k, -k\}$

The corresponding stabilisers are

$$Q_8, Q_8, \{1, i, -1, -i\}, \{1, j, -1, -j\}, \{1, k, -1, -k\}$$

6. $200 = 2^3 \times 5^2$ Sylow's theorem predicts subgroups of order 8 and 25.

A subgroup of order 8 is $\langle 25 \rangle = \{0, 25, 50, 75, 100, 125, 150, 175\}$ and a subgroup of order 25 is $\langle 8 \rangle$.

$12 = 2^2 \times 3$ Sylow's theorem predicts subgroups of order 4 and 3.

A subgroup of order 3 is $\{I, R^2, R^4\}$ and a subgroup of order 4 is of the form $\{I, R^3, M, N\}$ where M and N are a pair of reflections such that $MN = R^3$.

$6! = 720 = 2^4 \times 3^2 \times 5$ Sylow's theorem predicts subgroups of order 16, 9 and 5.

A subgroup of order 5 is $\langle (12345) \rangle$, a subgroup of order 9 is $\langle (123), (456) \rangle$.

Chapter 10

1. $\{0, 3\}$ and $\{0, 2, 4\}$ are subgroups of $\boldsymbol{Z}_6$. Since $\boldsymbol{Z}_6$ is abelian these are normal subgroups and hence $\boldsymbol{Z}_6$ is not simple.

Similarly $\{0, 5\}$ and $\{0, 2, 4, 6, 8\}$ are normal subgroups of $\boldsymbol{Z}_{10}$ and so $\boldsymbol{Z}_{10}$ is not simple.

2. A one element set has only a single permutation - the identity. This is even. Hence A_1 is trivial. A two element set has a pair of permutations - the identity and $(1\ 2)$. Of these, only the identity is even. Hence A_2 is trivial.

A_3 has three permutations: the identity $e, (1\ 2\ 3)$ and $(3\ 2\ 1)$. An isomorphism:

$A_3 \to \mathbf{Z}_3$ is $e \leftrightarrow 0, (1\ 2\ 3) \leftrightarrow 1$ and $(3\ 2\ 1) \leftrightarrow 2$. Hence $A_3 \cong \mathbf{Z}_3$.

3. S_4 is of order $4! = 24$.

Even: $(123), (132), (124), (142), (134), (143), (234), (243)$ $(12)(34), (13)(24), (14)(23), e.$

Odd: $(12), (13), (14), (23), (24), (34),$ $(1234), (1243), (1324), (1342), (1423), (1432).$

5. Suppose $\sigma, \pi \in A_n$. Then

$$\sigma = (a_1 b_1)(a_2 b_2)...(a_r b_r) \text{ and } \pi = (c_1 d_1)(c_2 d_2)...(c_s d_s)$$

where r and s are even. Clearly

$$\sigma \circ \pi = (a_1 b_1)(a_2 b_2)...(a_r b_r)(c_1 d_1)(c_2 d_2)...(c_s d_s)$$

is an even permutation, and $\sigma^{-1} = (a_r b_r)...(a_2 b_2)(a_1 b_1)$ is also even. Hence A_n is a subgroup of S_n.

Now suppose $\rho = (e_1 f_1)(e_2 f_2)...(e_t f_t) \in S_n$. Notice that t may be odd or even, but that

$$\rho \circ \sigma \circ \rho^{-1}$$
$$= (e_1 f_1)(e_2 f_2)...(e_t f_t)(a_1 b_1)(a_2 b_2)...(a_r b_r)(e_t f_t)...(e_2 f_2)(e_1 f_1) \in A_n$$

and so A_n is a normal subgroup of S_n.

6. (i) S_3 is of order 6.

Type	Number of elements	Order
e	1	1
(ab)	$\frac{3 \times 2}{2} = 3$	2
(abc)	$\frac{3 \times 2}{3} = 2$	3

The conjugacy class of (12) has at most three elements. We shall find the centraliser of (12):

Suppose $\pi = \begin{pmatrix} 1 & 2 & 3 \\ p & q & r \end{pmatrix}$. If $(12)\pi = \pi(12)$ then $\pi^{-1}(12)\pi = (12)$.

We have $\begin{pmatrix} p & q & r \\ 1 & 2 & 3 \end{pmatrix}(12)\begin{pmatrix} 1 & 2 & 3 \\ p & q & r \end{pmatrix} = (12)$ and so $(pq) = (12)$.

Hence either $\pi = e$ or $\pi = (12)$. Since the centraliser has two elements, the conjugacy class of (12) has three elements. By similar argument, the centraliser of (123) has three elements, and so its conjugacy class has two elements.

(ii) A_4 is of order 12.

Type	Number of elements	Order
e	1	1
(abc)	$\frac{4\times3\times2}{3} = 8$	3
$(ab)(cd)$	$\frac{4\times3\times2}{2\times2\times2} = 3$	2

The conjugacy class of (123) has at most eight elements. We shall find the centraliser of (123):

Suppose $\pi = \begin{pmatrix} 1 & 2 & 3 & 4 \\ p & q & r & s \end{pmatrix}$.

We have $\begin{pmatrix} p & q & r & s \\ 1 & 2 & 3 & 4 \end{pmatrix}(123)\begin{pmatrix} 1 & 2 & 3 & 4 \\ p & q & r & s \end{pmatrix} = (123)$ and so

$$(pqr) = (123).$$

Hence either $\pi = e, \pi = (123)$ or $\pi = (132)$. Since the centraliser has three elements, the conjugacy class of (123) has four elements. There is another conjugacy class of elements of this type, also having four elements.

By similar argument, the centraliser of $(12)(34)$ has four elements, and so its conjugacy class has three elements.

(iii) S_4 is of order 24.

Type	Number of elements	Order
e	1	1
(ab)	$\frac{4\times 3}{2}=6$	2
(abc)	$\frac{4\times 3\times 2}{3}=8$	3
$(ab)(cd)$	$\frac{4\times 3\times 2}{2\times 2\times 2}=3$	2
$(abcd)$	$\frac{4!}{4}=6$	4

$C(12)=\{e,(12),(34),(12)(34)\}$ and so the conjugacy class is of size 6.

$C(123)=\{e,(123),(132)\}$ and so the conjugacy class is of size 8.

The centraliser $C((12)(34))$ has 8 elements and so the conjugacy class is of size 3.

$C(1234)=\{e,(1234),(13)(24),(1423)\}$ and so the conjugacy class is of size 6.

7. Here is a sketch of the solution: A_6 is of order $\frac{6!}{2}=360$. Feasible orders for subgroups of A_6 are the divisors of 360:

$$2,3,4,5,6,8,9,10,12,15,18,20,24,30,36,40,45,60,90,120,180.$$

Type	Number of elements	Order
e	1	1
(abc)	$\frac{6\times 5\times 4}{3}=40$	3
$(ab)(cd)$	$\frac{6\times 5\times 4\times 3}{2\times 2\times 2}=45$	2
$(abc)(def)$	$\frac{6!}{3\times 3\times 2}=40$	3
$(abcde)$	$\frac{6!}{5}=144$	5
$(abcd)(ef)$	$\frac{6!}{4\times 2}=90$	4

Next you need to determine the sizes of the conjugacy classes. Finally observe that no sum of these, including the identity, gives a divisor of 360. *Good luck!*

8. Suppose that $M \in SL(n, F)$ so that $\det M = 1$.

For $N \in GL(n, F)$ we have

$$\det NMN^{-1} = \det N \times \det M \times \det N^{-1} = \det N \times \det N^{-1} = 1$$

so that $NMN^{-1} \in SL(n, F)$. Hence $SL(n, F) \lhd GL(n, F)$.

9. Observe that a scalar matrix commutes with *any* matrix.

If $M \in Z$ and $N \in GL(n, F)$ then

$NMN^{-1} = MNN^{-1} = M \in Z$ and so $Z \lhd GL(n, F)$.

10. (i) $(11^2 - 1) \times (11^2 - 11) = 120 \times 110 = 13200$

(ii) $(11^2 - 1) \times 11 = 120 \times 11 = 1320$

(iii) $1320 \div 2 = 660$

(iv) $(13^2 - 1) \times (13^2 - 13) = 168 \times 156 = 26208$

(v) $(13^2 - 1) \times 13 = 168 \times 13 = 2184$

(vi) $2184 \div 2 = 1092$

11. $|GL(4, p)| = (p^4 - 1)(p^4 - p)(p^4 - p^2)(p^4 - p^3)$

$|SL(4, p)| = (p^4 - 1)(p^4 - p)(p^4 - p^2)p^3$

12. Order 1: $\begin{pmatrix} 1 & 0 \\ 0 & 1 \end{pmatrix}$ Order 2: $\begin{pmatrix} 1 & 1 \\ 0 & 1 \end{pmatrix}, \begin{pmatrix} 1 & 0 \\ 1 & 1 \end{pmatrix}, \begin{pmatrix} 0 & 1 \\ 1 & 0 \end{pmatrix}$

Order 3: $\begin{pmatrix} 0 & 1 \\ 1 & 1 \end{pmatrix}, \begin{pmatrix} 1 & 1 \\ 1 & 0 \end{pmatrix}$

Chapter 11

1. (i) $\boldsymbol{Z}_6 \rhd \{0, 2, 4\} \rhd \{0\}$ and $\boldsymbol{Z}_6 \rhd \{0, 3\} \rhd \{0\}$

(ii) $\boldsymbol{Z}_8 \rhd \{0, 2, 4, 6\} \rhd \{0, 4\} \rhd \{0\}$

(iii) $\boldsymbol{Z}_{10} \rhd \{0, 2, 4, 6, 8\} \rhd \{0\}$ and $\boldsymbol{Z}_{10} \rhd \{0, 5\} \rhd \{0\}$.

(iv) $\quad \boldsymbol{Z}_{12} \rhd \{0,2,4,6,8,10\} \rhd \{0,4,8\} \rhd \{0\}$,
$\boldsymbol{Z}_{12} \rhd \{0,2,4,6,8,10\} \rhd \{0,6\} \rhd \{0\}$ and
$\boldsymbol{Z}_{12} \rhd \{0,3,6,9\} \rhd \{0,6\} \rhd \{0\}$.

2. D_4 has composition series

$$D_4 \rhd \{I,R,R^2,R^3\} \rhd \{I,R^2\} \rhd \{I\}.$$

$D_4/\{I,R,R^2,R^3\} \cong \boldsymbol{Z}_2$, $\{I,R,R^2,R^3\}/\{I,R^2\} \cong \boldsymbol{Z}_2$ and $\{I,R^2\}/\{I\} \cong \boldsymbol{Z}_2$ are all abelian and hence D_4 is soluble.

3. Q_8 has compositions series

$$Q_8 \rhd \{1,i,-1,-i\} \rhd \{1,-1\} \rhd \{1\}.$$

Each of the three composition factors is $\boldsymbol{Z}_2$, which is abelian, and hence Q_8 is soluble.

4. Subgroups H of $\boldsymbol{Z}$ such that $36\boldsymbol{Z} < H < \boldsymbol{Z}$ are

$$2\boldsymbol{Z}, 3\boldsymbol{Z}, 4\boldsymbol{Z}, 6\boldsymbol{Z}, 9\boldsymbol{Z}, 12\boldsymbol{Z} \text{ and } 18\boldsymbol{Z}.$$

The proper subgroups of $\boldsymbol{Z}_{36}$ are

$\{0,18\}, \{0,12,24\}, \{0,9,18,27\}, \{0,6,12,18,24,30\}$,

$\{0,4,8,12,16,20,24,28,32\}, \{0,3,6,9,12,15,18,21,24,27,30,33\}$

and $\{0,2,4,6,8,10,12,14,16,18,20,22,24,26,28,30,32,34\}$.

6. We know that if N and G/N are soluble then G is soluble. We prove the result by induction on n.

Basis: For $n = 1$ we have $G = G_1 \rhd G_0 = \{e\}$.

Now $G = G_1/G_0$, so if G_1/G_0 is soluble then G is soluble.

Inductive step: Suppose that the result is true for $n = k$, and consider

$$G = G_{k+1} \rhd G_k \rhd \ldots \rhd G_1 \rhd G_0 = \{e\}$$

By the inductive hypothesis G_k is soluble. By 11.12, since both G_k and G/G_k are soluble, G is soluble.

7. We have already shown that G_1H_1 is a subgroup of G in the proof of the second isomorphism theorem, 11.8. It only remains to show $G_1H_1 \lhd G$.

Suppose $g_1h_1 \in G_1H_1$ and $g \in G$. Then

$$g(g_1h_1)g^{-1} = (gg_1g^{-1})(gh_1g^{-1}) \in G_1H_1.$$

8. A composition series for $\boldsymbol{Z}_6$ is $\boldsymbol{Z}_6 \rhd \{0,2,4\} \rhd \{0\}$.

A composition series for S_3 is $S_3 \rhd A_3 \rhd \{e\}$.

In both cases the composition factors are $\boldsymbol{Z}_2$ and $\boldsymbol{Z}_3$.

9. $6\boldsymbol{Z}+9\boldsymbol{Z}=3\boldsymbol{Z}$ $\qquad$ $6\boldsymbol{Z}\cap 9\boldsymbol{Z}=18\boldsymbol{Z}$

$6\boldsymbol{Z}/18\boldsymbol{Z} \cong 3\boldsymbol{Z}/9\boldsymbol{Z}$ $\qquad$ $9\boldsymbol{Z}/18\boldsymbol{Z} \cong 3\boldsymbol{Z}/6\boldsymbol{Z}$

Chapter 12

1. (i) Take $p=2$:

$3 \not\equiv 0 \pmod 2$ but $8,-2 \equiv 0 \pmod 2$ and $-2 \not\equiv 0 \pmod 4$.

(ii) Take $p=3$:

$2 \not\equiv 0 \pmod 3$ but $6,-3,9,12 \equiv 0 \pmod 3$ and $12 \not\equiv 0 \pmod 9$.

(iii) Take $p=7$:

$4 \not\equiv 0 \pmod 7$ but $-14,7,21 \equiv 0 \pmod 7$ and $21 \not\equiv 0 \pmod{49}$.

(iv) We need to take $p=2$ because no other prime divides 8 and 4. But $4 \equiv 0 \pmod{2^2}$ so the criterion fails.

(v) We need to take $p=3$ because no other prime divides 6,-3, 9 and 12. But 3 also divides the coefficient 9 of the highest power so the criterion fails.

In the last two examples, failure of the criterion does *not* show that the polynomial fails to be irreducible. For example, if we find the roots of (iv) we see that this polynomial is irreducible over $\boldsymbol{Q}$.

2. Suppose that F is a subfield of $\boldsymbol{Q}$. Then we have $1 \in F$. For each positive integer m we have $m = 1+1+\ldots+1 \in F$. Since F contains additive inverses we have $m \in F$ for each integer m.

Since F is closed under division we have $\frac{m}{n} \in F$ for each pair of integers m and $n \neq 0$. Hence $F = \boldsymbol{Q}$.

3. Commutativity and associativity are clear. The additive identity is $0 + 0\sqrt{2}$, and the additive inverse of each element $a + b\sqrt{2}$ is $(-a) + (-b)\sqrt{2}$. Hence we have an abelian group.

$$
\begin{aligned}
(\lambda+\mu)\left(a+b\sqrt{2}\right) &= (\lambda+\mu)a + (\lambda+\mu)b\sqrt{2} \\
&= \lambda a + \mu a + \lambda b\sqrt{2} + \mu b\sqrt{2} \\
&= \lambda\left(a+b\sqrt{2}\right) + \mu\left(a+b\sqrt{2}\right).
\end{aligned}
$$

$$
\begin{aligned}
\lambda\left(\left(a+b\sqrt{2}\right)+\left(c+d\sqrt{2}\right)\right) &= \lambda\left((a+c)+(b+d)\sqrt{2}\right) \\
&= \lambda(a+c) + \lambda(b+d)\sqrt{2} \\
&= \lambda a + \lambda b\sqrt{2} + \lambda c + \lambda d\sqrt{2} \\
&= \lambda\left(a+b\sqrt{2}\right) + \lambda\left(c+d\sqrt{2}\right).
\end{aligned}
$$

$$\lambda(\mu(a+b\sqrt{2})) = \lambda(\mu a + \mu b\sqrt{2}) = \lambda\mu a + \lambda\mu b\sqrt{2} = (\lambda\mu)(a+b\sqrt{2}).$$

$$1(a+b\sqrt{2}) = 1a + 1b\sqrt{2} = a + b\sqrt{2}.$$

4. (i) The 8th roots of unity are:

$1\angle\frac{\pi}{4}, 1\angle\frac{\pi}{2} = i, 1\angle\frac{3\pi}{4}, 1\angle\pi = -1, 1\angle\frac{5\pi}{4}, 1\angle\frac{3\pi}{2} = -i, 1\angle\frac{7\pi}{4}$ and 1.

Of these, the primitive 8th roots of unity are:

$$1\angle\tfrac{\pi}{4}, 1\angle\tfrac{3\pi}{4}, 1\angle\tfrac{5\pi}{4} \text{ and } 1\angle\tfrac{7\pi}{4}.$$

(ii) The 7th roots of unity are:

$$1\angle\tfrac{2\pi}{7}, 1\angle\tfrac{4\pi}{7}, 1\angle\tfrac{6\pi}{7}, 1\angle\tfrac{8\pi}{7}, 1\angle\tfrac{10\pi}{7}, 1\angle\tfrac{12\pi}{7} \text{ and } 1.$$

Of these, all but the last are primitive 7th roots of unity.

5. (i) ω is a root of $x^8 - 1 = (x-1)(x+1)(x^2+1)(x^4+1)$.

The minimal polynomial of ω is $x^4 + 1$ of degree 4.

Hence $[\boldsymbol{Q}(\omega) : \boldsymbol{Q}] = 4$ and a basis for $\boldsymbol{Q}(\omega)$ is $\{1, \omega, \omega^2, \omega^3\}$.

(ii) ω is a root of $x^7 - 1 = (x-1)(x^6 + x^5 + x^4 + x^3 + x^2 + x + 1)$.

The minimal polynomial of ω is $x^6 + x^5 + x^4 + x^3 + x^2 + x + 1$ of degree 6.

Hence $[\boldsymbol{Q}(\omega) : \boldsymbol{Q}] = 6$ and a basis for $\boldsymbol{Q}(\omega)$ is

$$\{1, \omega, \omega^2, \omega^3, \omega^4, \omega^5\}.$$

6.(i) $x^2 - 7 = (x - \sqrt{7})(x + \sqrt{7})$ so the splitting field is $\boldsymbol{Q}(\sqrt{7})$.

(ii) $x^2 - 9 = (x-3)(x+3)$ so the splitting field is $\boldsymbol{Q}$.

(iii) $x^3 - 4 = (x - \sqrt[3]{4})(x - \omega\sqrt[3]{4})(x - \omega^2\sqrt[3]{4})$ so the splitting field is

$\boldsymbol{Q}(\omega, \sqrt[3]{4})$ where ω is a primitive cube root of unity.

(iv) $x^2 - x - 1$ has roots $x = \frac{1}{2}(1 \pm \sqrt{5})$ so the splitting field is $\boldsymbol{Q}(\sqrt{5})$.

(v) $x^2 + 1 = (x+i)(x-i)$ so the splitting field is $\boldsymbol{Q}(i)$.

(vi)
$$\begin{aligned} x^4 - 2 &= (x^2 - \sqrt{2})(x^2 + \sqrt{2}) \\ &= (x - \sqrt[4]{2})(x + \sqrt[4]{2})(x - \sqrt[4]{2}\,i)(x + \sqrt[4]{2}\,i) \end{aligned}$$

so the splitting field is $\boldsymbol{Q}(\sqrt[4]{2}, i)$.

The dimensions of these splitting fields as vector spaces over $\boldsymbol{Q}$ are

(i) 2 (ii) 1

(iii) $\boldsymbol{Q}(\omega, \sqrt[3]{4}) \supset \boldsymbol{Q}(\omega) \supset \boldsymbol{Q}$

$\sqrt[3]{4}$ has minimum polynomial $x^3 - 4$ of degree 3,

ω has minimum polynomial $x^2 + x + 1$ of degree 2,

and so $[\boldsymbol{Q}(\omega, \sqrt[3]{4}) : \boldsymbol{Q}] = 6$.

(iv) 2 (v) 2

(vi) $\boldsymbol{Q}(i, \sqrt[4]{2}) \supset \boldsymbol{Q}(i) \supset \boldsymbol{Q}$

$\sqrt[4]{2}$ has minimum polynomial $x^4 - 2$ of degree 4, i has minimum polynomial $x^2 + 1$ of degree 2, and so $[\boldsymbol{Q}(i, \sqrt[4]{2}) : \boldsymbol{Q}] = 8$.

7. $\sqrt{2}, \sqrt{3} \in \boldsymbol{Q}(\sqrt{2}, \sqrt{3})$ and hence $\sqrt{2}+\sqrt{3} \in \boldsymbol{Q}(\sqrt{2}, \sqrt{3})$. Now since $\boldsymbol{Q}(\sqrt{2}+\sqrt{3})$ is the smallest extension of $\boldsymbol{Q}$ containing $\sqrt{2}+\sqrt{3}$ we have

$$\boldsymbol{Q}(\sqrt{2}+\sqrt{3}) \subseteq \boldsymbol{Q}(\sqrt{2}, \sqrt{3}).$$

$\sqrt{2}+\sqrt{3} \in \boldsymbol{Q}(\sqrt{2}+\sqrt{3})$ and hence

$$(\sqrt{2}+\sqrt{3})^2 = 5+2\sqrt{6} \in \boldsymbol{Q}(\sqrt{2}+\sqrt{3})$$

So we have $\sqrt{6} \in \boldsymbol{Q}(\sqrt{2}+\sqrt{3})$ and so

$$\sqrt{6}(\sqrt{2}+\sqrt{3}) = 3\sqrt{2}+2\sqrt{3} \in \boldsymbol{Q}(\sqrt{2}+\sqrt{3}).$$

It follows that $\sqrt{2}, \sqrt{3} \in \boldsymbol{Q}(\sqrt{2}+\sqrt{3})$ (why?)

Now since $\boldsymbol{Q}(\sqrt{2}, \sqrt{3})$ is the smallest extension of $\boldsymbol{Q}$ containing $\sqrt{2}$ and $\sqrt{3}$ we have

$$\boldsymbol{Q}(\sqrt{2}, \sqrt{3}) \subseteq \boldsymbol{Q}(\sqrt{2}+\sqrt{3}).$$

Hence the result.

8. $[\boldsymbol{Q}(\sqrt{3}, \sqrt[3]{4}) : \boldsymbol{Q}] = 6$ with basis

$$\left\{1, 4^{\frac{1}{3}}, 4^{\frac{2}{3}}, \sqrt{3}, \sqrt{3}\cdot 4^{\frac{1}{3}}, \sqrt{3}\cdot 4^{\frac{2}{3}}\right\}$$

Chapter 13

1. We have $0+0=0$ and so $\sigma(0+0)=\sigma(0)$.

Hence $\sigma(0)+\sigma(0)=\sigma(0)+0$ and by cancellation $\sigma(0)=0$.

We have $\sigma(a)+\sigma(-a)=\sigma(a+(-a))=\sigma(0)=0$

By uniqueness of inverses, $\sigma(-a)=-\sigma(a)$.

We have $\sigma(a)\cdot\sigma(a^{-1})=\sigma(a\cdot a^{-1})=\sigma(1)=1$

By uniqueness of inverses, $\sigma(a^{-1})=\sigma(a)^{-1}$.

2. (i) There are two automorphisms of $\boldsymbol{Q}(\sqrt{3})$ that fix $\boldsymbol{Q}$:

$$a+b\sqrt{3} \mapsto a+b\sqrt{3}$$

$$a+b\sqrt{3} \mapsto a-b\sqrt{3}$$

Hence $Gal(\boldsymbol{Q}(\sqrt{3})/\boldsymbol{Q}) \cong \boldsymbol{Z}_2$

(ii) There are four automorphisms of $\boldsymbol{Q}(\sqrt{3}, \sqrt{5})$ that fix $\boldsymbol{Q}$:

$$a + b\sqrt{3} + c\sqrt{5} + d\sqrt{15} \mapsto a + b\sqrt{3} + c\sqrt{5} + d\sqrt{15}$$

$$a + b\sqrt{3} + c\sqrt{5} + d\sqrt{15} \mapsto a - b\sqrt{3} + c\sqrt{5} - d\sqrt{15}$$

$$a + b\sqrt{3} + c\sqrt{5} + d\sqrt{15} \mapsto a + b\sqrt{3} - c\sqrt{5} - d\sqrt{15}$$

$$a + b\sqrt{3} + c\sqrt{5} + d\sqrt{15} \mapsto a - b\sqrt{3} - c\sqrt{5} + d\sqrt{15}$$

Each non-identity automorphism is of order 2 and hence

$$Gal(\boldsymbol{Q}(\sqrt{3}, \sqrt{5})/\boldsymbol{Q}) \cong \boldsymbol{Z}_2 \times \boldsymbol{Z}_2.$$

(iii) There are 8 automorphisms of $\boldsymbol{Q}(\sqrt{2}, \sqrt{3}, \sqrt{5})$ that fix $\boldsymbol{Q}$:

Following a similar pattern to the last two solutions, these eight are specified by $\sqrt{2} \to \pm\sqrt{2}$, $\sqrt{3} \to \pm\sqrt{3}$, $\sqrt{5} \to \pm\sqrt{5}$

Hence $Gal(\boldsymbol{Q}(\sqrt{2}, \sqrt{3}, \sqrt{5})/\boldsymbol{Q}) \cong \boldsymbol{Z}_2 \times \boldsymbol{Z}_2 \times \boldsymbol{Z}_2$.

3. $\phi(\sqrt{2})\phi(\sqrt{2}) = \sqrt{3}\sqrt{3} = 3$, whereas $\phi(\sqrt{2}\sqrt{2}) = \phi(2) = 2$, so ϕ is not an isomorphism.

4. (i) $[\boldsymbol{Q}(\sqrt{7}) : \boldsymbol{Q}] = 2$ hence $Gal(\boldsymbol{Q}(\sqrt{7})/\boldsymbol{Q}) \cong \boldsymbol{Z}_2$.

(ii) $Gal(\boldsymbol{Q}/\boldsymbol{Q}) = \{e\}$.

(iii) $[\boldsymbol{Q}(\omega, \sqrt[3]{4}) : \boldsymbol{Q}] = 6$ so the order of the Galois group is 6. The polynomial is of degree 3, so the Galois group is a subgroup of S_3.

Hence $Gal(\boldsymbol{Q}(\omega, \sqrt[3]{4})/\boldsymbol{Q}) \cong S_3$.

(iv) $Gal(\boldsymbol{Q}(\sqrt{5})/\boldsymbol{Q}) \cong \boldsymbol{Z}_2$

(v) $Gal(\boldsymbol{Q}(i)/\boldsymbol{Q}) \cong \boldsymbol{Z}_2$

(vi) $[\boldsymbol{Q}(i, \sqrt[4]{2}) : \boldsymbol{Q}] = 8$ so the order of the Galois group is 8. The polynomial is of degree 4, so the Galois group is a subgroup of S_4.

There are several groups of order 8. The answer is

$$Gal(\boldsymbol{Q}(i, \sqrt[4]{2})/\boldsymbol{Q}) \cong D_4$$

How do we know that D_4 is the correct group of order 8? Could S_4 have a subgroup isomorphic to Q_8, for example?

Chapter 14

2. Suppose that a ring has two unity elements e_1 and e_2. Then

$e_1 \cdot e_2 = e_1$ because e_2 is a unity

$e_1 \cdot e_2 = e_2$ because e_1 is a unity.

Hence $e_1 = e_2$ and so there is only one unity.

3. (i) $0+0=0$ and so $(0+0)\cdot a = 0\cdot a$, so that

$$0\cdot a + 0\cdot a = 0\cdot a$$

Hence $0\cdot a + 0\cdot a = 0\cdot a + 0$ and so by additive cancellation

$$a\cdot 0 = 0.$$

(iii) By definition $-(a\cdot b)$ is the additive inverse for $a\cdot b$

But $a\cdot b + a\cdot(-b) = a\cdot(b + -b) = a\cdot 0 = 0$ by (i)

and $a\cdot(-b) + a\cdot b = a\cdot(-b+b) = a\cdot 0 = 0$ also by (i).

Hence $a\cdot(-b)$ is also the additive inverse for $a\cdot b$.

By uniqueness of additive inverses we have $-(a\cdot b) = a\cdot(-b)$.

4. Suppose $a+c=b+c$. Then we have

$$(a+c)+(-c) = (b+c)+(-c)$$

and by associativity $a+(c+(-c)) = b+(c+(-c))$

Then $a+0=b+0$ so that $a=b$.

Now suppose $a+x=b$, then we have $-a+(a+x) = -a+b$, and by associativity $(-a+a)+x = -a+b$.

Hence $0+x = -a+b$ so that $x = -a+b$.

Suppose that there are *two* solutions x_1 and x_2 so that

$$a+x_1=b$$
$$a+x_2=b.$$

Then $a+x_1=a+x_2$ and by additive cancellation $x_1=x_2$. Hence the solution is unique.

5. $a\cdot a=a$ for all $a\in R$.

(i) $(a+a)^2=(a+a)\cdot(a+a)=a+a$

But also
$$\begin{aligned}(a+a)^2 &= (a+a)\cdot(a+a)=a\cdot(a+a)+a\cdot(a+a)\\ &= a\cdot a+a\cdot a+a\cdot a+a\cdot a=a+a+a+a\end{aligned}$$

Now $a+a+a+a=a+a$, by cancellation $a+a=0$, and so $a=-a$.

(ii) $(a+b)^2=(a+b)\cdot(a+b)=a+b$

But also
$$\begin{aligned}(a+b)^2 &= (a+b)\cdot(a+b)\\ &= a\cdot a+a\cdot b+b\cdot a+b\cdot b\\ &= a+a\cdot b+b\cdot a+b\end{aligned}$$

Hence $a+b=a+a\cdot b+b\cdot a+b$, and by cancellation and (i) above $a\cdot b=b\cdot a$.

6.
$$\begin{aligned}(a+b)+(c+d) &= (a+(b+(c+d)) \text{ by associativity}\\ &= a+((b+c)+d) \text{ by associativity}\\ &= a+((c+b)+d) \text{ by commutativity}\\ &= a+(c+(b+d)) \text{ by associativity}\\ &= (a+c)+(b+d) \text{ by associativity}\end{aligned}$$

7. Commutativity of addition is verified like this:

$$\begin{aligned}(x_1,y_1)+(x_2,y_2)&=(x_1+x_2,y_1+y_2)\\ &=(x_2+x_1,y_2+y_1)=(x_2,y_2)+(x_1,y_1)\end{aligned}$$

Verification of the other axioms is similar.

8.
$$\boldsymbol{Z}_3\oplus\boldsymbol{Z}_4=\left\{\begin{array}{c}(0,0),(0,1),(0,2),(0,3),(1,0),(1,1),(1,2),(1,3),\\ (2,0),(2,1),(2,2),(2,3)\end{array}\right\}$$

char $\boldsymbol{Z}_3\oplus\boldsymbol{Z}_4=12$.

Chapter 15

1. (a) If $a, b \in \boldsymbol{Z}$ then $a+b,\ a \cdot b,\ -a \in \boldsymbol{Z}$.

If $\frac{a}{b}, \frac{a'}{b'} \in \boldsymbol{Q}$ then $\frac{a}{b} + \frac{a'}{b'} = \frac{ab' + a'b}{bb'} \in \boldsymbol{Q}$,

$\frac{a}{b} \times \frac{a'}{b'} = \frac{aa'}{bb'} \in \boldsymbol{Q}$ and $-\frac{a}{b} = \frac{-a}{b} \in \boldsymbol{Q}$.

(b) If A and B are a pair of 2×2 matrices with integer entries, then $A+B, AB$ and $-A$ are also 2×2 matrices with integer entries.

The reasoning for $Mat(2, \boldsymbol{Q})$ is similar.

(c) If $p(x)$ and $q(x)$ are a pair of polynomials with integer coefficients, then $p(x)+q(x), p(x)q(x)$ and $-p(x)$ are also polynomials with integer coefficients.

The reasoning for $\boldsymbol{Q}[x]$ is similar.

(d) Clearly $n\boldsymbol{Z} \subset \boldsymbol{Z}$. Suppose $a, b \in n\boldsymbol{Z}$.

Then $a = rn$ and $b = sn$ for some $r, s \in \boldsymbol{Z}$.

$$\begin{aligned} a+b &= rn+sn = (r+s)n \in n\boldsymbol{Z} \\ a \cdot b &= rn \cdot sn = (rsn)n \in n\boldsymbol{Z} \\ -a &= -rn = (-r)n \in n\boldsymbol{Z}. \end{aligned}$$

(f) $$-\begin{pmatrix} a & b & 0 \\ c & d & 0 \\ 0 & 0 & e \end{pmatrix} = \begin{pmatrix} -a & -b & 0 \\ -c & -d & 0 \\ 0 & 0 & -e \end{pmatrix} \in S$$

$$\begin{pmatrix} a_1 & b_1 & 0 \\ c_1 & d_1 & 0 \\ 0 & 0 & e_1 \end{pmatrix} + \begin{pmatrix} a_2 & b_2 & 0 \\ c_2 & d_2 & 0 \\ 0 & 0 & e_2 \end{pmatrix} = \begin{pmatrix} a_1+a_2 & b_1+b_2 & 0 \\ c_1+c_2 & d_1+d_2 & 0 \\ 0 & 0 & e_1+e_2 \end{pmatrix} \in S$$

$$\begin{pmatrix} a_1 & b_1 & 0 \\ c_1 & d_1 & 0 \\ 0 & 0 & e_1 \end{pmatrix} \cdot \begin{pmatrix} a_2 & b_2 & 0 \\ c_2 & d_2 & 0 \\ 0 & 0 & e_2 \end{pmatrix} = \begin{pmatrix} a_1a_2+b_1c_2 & a_1b_2+b_1d_2 & 0 \\ c_1a_2+d_1c_2 & c_1b_2+d_1d_2 & 0 \\ 0 & 0 & e_1e_2 \end{pmatrix} \in S$$

$\mathbf{Z}_3$ is not a subring of $\mathbf{Z}_4$ because the binary operations in the two rings are different. For example, in $\mathbf{Z}_3$ we have $2+2=1$, whereas in $\mathbf{Z}_4$ we have $2+2=0$.

2. (i) $$\begin{pmatrix} x & y \\ -y & x \end{pmatrix} + \begin{pmatrix} x' & y' \\ -y' & x' \end{pmatrix} = \begin{pmatrix} x+x' & y+y' \\ -(y+y') & x+x' \end{pmatrix} \in S$$

$$\begin{pmatrix} x & y \\ -y & x \end{pmatrix} \begin{pmatrix} x' & y' \\ -y' & x' \end{pmatrix} = \begin{pmatrix} xx'-yy' & xy'+x'y \\ -(xy'+x'y) & xx'-yy' \end{pmatrix} \in S$$

$$-\begin{pmatrix} x & y \\ -y & x \end{pmatrix} = \begin{pmatrix} -x & -y \\ y & -x \end{pmatrix} \in S.$$

(ii) is similar, but requires thinking about the complex conjugate.

3. (i) $$S = \left\{ \begin{pmatrix} a & b \\ 0 & c \end{pmatrix} : a, b, c \in \mathbf{Z} \right\}$$

$$\begin{pmatrix} a_1 & b_1 \\ 0 & c_1 \end{pmatrix} + \begin{pmatrix} a_2 & b_2 \\ 0 & c_2 \end{pmatrix} = \begin{pmatrix} a_1+a_2 & b_1+b_2 \\ 0 & c_1+c_2 \end{pmatrix} \in S$$

$$\begin{pmatrix} a_1 & b_1 \\ 0 & c_1 \end{pmatrix} \begin{pmatrix} a_2 & b_2 \\ 0 & c_2 \end{pmatrix} = \begin{pmatrix} a_1a_2 & a_1b_2+b_1c_2 \\ 0 & c_1c_2 \end{pmatrix} \in S$$

$$-\begin{pmatrix} a_1 & b_1 \\ 0 & c_1 \end{pmatrix} = \begin{pmatrix} -a_1 & -b_1 \\ 0 & -c_1 \end{pmatrix} \in S$$

Hence S is a subring of $Mat(2, \mathbf{Z})$.

(ii) $S = \{A \in M_2(\mathbf{Z}) : \det A = 1\}$.

Clearly $I = \begin{pmatrix} 1 & 0 \\ 0 & 1 \end{pmatrix} \in S$, but $I + I = \begin{pmatrix} 2 & 0 \\ 0 & 2 \end{pmatrix} \notin S$ since it has determinant 4. Hence S is *not* a subring of $Mat(2, \mathbf{Z})$.

(iii) For given $B \in Mat(2, \mathbf{Z})$ let

$$S = \{A \in Mat(2, \mathbf{Z}) : AB = BA\}$$

If $A_1, A_2 \in S$ then $A_1B = BA_1$ and $A_2B = BA_2$.

Then $(A_1 + A_2)B = A_1B + A_2B = BA_1 + BA_2 = B(A_1 + A_2)$ so that $A_1 + A_2 \in S$.

$(A_1A_2)B = A_1(A_2B) = A_1(BA_2) = (A_1B)A_2 = (BA_1)A_2 = B(A_1A_2)$ so that $A_1A_2 \in S$

Finally $(-A_1)B = -(A_1B) = -(BA_1) = B(-A_1)$ so that $-A_1 \in S$.

Hence S is a subring of $Mat(2, \boldsymbol{Z})$.

4. (i) $\boldsymbol{R}$ is *not* an ideal in $\boldsymbol{R}[x]$, for example $1 \in \boldsymbol{R}$ and $x \in \boldsymbol{R}[x]$ but $1 \cdot x \notin \boldsymbol{R}$.

(ii) $I = \{a_0 + a_1x + a_2x^2 + ... + a_nx^n : a_0 + a_1 + a_2 + ... + a_n = 0\}$

Notice that $p(x) \in I$ iff $\phi_1(p(x)) = 0$ iff $p(x) \in \ker \phi_1$.

This is an ideal in $\boldsymbol{R}[x]$.

(iii) $\boldsymbol{Q}[x]$ is not an ideal in $\boldsymbol{Z}[x]$, for example

$\sqrt{2}\,x \in \boldsymbol{R}[x]$, $x \in \boldsymbol{Q}[x]$ but $\sqrt{2}\,x.x = \sqrt{2}\,x^2 \notin \boldsymbol{Q}[x]$

(iv) $I = \{a_0 + a_1x + a_2x^2 + ... + a_nx^n : a_0 = 0\}$

Notice that $p(x) \in I$ iff $\phi_0(p(x)) = 0$ iff $p(x) \in \ker \phi_0$.

This is an ideal in $\boldsymbol{R}[x]$.

(v) $\langle x-2 \rangle$ is an ideal in $\boldsymbol{R}[x]$: If $p(x), q(x) \in \langle x-2 \rangle$ then $p(x) = (x-2)a(x)$ and $q(x) = (x-2)b(x)$ for some $a(x), b(x) \in \boldsymbol{R}[x]$.
We have $p(x) + q(x) = (x-2)(a(x) + b(x)) \in \langle x-2 \rangle$ and $-p(x) = -(x-2)a(x) \in \langle x-2 \rangle$

Finally if $r(x) \in \boldsymbol{R}[x]$ then

$$p(x)r(x) = r(x)p(x) = (x-2)a(x)r(x) \in \langle x-2 \rangle$$

(vi) Any ring is an ideal in itself, so $\boldsymbol{R}[x]$ is an ideal in $\boldsymbol{R}[x]$.

5. $\{0\}, \{0, 10\}, \{0, 5, 10, 15\}, \{0, 4, 8, 12, 16\}$, $\{0, 2, 4, 6, 8, 10, 12, 14, 16, 18\}$.

6. If $1 \in I$ and $r \in R$ then $r = r \cdot 1 \in I$.

Hence any $r \in R$ is also in I, so that $I = R$.

7. Let $x = a + b, y = a' + b'$, where $a, a' \in I$ and $b, b' \in J$.

$x + y = (a + b) + (a' + b') = (a + a') + (b + b') \in I + J$

$-x = -(a + b) = (-a) + (-b) \in I + J$

$rx = r(a + b) = ra + rb \in I + J$

$xr = (a + b)r = ar + br \in I + J$

Hence $I + J$ is an ideal.

8. (a) The mapping is *not* a homomorphism, for example

$f(2 \times 2) = f(4) = -4$ whereas $f(2) \times f(2) = -2 \times -2 = 4$.

(b) The mapping is a homomorphism:

$$\begin{aligned} f((a_1, b_1) + (a_2, b_2)) &= f(a_1 + a_2, b_1 + b_2) = (a_1 + a_2, 0) \\ &= = (a_1, 0) + (a_2, 0) \\ &= f(a_1, b_1) + f(a_2, b_2). \\ f((a_1, b_1) \cdot (a_2, b_2)) &= f(a_1 a_2, b_1 b_2) = (a_1 a_2, 0) = (a_1, 0) \cdot (a_2, 0) \\ &= f(a_1 b_1) \cdot f(a_2 b_2). \end{aligned}$$

(c) The mapping is *not* a homomorphism, for example

$f((0, 1) + (0, 1)) = f(0, 2) = (0, 1)$ whereas
$f(0, 1) + f(0, 1) = (0, 1) + (0, 1) = (0, 2)$.

(d) The mapping is a homomorphism:

$$f((a, b) + (a', b')) = f(a + a', b + b') = \begin{pmatrix} a + a' & 0 \\ 0 & b + b' \end{pmatrix}$$

$$= \begin{pmatrix} a & 0 \\ 0 & b \end{pmatrix} + \begin{pmatrix} a' & 0 \\ 0 & b' \end{pmatrix} = f(a, b) + f(a', b')$$

$$f((a,b)\cdot(a',b')) = f(aa',bb') = \begin{pmatrix} aa' & 0 \\ 0 & bb' \end{pmatrix}$$

$$= \begin{pmatrix} a & 0 \\ 0 & b \end{pmatrix}\begin{pmatrix} a' & 0 \\ 0 & b' \end{pmatrix} = f(a,b)\cdot f(a',b')$$

(e) The mapping is a homomorphism:

$$f((a+bi)+(a'+b'i)) = f((a+a')+(b+b')i) = (a+a')-(b+b')i$$
$$= (a-bi)+(a'-b'i) = f(a+bi)+f(a'+b'i)$$

$$f((a+bi)(a'+b'i)) = f((aa'-bb')+(ab'+a'b)i)$$
$$= (aa'-bb')-(ab'+a'b)i$$
$$= (a-bi)(a'-b'i) = f(a+bi)f(a'+b'i)$$

(f) The mapping is *not* a homomorphism, for example

$f(1\times 1) = f(1) = i$ whereas $f(1)f(1) = i\times i = -1$.

9. $f: R\to S$ is a monomorphism iff $\ker f = \{0\}$

Proof: "$\Rightarrow$" Because f is a homomorphism $f(0) = 0$.

Suppose $a\in \ker f$ so that $f(a) = 0$.

Because f is a monomorphism, if $f(a) = f(0)$ then $a = 0$.

Hence $\ker f = \{0\}$.

"$\Leftarrow$" Suppose that we have $f(a) = f(b)$. Then $f(a)-f(b) = 0$.

Because f is a homomorphism $f(a)-f(b) = f(a-b) = 0$.

Hence $a-b\in \ker f$. But $\ker f = \{0\}$ so we must have $a-b = 0$ so that $a = b$. Hence f is a monomorphism.

Chapter 16

1. (i)
$$\begin{aligned}((a+I)(b+I))(c+I) &= (ab+I)(c+I) = (ab)c+I \\ &= a(bc)+I = (a+I)(bc+I) \\ &= (a+I)((b+I)(c+I))\end{aligned}$$

(ii)
$$\begin{aligned}((a+I)+(b+I))(c+I) &= (a+b+I)(c+I) = (a+b)c+I \\ &= ac+bc+I = (ac+I)+(bc+I) \\ &= (a+I)(c+I)+(b+I)(c+I)\end{aligned}$$

2. (i) R/I has a unity $1+I$ if and only if R has a unity 1. Then

$$\begin{aligned}(1+I)\cdot(a+I) &= 1a+I = a+I \\ \text{and } (a+I)\cdot(1+I) &= a1+I = a+I.\end{aligned}$$

(ii) R/I is commutative iff R is commutative:

$$(a+I)(b+I) = ab+I = ba+I = (b+I)(a+I).$$

3. (i) $I = \{0\}$.

$$p(x) \sim q(x) \text{ iff } p(x) - q(x) \in \{0\} \text{ iff } p(x) = q(x).$$

Each distinct polynomial is in its own singleton equivalence class.

$$\mathbf{Z}[x]/\{0\} \cong \mathbf{Z}[x].$$

(ii) $I = \{2a_0 + a_1x + a_2x^2 + \ldots + a_nx^n : a_0, a_1, \ldots, a_n \in \mathbf{Z}\}$

A pair of polynomials are equivalent if their difference has an even constant; this is the case if either both constants are even or both constants are odd. There are two equivalence classes: one comprising all polynomials with even constants, the other comprising all polynomials with odd constants.

Define $f: \mathbf{Z}[x] \to \mathbf{Z}_2$ by $f(p_0 + p_1x + p_2x^2 + \ldots + p_nx^n) = [p_0]$, where $[p_0]$ is the congruence class of p_0 modulo 2.

Then $\ker f = I$ and $\operatorname{im} f = \mathbf{Z}_2$.

Hence by the first isomorphism theorem 16.5, $\mathbf{Z}[x]/I \cong \mathbf{Z}_2$.

(iii) $I = 2\boldsymbol{Z}[x]$

$$p(x) \sim q(x) \text{ iff } p_0 - q_0, p_1 - q_1, ..., p_n - q_n \in 2\boldsymbol{Z}.$$

Define $f: \boldsymbol{Z}[x] \to \boldsymbol{Z}_2[x]$ by

$$f(p_0 + p_1x + p_2x^2 + ... + p_nx^n) = [p_0] + [p_1]x + [p_2]x^2 + ... + [p_n]x^n$$

Then $\ker f = 2\boldsymbol{Z}[x]$ and $\operatorname{im} f = \boldsymbol{Z}_2[x]$.

Hence by the first isomorphism theorem $\boldsymbol{Z}[x]/2\boldsymbol{Z}[x] \cong \boldsymbol{Z}_2[x]$.

(iv) $I = \langle x \rangle$

$p(x) \sim q(x)$ iff $p(x) - q(x) \in \langle x \rangle$ iff $p(x)$ and $q(x)$ have the same constant.

There is one equivalence class for each integer n, comprising all polynomials with constant n.

Define $f: \boldsymbol{Z}[x] \to \boldsymbol{Z}$ by $f(p(x)) = \phi_0(p(x))$.

Then $\ker f = \langle x \rangle$ and $\operatorname{im} f = \boldsymbol{Z}$.

Hence by the first isomorphism theorem $\boldsymbol{Z}[x]/\langle x \rangle \cong \boldsymbol{Z}$.

(v) $I = \langle x^2 + 1 \rangle$. The evaluation $\phi_i : \boldsymbol{Z}[x] \to \boldsymbol{Z}[i]$ has $\ker \phi_i = \langle x^2 + 1 \rangle$ and $\operatorname{im} \phi_i = \boldsymbol{Z}[i]$.

Hence $\boldsymbol{Z}[x]/\langle x^2 + 1 \rangle \cong \boldsymbol{Z}[i]$.

(vi) $I = \boldsymbol{Z}[x]$

$$p(x) \sim q(x) \text{ for all } p(x), q(x) \in \boldsymbol{Z}[x].$$

There is one equivalence class comprising all polynomials in $\boldsymbol{Z}[x]$.

$$\boldsymbol{Z}[x]/\boldsymbol{Z}[x] \cong \{0\}.$$

Chapter 17

1. Divisors of zero in $\boldsymbol{Z}_8$ are 2, 4 and 6. $\boldsymbol{Z}_{11}$ has no divisors of zero because 11 is prime.

2. Suppose that $a \neq 0$ has multiplicative inverses a^{-1} and a^*. Then $a^{-1} = a^{-1} \cdot 1 = a^{-1} \cdot (a \cdot a^*) = (a^{-1} \cdot a) \cdot a^* = 1 \cdot a^* = a^*$

Hence $a^{-1} = a^*$ so that the inverse is unique.

3. $x^2 - 1 = (x+1)(x-1)$, so that for example $\langle x^2 - 1 \rangle$ lies inside the ideal $\langle x-1 \rangle$ and so $\langle x^2 - 1 \rangle$ is not maximal.

Hence $\boldsymbol{Z}[x]/\langle x^2 - 1 \rangle$ is not a field.

4. $\boldsymbol{Z}[x]/\langle x-3 \rangle \cong \boldsymbol{Z}$ is an integral domain but *not* a field.

Hence the ideal $\langle x-3 \rangle$ in $\boldsymbol{Z}[x]$ is prime but *not* maximal.

5. Maximal ideals in $\boldsymbol{Z}_{30}$:

$\{0,2,4,6,8,10,12,14,16,18,20,22,24,26,28\}$,

$\{0,3,6,9,12,15,18,21,24,27\}$, $\{0,5,10,15,20,25\}$.

6. $J = \{b + ra : b \in I, r \in R\}$

(i) $j_1, j_2 \in J$ so that $j_1 = b_1 + r_1 a,\ j_2 = b_2 + r_2 a$ for some $b_1, b_2 \in I,\ r_1, r_2 \in R$. Then

$$\begin{aligned} j_1 + j_2 &= (b_1 + r_1 a) + (b_2 + r_2 a) \\ &= (b_1 + b_2) + (r_1 + r_2)a \in J \end{aligned}$$

since $b_1 + b_2 \in I$ and $r_1 + r_2 \in R$.

(ii) $j \in J$ and so $j = b + ra$ for some $b \in I,\ r \in R$

Then $-j = (-b) + (-r)a \in J$ since $-b \in I$ and $-r \in R$.

(iii) $j \in J,\ s \in R$ so that $j = b + ra$ for some $b \in I, r \in R$.

Then $js = bs + (rs)a \in J$ since $bs \in I$ and $rs \in R$.

Hence J is an ideal in R.

7. Suppose $a \in R, a \neq 0$. Since R has no proper ideals we have $R = \langle a \rangle$ and so $1 \in \langle a \rangle$.

Hence $1 = xa$ for some $x \in R$

x is the multiplicative inverse of a, and so R is a field.

Chapter 18

1. (i) and (ii) are not irreducible: in each case 1 is a root.

   (iii) is irreducible, since neither 0 or 1 is a root.

   (iv) is not irreducible; although neither 0 or 1 is a root, the polynomial factorises as $x^4 + x^2 + 1 = (x^2 + x + 1)^2$.

2. The field of nine elements, F_9:

$+$	0	1	2	x	$x+1$	$x+2$	$2x$	$2x+1$	$2x+2$
0	0	1	2	x	$x+1$	$x+2$	$2x$	$2x+1$	$2x+2$
1	1	2	0	$x+1$	$x+2$	x	$2x+1$	$2x+2$	$2x$
2	2	0	1	$x+2$	x	$x+1$	$2x+2$	$2x$	$2x+1$
x	x	$x+1$	$x+2$	$2x$	$2x+1$	$2x+2$	0	1	2
$x+1$	$x+1$	$x+2$	x	$2x+1$	$2x+2$	$2x$	1	2	0
$x+2$	$x+2$	x	$x+1$	$2x+2$	$2x$	$2x+1$	2	0	1
$2x$	$2x$	$2x+1$	$2x+2$	0	1	2	x	$x+1$	$x+2$
$2x+1$	$2x+1$	$2x+2$	$2x$	1	2	0	$x+1$	$x+2$	x
$2x+2$	$2x+2$	$2x$	$2x+1$	2	0	1	$x+2$	x	$x+1$

$\cdot$	0	1	2	x	$x+1$	$x+2$	$2x$	$2x+1$	$2x+2$
0	0	0	0	0	0	0	0	0	0
1	0	1	2	x	$x+1$	$x+2$	$2x$	$2x+1$	$2x+2$
2	0	2	1	$2x$	$2x+2$	$2x+1$	x	$x+2$	$x+1$
x	0	x	$2x$	2	$x+2$	$2x+2$	1	$x+1$	$2x+1$
$x+1$	0	$x+1$	$2x+2$	$x+2$	$2x$	1	$2x+1$	2	x
$x+2$	0	$x+2$	$2x+1$	$2x+2$	1	x	$x+1$	$2x$	2
$2x$	0	$2x$	x	1	$2x+1$	$x+1$	2	$2x+2$	$x+2$
$2x+1$	0	$2x+1$	$x+2$	$x+1$	2	$2x$	$2x+2$	x	1
$2x+2$	0	$2x+2$	$x+1$	$2x+1$	x	2	$x+2$	1	$2x$

3. The field of eight elements, F_8:

$+$	0	1	x	$x+1$	x^2	x^2+1	x^2+x	x^2+x+1
0	0	1	x	$x+1$	x^2	x^2+1	x^2+x	x^2+x+1
1	1	0	$x+1$	x	x^2+1	x^2	x^2+x+1	x^2+x
x	x	$x+1$	0	1	x^2+x	x^2+x+1	x^2	x^2+1
$x+1$	$x+1$	x	1	0	x^2+x+1	x^2+x	x^2+1	x^2
x^2	x^2	x^2+1	x^2+x	x^2+x+1	0	1	x	$x+1$
x^2+1	x^2+1	x^2	x^2+x+1	x^2+x	1	0	$x+1$	x
x^2+x	x^2+x	x^2+x+1	x^2	x^2+1	x	$x+1$	0	1
x^2+x+1	x^2+x+1	x^2+x	x^2+1	x^2	$x+1$	x	1	0

$\cdot$	0	1	x	$x+1$	x^2	x^2+1	x^2+x	x^2+x+1
0	0	0	0	0	0	0	0	0
1	0	1	x	$x+1$	x^2	x^2+1	x^2+x	x^2+x+1
x	0	x	x^2	x^2+x	$x+1$	1	x^2+x+1	x^2+1
$x+1$	0	$x+1$	x^2+x	x^2+1	x^2+x+1	x^2	1	x
x^2	0	x^2	$x+1$	x^2+x+1	x^2+x	x	x^2+1	1
x^2+1	0	x^2+1	1	x^2	x	x^2+x+1	$x+1$	x^2+x
x^2+x	0	x^2+x	x^2+x+1	1	x^2+1	$x+1$	x	x^2
x^2+x+1	0	x^2+x+1	x^2+1	x	1	x^2+x	x^2	$x+1$

4. We may choose $\gamma = x+1$ as a generator for the multiplicative group of F_9:

γ^0	γ^1	γ^2	γ^3	γ^4	γ^5	γ^6	γ^7
1	$x+1$	$2x$	$2x+1$	2	$2x+2$	x	$x+2$

5. We may choose x as a generator for the multiplicative group of F_8:

x^0	x^1	x^2	x^3	x^4	x^5	x^6
1	x	x^2	$x+1$	x^2+x	x^2+x+1	x^2+1

6. Frobenius automorphisms:

a	0	1	x	$x+1$	x^2	x^2+1	x^2+x	x^2+x+1
a^2	0	1	x^2	x^2+1	x^2+x	x^2+x+1	x	$x+1$

a	0	1	2	x	$x+1$	$x+2$	$2x$	$2x+1$	$2x+2$
a^3	0	1	2	$2x$	$2x+1$	$2x+2$	x	$x+1$	$x+2$

Chapter 19

1. (i) The units in $\boldsymbol{Z}_{10}$ are 1, 3, 5, 7 and 9.

2, 4, 6 and 8 are associates.

(ii) The units in $\boldsymbol{Z}_{12}$ are 1, 5, 7 and 11.

2 and 10 are associates, likewise 3 and 9, and also 4 and 8.

(iii) The units in $\boldsymbol{Z}_{14}$ are 1, 3, 5, 9, 11 and 13.

2, 4, 6, 8, 10 and 12 are associates.

(iv) The units in $\boldsymbol{Z}_{15}$ are 1, 2, 4, 7, 8, 11, 13 and 14.

3, 6, 9 and 12 are associates. 5 and 10 are associates.

2. (i)

$$\begin{aligned}
N((a+bi)(c+di)) &= N((ac-bd)+(ad+bc)i) \\
&= (ac-bd)^2+(ad+bc)^2 \\
&= a^2c^2-2abcd+b^2d^2+a^2d^2+2abcd+b^2c^2 \\
&= a^2c^2+b^2d^2+a^2d^2+b^2c^2 \\
&= (a^2+b^2)(c^2+d^2) = N(a+bi)N(c+di)
\end{aligned}$$

(ii)

$$N\left(\left(a+b\sqrt{-3}\right)\left(c+d\sqrt{-3}\right)\right)=N\left((ac-3bd)+(ad+bc)\sqrt{-3}\right)$$
$$=(ac-3bd)^2+3(ad+bc)^2$$
$$=a^2c^2-6abcd+9b^2d^2+3a^2d^2$$
$$+6abcd+3b^2c^2$$
$$=a^2c^2+9b^2d^2+3a^2d^2+3b^2c^2$$
$$=(a^2+3b^2)(c^2+3d^2)$$
$$=N\left(a+b\sqrt{-3}\right)N\left(c+d\sqrt{-3}\right)$$

3. $5=(1+2i)(1-2i)$, so is not irreducible.

$13=(2+3i)(2-3i)$, so is not irreducible.

Suppose $7=(a+bi)(c+di)$ so that $N(7)=N(a+bi)N(c+di)$.

Then $49=(a^2+b^2)(c^2+d^2)$. But for all integers $a^2+b^2\neq 7$.

So we must have $a^2+b^2=1$ or $c^2+d^2=1$.

Hence one of the two factors is a unit, and so 7 is irreducible.

The story is similar for 11, which is also irreducible.

4. (i) We use the norm $N:\boldsymbol{Z}[\sqrt{-5}\,]\to\boldsymbol{Z}$ given by

$$N(a+b\sqrt{-5}\,)=a^2+5b^2.$$

If α is a unit then $N(\alpha)=1$. Hence $\alpha=\pm 1$.

(ii) Suppose that $1+\sqrt{-5}=(a+b\sqrt{-5}\,)(c+d\sqrt{-5}\,)$ for some $a,b,c,d\in\boldsymbol{Z}$.

Then $N(1+\sqrt{-5}\,)=N(a+b\sqrt{-5}\,)N(c+d\sqrt{-5}\,)$ so that $6=(a^2+5b^2)(c^2+5d^2)$ in $\boldsymbol{Z}$.

But $a^2+5b^2, c^2+5d^2\notin\{2,3\}$ for any $a,b,c,d\in\boldsymbol{Z}$.

So we have either $a^2+5b^2=1$, in which case $a+b\sqrt{-5}=\pm 1$ is a unit, or else $c^2+5d^2=1$, in which case $c+d\sqrt{-5}=\pm 1$ is a unit.

Hence the element $1+\sqrt{-5}$ is irreducible.

(iii) Suppose $1+\sqrt{-5}$ is prime.

$(1+\sqrt{-5})(1-\sqrt{-5})=6$ and so $(1+\sqrt{-5})|6$.

Since $6=2\times 3$ we must have $(1+\sqrt{-5})\,|\,2$ or $1+\sqrt{-5})\,|\,3$.

If the first of these then $(1+\sqrt{-5})(a+b\sqrt{-5})=2$ for some $a,b\in \boldsymbol{Z}$ so that

$$\left.\begin{array}{rcl} a-5b & = & 2 \\ a+b & = & 0 \end{array}\right\}$$

But the solutions $a=\frac{1}{3},\ b=-\frac{1}{3}$ are not in $\boldsymbol{Z}$.

If the second of these, then $(1+\sqrt{-5})(a+b\sqrt{-5})=3$ for some $a,b\in \boldsymbol{Z}$ so that

$$\left.\begin{array}{rcl} a-5b & = & 3 \\ a+b & = & 0 \end{array}\right\}$$

But again the solutions $a=\frac{1}{2},\ b=-\frac{1}{2}$ are not in $\boldsymbol{Z}$.

Hence $1+\sqrt{-5}$ is not prime in $\boldsymbol{Z}[\sqrt{-3}\,]$.

(iv) The element $6\in \boldsymbol{Z}[\sqrt{-5}\,]$ may be factorised as 2×3or as $(1+\sqrt{-5})(1-\sqrt{-5})$, and so $\boldsymbol{Z}[\sqrt{-5}\,]$ is *not* a UFD.

5. Because $\boldsymbol{Z}[\sqrt{-2}\,]$ is a Euclidean domain, it is a PID and hence also a UFD.

Chapter 20

2. (i) $\dim \boldsymbol{R}^4=4$ (ii) $\dim Mat(3,\boldsymbol{R})=3\times 3=9$

(iii) $\dim Mat(2,\boldsymbol{C})=2\times 2\times 2=8$

3.

$$\left(\begin{pmatrix} x_1 \\ y_1 \end{pmatrix} + \begin{pmatrix} x_2 \\ y_2 \end{pmatrix}\right) \cdot \begin{pmatrix} x_3 \\ y_3 \end{pmatrix} = \begin{pmatrix} x_1 + x_2 \\ y_1 + y_2 \end{pmatrix} \cdot \begin{pmatrix} x_3 \\ y_3 \end{pmatrix} = \begin{pmatrix} (x_1 + x_2)x_3 \\ (y_1 + y_2)y_3 \end{pmatrix}$$

$$= \begin{pmatrix} x_1x_3 + x_2x_3 \\ y_1y_3 + y_2y_3 \end{pmatrix}$$

$$= \begin{pmatrix} x_1x_3 \\ y_1y_3 \end{pmatrix} + \begin{pmatrix} x_2x_3 \\ y_2y_3 \end{pmatrix}$$

$$= \begin{pmatrix} x_1 \\ y_1 \end{pmatrix} \cdot \begin{pmatrix} x_3 \\ y_3 \end{pmatrix} + \begin{pmatrix} x_2 \\ y_2 \end{pmatrix} \cdot \begin{pmatrix} x_3 \\ y_3 \end{pmatrix}$$

Similarly $\begin{pmatrix} x_1 \\ y_1 \end{pmatrix} \cdot \left(\begin{pmatrix} x_2 \\ y_2 \end{pmatrix} + \begin{pmatrix} x_3 \\ y_3 \end{pmatrix}\right) = \begin{pmatrix} x_1 \\ y_1 \end{pmatrix} \cdot \begin{pmatrix} x_2 \\ y_2 \end{pmatrix} + \begin{pmatrix} x_1 \\ y_1 \end{pmatrix} \cdot \begin{pmatrix} x_3 \\ y_3 \end{pmatrix}$

$$\left(\lambda\begin{pmatrix} x_1 \\ y_1 \end{pmatrix}\right) \cdot \begin{pmatrix} x_2 \\ y_2 \end{pmatrix} = \begin{pmatrix} \lambda x_1 \\ \lambda y_1 \end{pmatrix} \cdot \begin{pmatrix} x_2 \\ y_2 \end{pmatrix} = \begin{pmatrix} \lambda x_1x_2 \\ \lambda y_1y_2 \end{pmatrix}$$

$$= \lambda\begin{pmatrix} x_1x_2 \\ y_1y_2 \end{pmatrix} = \lambda\left(\begin{pmatrix} x_1 \\ y_1 \end{pmatrix} \cdot \begin{pmatrix} x_2 \\ y_2 \end{pmatrix}\right)$$

4. All four are subspaces.

(i) $\begin{pmatrix} x & 0 \\ 0 & y \end{pmatrix}\begin{pmatrix} x' & 0 \\ 0 & y' \end{pmatrix} = \begin{pmatrix} xx' & 0 \\ 0 & yy' \end{pmatrix}$, so this is a subalgebra.

(ii) $\begin{pmatrix} 0 & 1 \\ 1 & 0 \end{pmatrix}\begin{pmatrix} 0 & 1 \\ 1 & 0 \end{pmatrix} = \begin{pmatrix} 1 & 0 \\ 0 & 1 \end{pmatrix}$, so this is *not* a subalgebra.

(iii) This is a subalgebra.

(iv) $\begin{pmatrix} 1 & 1 \\ 1 & -1 \end{pmatrix}\begin{pmatrix} 1 & 1 \\ 1 & -1 \end{pmatrix} = \begin{pmatrix} 2 & 0 \\ 0 & 2 \end{pmatrix}$, so this is *not* a subalgebra.

5. Suppose

$$\lambda_1\begin{pmatrix} 1 & 0 \\ 0 & 1 \end{pmatrix} + \lambda_2\begin{pmatrix} 1 & 0 \\ 0 & -1 \end{pmatrix} + \lambda_3\begin{pmatrix} 0 & 1 \\ 1 & 0 \end{pmatrix} + \lambda_4\begin{pmatrix} 0 & 1 \\ -1 & 0 \end{pmatrix} = \begin{pmatrix} 0 & 0 \\ 0 & 0 \end{pmatrix}$$

We have $\lambda_1 + \lambda_2 = 0,\ \lambda_1 - \lambda_2 = 0$ so that $\lambda_1 = \lambda_2 = 0$.

Similarly $\lambda_3 + \lambda_4 = 0, \lambda_3 - \lambda_4 = 0$ so that $\lambda_3 = \lambda_4 = 0$.

Hence the matrices are linearly independent.

The show that the matrices span, observe that

$$\begin{pmatrix} a & b \\ c & d \end{pmatrix} = \left(\frac{1}{2}a + \frac{1}{2}d\right)\begin{pmatrix} 1 & 0 \\ 0 & 1 \end{pmatrix} + \left(\frac{1}{2}a - \frac{1}{2}d\right)\begin{pmatrix} 1 & 0 \\ 0 & -1 \end{pmatrix}$$
$$+ \left(\frac{1}{2}b + \frac{1}{2}c\right)\begin{pmatrix} 0 & 1 \\ 1 & 0 \end{pmatrix} + \left(\frac{1}{2}b - \frac{1}{2}c\right)\begin{pmatrix} 0 & 1 \\ -1 & 0 \end{pmatrix}$$

6. First we show that the mapping is a linear transformation:

$$f\left(\begin{pmatrix} w \\ x \\ y \\ z \end{pmatrix} + \begin{pmatrix} w' \\ x' \\ y' \\ z' \end{pmatrix}\right) = f\begin{pmatrix} w+w' \\ x+x' \\ y+y' \\ z+z' \end{pmatrix} = \begin{pmatrix} w+w' & x+x' \\ y+y' & z+z' \end{pmatrix}$$

$$= \begin{pmatrix} w & x \\ y & z \end{pmatrix} + \begin{pmatrix} w' & x' \\ y' & z' \end{pmatrix} = f\begin{pmatrix} w \\ x \\ y \\ z \end{pmatrix} + f\begin{pmatrix} w' \\ x' \\ y' \\ z' \end{pmatrix}$$

$$f\left(\lambda\begin{pmatrix} w \\ x \\ y \\ z \end{pmatrix}\right) = f\begin{pmatrix} \lambda w \\ \lambda x \\ \lambda y \\ \lambda z \end{pmatrix} = \begin{pmatrix} \lambda w & \lambda x \\ \lambda y & \lambda z \end{pmatrix} = \lambda\begin{pmatrix} w & x \\ y & z \end{pmatrix} = \lambda f\begin{pmatrix} w \\ x \\ y \\ z \end{pmatrix}$$

The mapping is an isomorphism, since it has an inverse and so is bijective.

7. (i) $\{(1,0),(0,1)\}$

(ii) $\{(1,0),(i,0),(0,1),(0,i)\}$

(iii)

$$\left\{\left(\begin{pmatrix}1\\0\end{pmatrix},\begin{pmatrix}0\\0\\0\end{pmatrix}\right),\left(\begin{pmatrix}0\\1\end{pmatrix},\begin{pmatrix}0\\0\\0\end{pmatrix}\right),\left(\begin{pmatrix}0\\0\end{pmatrix},\begin{pmatrix}1\\0\\0\end{pmatrix}\right),\left(\begin{pmatrix}0\\0\end{pmatrix},\begin{pmatrix}0\\1\\0\end{pmatrix}\right),\left(\begin{pmatrix}0\\0\end{pmatrix},\begin{pmatrix}0\\0\\1\end{pmatrix}\right)\right\}$$

Chapter 21

1. $\Lambda \boldsymbol{R}^4$ has basis

$$\{1, e_1, e_2, e_3, e_4, e_1\wedge e_2, e_1\wedge e_3, e_1\wedge e_4, e_2\wedge e_3, e_2\wedge e_4, e_3\wedge e_4, \\ e_2\wedge e_3\wedge e_4, e_1\wedge e_3\wedge e_4, e_1\wedge e_2\wedge e_4, e_1\wedge e_2\wedge e_3, e_1\wedge e_2\wedge e_3\wedge e_4\}$$

2. (i) $8e_1\wedge e_2 + 5e_2\wedge e_1 = 8e_1\wedge e_2 - 5e_1\wedge e_2 = 3e_1\wedge e_2$

(ii)

$$\begin{aligned} &6e_1\wedge e_3\wedge e_2 + 7e_3\wedge e_2\wedge e_1 + 10e_2\wedge e_3\wedge e_1 \\ &= -6e_1\wedge e_2\wedge e_3 - 7e_1\wedge e_2\wedge e_3 + 10e_1\wedge e_2\wedge e_3 \\ &= -3e_1\wedge e_2\wedge e_3 \end{aligned}$$

(iii)

$$\begin{aligned} &e_1\wedge e_2\wedge e_3\wedge (e_1 + e_2 + e_3) \\ &= e_1\wedge e_2\wedge e_3\wedge e_1 + e_1\wedge e_2\wedge e_3\wedge e_2 + e_1\wedge e_2\wedge e_3\wedge e_3 \\ &= \mathbf{0} + \mathbf{0} + \mathbf{0} = \mathbf{0}. \end{aligned}$$

3. (i)

$(e_1 + 3e_2)\wedge(4e_1 - 5e_2) = -5e_1\wedge e_2 + 12e_2\wedge e_1 = -17e_1\wedge e_2$

(ii) $-15e_1\wedge e_2 - 8e_1\wedge e_3 + 26e_2\wedge e_3$ (iii) 0

(iv)
$$\begin{aligned} (e_1 + e_2\wedge e_3)\wedge(e_1 + e_2\wedge e_3) &= e_1\wedge e_2\wedge e_3 + e_2\wedge e_3\wedge e_1 \\ &= 2e_1\wedge e_2\wedge e_3 \end{aligned}$$

(v) $-5e_1\wedge e_3\wedge e_4 + 6e_2\wedge e_4\wedge e_1$

4. Let $v = xe_1 + ye_2 + ze_3$ and $u = x'e_1 + y'e_2 + z'e_3$.

$$\begin{aligned} v \wedge v &= (xe_1 + ye_2 + ze_3) \wedge (xe_1 + ye_2 + ze_3) \\ &= xye_1 \wedge e_2 + xze_1 \wedge e_3 - xye_1 \wedge e_2 + yze_2 \wedge e_3 \\ &\quad - xze_1 \wedge e_3 - yze_2 \wedge e_3 \\ &= \mathbf{0}. \end{aligned}$$

$$\begin{aligned} v \wedge u &= (xe_1 + ye_2 + ze_3) \wedge (x'e_1 + y'e_2 + z'e_3) \\ &= (xy' - x'y)e_1 \wedge e_2 + (xz' - x'z)e_1 \wedge e_3 + (yz' - y'z)e_2 \wedge e_3 \\ &= -(x'y - xy')e_1 \wedge e_2 - (x'z - xz')e_1 \wedge e_3 - (y'z - yz')e_2 \wedge e_3 \\ &= -(x'e_1 + y'e_2 + z'e_3) \wedge (xe_1 + ye_2 + ze_3) \\ &= -u \wedge v. \end{aligned}$$

5. (i)

$$\begin{aligned} &(e_1 + 2e_3 + e_4) \wedge (e_1 - e_2) \wedge (e_2 + 2e_3 + e_4) \\ &= (e_1 + 2e_3 + e_4) \\ &\quad \wedge (e_1 \wedge e_2 + 2e_1 \wedge e_3 + e_1 \wedge e_4 - 2e_2 \wedge e_3 - e_2 \wedge e_4) \\ &= -2e_1 \wedge e_2 \wedge e_3 - e_1 \wedge e_2 \wedge e_4 + 2e_3 \wedge e_1 \wedge e_2 + 2e_3 \wedge e_1 \wedge e_4 \\ &\quad - 2e_3 \wedge e_2 \wedge e_4 + e_4 \wedge e_1 \wedge e_2 + e_4 \wedge e_1 \wedge e_2 + 2e_4 \wedge e_1 \wedge e_3 \\ &\quad - 2e_4 \wedge e_2 \wedge e_3 \\ &= \mathbf{0}. \end{aligned}$$

Hence the vectors are linearly *dependent*.

(ii)

$$
\begin{aligned}
&(e_1 + 2e_3 + e_4) \wedge (e_1 - e_2) \wedge (e_2 + 3e_4) \\
&= (e_1 + 2e_3 + e_4) \wedge (e_1 \wedge e_2 + 3e_1 \wedge e_4 - 3e_2 \wedge e_4) \\
&= -3e_1 \wedge e_2 \wedge e_4 + 2e_3 \wedge e_1 \wedge e_2 + 6e_3 \wedge e_1 \wedge e_4 \\
&\quad - 6e_3 \wedge e_2 \wedge e_4 + e_4 \wedge e_1 \wedge e_2 \\
&= -2e_1 \wedge e_2 \wedge e_4 + 2e_1 \wedge e_2 \wedge e_3 - 6e_1 \wedge e_3 \wedge e_4 + 6e_2 \wedge e_3 \wedge e_4 \\
&\neq \mathbf{0}.
\end{aligned}
$$

Hence the vectors are linearly *independent.*

6. (i) $(3e_1 + 4e_2) \wedge (5e_1 + 8e_2) = 24e_1 \wedge e_2 + 20e_2 \wedge e_1 = 4e_1 \wedge e_2$

so determinant is 4.

(ii) $(2e_1 + 3e_2) \wedge (4e_2 - e_3) \wedge (e_1 - e_2 + 2e_3) = 11e_1 \wedge e_2 \wedge e_3$

so determinant is 11.

(iii) $(2e_1 - 3e_2) \wedge (4e_2 - e_3) \wedge (2e_1 + e_2 - e_3) = 0e_1 \wedge e_2 \wedge e_3$

so determinant is 0.

7. For a 2×2 matrix $A = \begin{pmatrix} a & b \\ c & d \end{pmatrix}$ we have

$$
A^2 = \begin{pmatrix} a^2 + bc & (a+d)b \\ (a+d)c & bc + d^2 \end{pmatrix}.
$$

Hence

$$
\begin{aligned}
\tfrac{1}{2}((\operatorname{tr} A)^2 - \operatorname{tr} A^2) &= \tfrac{1}{2}((a+d)^2 - (a^2 + 2bc + d^2)) \\
&= ad - bc = \det A.
\end{aligned}
$$

More elegantly, using exterior products:

$$(\text{tr } A)^2 e_1 \wedge e_2 = A^{(1)} \circ A^{(1)}(e_1 \wedge e_2)$$
$$= A^{(1)}(Ae_1 \wedge e_2 + e_1 \wedge Ae_2)$$
$$= A^2 e_1 \wedge e_2 + Ae_1 \wedge Ae_2 + Ae_1 \wedge Ae_2 + e_1 \wedge A^2 e_2$$
$$= (\text{tr } A^2) e_1 \wedge e_2 + 2 \det A(e_1 \wedge e_2).$$

Hence the result.

8. tr $A = 7$ and $\det A = 12$.

$$Ae_1 \wedge Ae_2 \wedge e_3 + Ae_1 \wedge e_2 \wedge Ae_3 + e_1 \wedge Ae_2 \wedge Ae_3$$
$$= 16 e_1 \wedge e_2 \wedge e_3.$$

Hence the characteristic equation is $x^3 - 7x^2 + 16x - 12 = 0$.

Chapter 22

1. (ii)

$$(u \times v) \times w + (v \times w) \times u + (w \times u) \times v$$
$$= (u.w)v - (v.w)u + (v.u)w - (w.u)v + (w.v)u - (u.v)w$$
$$= \mathbf{0}, \text{ cancelling terms in pairs.}$$

2. $[x+y, x+y] = [x,x] + [x,y] + [y,x] + [y,y] = [x,y] + [y,x]$.

But $[x+y, x+y] = \mathbf{0}$ and so $[x,y] + [y,x] = \mathbf{0}$, hence the result.

3.

$$[[A,B],C] = [AB - BA, C]$$
$$= (AB - BA)C - C(AB - BA) = ABC - BAC - CAB + CBA.$$

Similarly, $[[B,C],A] = BCA - CBA - ABC + ACB$

and $[[C,A],B] = CAB - ACB - BCA + BAC$.

Now add these three together and observe that the terms cancel in pairs, giving $\mathbf{0}$.

4. Suppose $x, y \in Z(L)$ so that

$$[x,z]=0 \text{ and } [y,z]=\mathbf{0} \text{ for all } z \in L.$$

In particular, $[x,y]=\mathbf{0}$. Hence $[[x,y],z]=[\mathbf{0},z]=\mathbf{0}$ and hence $[x,y] \in Z(L)$

5. $\operatorname{tr}[A,B]=\operatorname{tr}(AB-BA)=\operatorname{tr} AB-\operatorname{tr} BA=0.$

Hence $[A,B] \in sl(n,\boldsymbol{R})$ for all $A,B \in gl(n,\boldsymbol{R})$.

6. If A and B are skew-hermitian then

$(A+B)^* = A^*+B^* = (-A)+(-B) = -(A+B)$ so that $A+B$ is skew-hermitian.

$(\lambda A)^* = \lambda A^* = \lambda(-A) = -\lambda A$ so that λA is skew-hermitian.

Finally,

$$\begin{aligned}[A,B]^* &= (AB-BA)^* = (AB)^*-(BA)^* \\ &= B^*A^*-A^*B^* = (-B)(-A)-(-A)(-B) \\ &= BA-AB = -(AB-BA) \\ &= -[A,B]\end{aligned}$$

so that $[A,B]$ is also skew-hermitian.

9. $\dim ut(n,\boldsymbol{R}) = \frac{1}{2}n(n+1).$

The centre of $ut(n,\boldsymbol{R})$ is the algebra of scalar matrices.

10. $\dim sut(n,\boldsymbol{R}) = \frac{1}{2}n(n-1).$

Chapter 23

1. If A is orthogonal then $AA^T = I$ so that $(\det A)^2 = \det A \det A^T = \det(AA^T) = \det I = 1.$

Hence $\det A = \pm 1$.

If A is unitary then $AA^* = I$. Then

$$|\det A|^2 = \det A\ \overline{\det A} = \det A \det A^*$$
$$= \det AA^* = \det I = 1$$

Hence $|\det A| = 1$.

2. Mimic the procedure followed for the real (orthogonal) case.

6. $SO(6)$: Every matrix in the centre is of the form

$$A = \begin{pmatrix} a & 0 & 0 & 0 & 0 & 0 \\ 0 & a & 0 & 0 & 0 & 0 \\ 0 & 0 & a & 0 & 0 & 0 \\ 0 & 0 & 0 & a & 0 & 0 \\ 0 & 0 & 0 & 0 & a & 0 \\ 0 & 0 & 0 & 0 & 0 & a \end{pmatrix}$$

where $a \in \mathbf{R}$. Since $\det A = 1$ we have $a^6 = 1$ and so $a = \pm 1$.

Hence $Z(SO(6)) = \{-I, I\}$.

$SU(4)$: Every matrix in the centre is of the form

$$A = \begin{pmatrix} a & 0 & 0 & 0 \\ 0 & a & 0 & 0 \\ 0 & 0 & a & 0 \\ 0 & 0 & 0 & a \end{pmatrix}$$

where $a \in \mathbf{C}$.

Since $\det A = 1$ we have $a^4 = 1$ and so $a \in \{i, -1, -i, 1\}$.

Hence $Z(SU(4)) = \{iI, -I, -iI, I\}$.

It follows that $SO(6) \not\cong SU(4)$, because the centres are different.

The matrices in the centre of $SO(9)$ are scalar matrices aI, with $a \in \mathbf{R}$. Since the determinant is 1, we have $a^9 = 1$, hence $a = 1$ so that $Z(SO(9)) = \{I\}$.

The matrices in the centre of $U(6)$ are scalar matrices of the form aI, with $a \in \boldsymbol{C}$.

Because the matrices are unitary we have $|a^6| = 1$ and so $|a| = 1$.

Hence $Z(U(6)) \cong \{a \in \boldsymbol{C} : |a| = 1\} = U(1)$.

It follows that $SO(9) \not\cong U(6)$, because the centres are different.

Chapter 24

1.

$$\begin{aligned}
&\| (x_1 + y_1 i)(x_2 + y_2 i) \| \\
&= \| (x_1 x_2 - y_1 y_2) + (x_1 y_2 + x_2 y_1) i \| \\
&= \sqrt{(x_1 x_2 - y_1 y_2)^2 + (x_1 y_2 + x_2 y_1)^2} \\
&= \sqrt{x_1^2 x_2^2 - 2 x_1 x_2 y_1 y_2 + y_1^2 y_2^2 + x_1^2 y_2^2 + 2 x_1 x_2 y_1 y_2 + x_2^2 y_1^2} \\
&= \sqrt{x_1^2 x_2^2 + y_1^2 y_2^2 + x_1^2 y_2^2 + x_2^2 y_1^2} \\
&= \sqrt{(x_1^2 + y_1^2)(x_2^2 + y_2^2)} \\
&= \sqrt{x_1^2 + y_1^2}\,\sqrt{x_2^2 + y_2^2} \\
&= \| x_1 + y_1 i \| \cdot \| x_2 + y_2 i \|
\end{aligned}$$

2. $\|zx\|^2 = \|z\|^2 \cdot \|x\|^2$, and so $\langle zx, zx \rangle = \langle z, z \rangle \langle x, x \rangle$.

Replacing x by $x + y$ we get $\langle z(x+y), z(x+y) \rangle = \langle z, z \rangle \langle x+y, x+y \rangle$.

Hence $\langle zx, zx \rangle + 2\langle zx, zy \rangle + \langle zy, zy \rangle = \langle z, z \rangle \{ \langle x, x \rangle + 2\langle x, y \rangle + \langle y, y \rangle \}$ and so

$$\begin{aligned}
&\langle z, z \rangle \langle x, x \rangle + 2\langle zx, zy \rangle + \langle z, z \rangle \langle y, y \rangle \\
&= \langle z, z \rangle \langle x, x \rangle + 2\langle z, z \rangle \langle x, y \rangle + \langle z, z \rangle \langle y, y \rangle
\end{aligned}$$

By cancellation we get $\langle zx, zy \rangle = \langle z, z \rangle \langle x, y \rangle$.

4. (i) $4+3i-j+3k$ (ii) $2-5i+5j-k$

(iii) $11+18i-j+2k$ (iv) $11+4i-13j+12k$

(v) $\sqrt{15}$ (vi) $\frac{1}{15}(3+i-2j-k)=\frac{1}{5}+\frac{1}{15}i-\frac{2}{15}j-\frac{1}{15}k$

(vii) $\sqrt{30}$ (viii) $\frac{1}{30}(1-4i+3j-2k)=\frac{1}{30}-\frac{2}{15}i+\frac{1}{10}j-\frac{1}{15}k$

5. (i) $\begin{pmatrix} 0 & -1 & 0 & 0 \\ 1 & 0 & 0 & 0 \\ 0 & 0 & 0 & -1 \\ 0 & 0 & 1 & 0 \end{pmatrix}$ (ii) $\begin{pmatrix} 0 & 0 & -2 & -3 \\ 0 & 0 & -3 & 2 \\ 2 & 3 & 0 & 0 \\ 3 & -2 & 0 & 0 \end{pmatrix}$

(iii) $\begin{pmatrix} 4 & -3 & -2 & -1 \\ 3 & 4 & -1 & 2 \\ 2 & 1 & 4 & -3 \\ 1 & -2 & 3 & 4 \end{pmatrix}$

Chapter 25

1. $\dim \boldsymbol{R}^3 \otimes \boldsymbol{R}^4 = 3\times 4 = 12$ $\dim \boldsymbol{C}\otimes \boldsymbol{C} = 2\times 2 = 4$

$\dim Mat(2,\boldsymbol{R}) \otimes Mat(3,\boldsymbol{R}) = 4\times 9 = 36$

$\dim Mat(2,\boldsymbol{H}) \otimes Mat(3,\boldsymbol{C}) = 16\times 18 = 288$

3. $\{u_1 \otimes v_1, u_1 \otimes v_2, u_2 \otimes v_1, u_2 \otimes v_2, u_3 \otimes v_1, u_3 \otimes v_2\}$

4. Begin with the standard basis for $\boldsymbol{C}\otimes\boldsymbol{C}$:

$$\{1\otimes 1, 1\otimes i, i\otimes 1, i\otimes i\}$$

An isomorphism must correlate the unity of one algebra with the unity of the other, so our isomorphism must map $1\otimes 1 \leftrightarrow (1,1)$.

Now notice that $1\otimes i \cdot 1\otimes i = 1\otimes -1 = -(1\otimes 1)$. So we need a second basis element for $\boldsymbol{C}\oplus\boldsymbol{C}$ that squares to make minus the unity. We may take $1\otimes i \leftrightarrow (i,i)$, since

$$(i,i)\cdot(i,i) = (-1,-1) = -(1,1).$$

Similarly $i\otimes 1\cdot i\otimes 1=-1\otimes 1=-(1\otimes 1)$. So we need another basis element for $\boldsymbol{C}\oplus\boldsymbol{C}$ that squares to make minus the unity. We may take $i\otimes 1\leftrightarrow(-i,i)$, since

$$(-i,i)\cdot(-i,i)=(-1,-1)=-(1,1).$$

Finally notice the relationship between the last three basis elements for $\boldsymbol{C}\otimes\boldsymbol{C}$: $\quad 1\otimes i\cdot i\otimes 1=i\otimes i$.

This suggests the final basis element for $\boldsymbol{C}\oplus\boldsymbol{C}$:

$$(i,i)\cdot(-i,i)=(1,-1).$$

It is routine to check that $\{(1,1),(i,i),(-i,i),(1,-1)\}$ is indeed a basis for $\boldsymbol{C}\oplus\boldsymbol{C}$, although not of course the standard one.

5.

$$\begin{aligned}
Cl(7)&=Cl'(5)\otimes Cl(2)\\
&=(Mat(2,\boldsymbol{H})\oplus Mat(2,\boldsymbol{H}))\otimes\boldsymbol{H}\\
&=(Mat(2,\boldsymbol{R})\otimes\boldsymbol{H}\oplus Mat(2,\boldsymbol{R})\otimes H)\otimes\boldsymbol{H}\\
&=Mat(2,\boldsymbol{R})\otimes\boldsymbol{H}\otimes\boldsymbol{H}\oplus Mat(2,\boldsymbol{R})\otimes\boldsymbol{H}\otimes\boldsymbol{H}\\
&=Mat(2,\boldsymbol{R})\otimes Mat(4,\boldsymbol{R})\oplus Mat(2,\boldsymbol{R})\otimes Mat(4,\boldsymbol{R})\\
&=Mat(8,\boldsymbol{R})\oplus Mat(8,\boldsymbol{R})
\end{aligned}$$

$$\begin{aligned}
Cl'(7)&=Cl(5)\otimes Cl'(2)=Mat(2,\boldsymbol{C})\otimes Mat(2,\boldsymbol{R})\\
&=\boldsymbol{C}\otimes Mat(2,\boldsymbol{R})\otimes Mat(2,\boldsymbol{R})=\boldsymbol{C}\otimes Mat(4,\boldsymbol{R})\\
&=Mat(4,\boldsymbol{C})
\end{aligned}$$

$$\begin{aligned}
Cl(8)&=Cl'(6)\otimes Cl(2)=Mat(4,\boldsymbol{H})\otimes\boldsymbol{H}=Mat(4,\boldsymbol{R})\otimes\boldsymbol{H}\otimes\boldsymbol{H}\\
&=Mat(4,\boldsymbol{R})\otimes Mat(4,\boldsymbol{R})=Mat(16,\boldsymbol{R})
\end{aligned}$$

$$Cl'(8)=Cl(6)\otimes Cl'(2)=Mat(8,\boldsymbol{R})\otimes Mat(2,\boldsymbol{R})=Mat(16,\boldsymbol{R})$$

6.

$$Cl(24) = Mat(16, Cl(16)) = Mat(256, Cl(8)) = Mat(4096, \boldsymbol{R})$$

$$Cl(25) = Mat(4096, \boldsymbol{C})$$

$$Cl(26) = Mat(4096, \boldsymbol{H})$$

$$Cl(27) = Mat(4096, \boldsymbol{H} \oplus \boldsymbol{H}) = Mat(4096, \boldsymbol{H}) \oplus Mat(4096, \boldsymbol{H}).$$

7. To prove: $Cl'(n+8) = Mat(16, Cl'(n))$

$$\begin{aligned} Cl'(n+8) &= Cl(n+6) \otimes Mat(2, \boldsymbol{R}) \\ &= Cl'(n+4) \otimes Mat(2, \boldsymbol{R}) \otimes \boldsymbol{H} \\ &= Cl(n+2) \otimes Mat(2, \boldsymbol{R}) \otimes Mat(2, \boldsymbol{R}) \otimes \boldsymbol{H} \\ &= Cl'(n) \otimes Mat(2, \boldsymbol{R}) \otimes Mat(2, \boldsymbol{R}) \otimes \boldsymbol{H} \otimes \boldsymbol{H} \\ &= Cl'(n) \otimes Mat(4, \boldsymbol{R}) \otimes Mat(4, \boldsymbol{R}) \\ &= Cl'(n) \otimes Mat(16, \boldsymbol{R}) \\ &= Mat(16, Cl'(n)) \end{aligned}$$

Printed in Great Britain
by Amazon

48857764R00220